Serverless ETL and Analytics with AWS Glue

Your comprehensive reference guide to learning about AWS Glue and its features

Vishal Pathak

Subramanya Vajiraya

Noritaka Sekiyama

Tomohiro Tanaka

Albert Quiroga

Ishan Gaur

BIRMINGHAM—MUMBAI

Serverless ETL and Analytics with AWS Glue

Publishing Product Manager: Reshma Raman
Senior Editor: Tazeen Shaikh
Content Development Editor: Sean Lobo
Technical Editor: Devanshi Ayare
Copy Editor: Safis Editing
Project Coordinator: Farheen Fathima
Proofreader: Safis Editing
Indexer: Pratik Shirodkar
Production Designer: Jyoti Chauhan
Marketing Coordinator: Nivedita Singh

First published: August 2022

Production reference: 1220722

Published by Packt Publishing Ltd.
Livery Place
35 Livery Street
Birmingham
B3 2PB, UK.

ISBN 978-1-80056-498-5

www.packt.com

Contributors

About the authors

Vishal Pathak is a Data Lab Solutions Architect at AWS. Vishal works with customers on their use cases, architects solutions to solve their business problems, and helps them build scalable prototypes. Prior to his journey in AWS, Vishal helped customers implement business intelligence, data warehouse, and data lake projects in the US and Australia.

Subramanya Vajiraya is a Big data Cloud Engineer at AWS Sydney specializing in AWS Glue. He obtained his Bachelor of Engineering degree specializing in Information Science & Engineering from NMAM Institute of Technology, Nitte, KA, India (Visvesvaraya Technological University, Belgaum) in 2015 and obtained his Master of Information Technology degree specialized in Internetworking from the University of New South Wales, Sydney, Australia in 2017. He is passionate about helping customers solve challenging technical issues related to their ETL workload and implementing scalable data integration and analytics pipelines on AWS.

Noritaka Sekiyama is a Senior Big Data Architect on the AWS Glue and AWS Lake Formation team. He has 11 years of experience working in the software industry. Based in Tokyo, Japan, he is responsible for implementing software artifacts, building libraries, troubleshooting complex issues and helping guide customer architectures.

Tomohiro Tanaka is a senior cloud support engineer at AWS. He works to help customers solve their issues and build data lakes across AWS Glue, AWS IoT, and big data technologies such Apache Spark, Hadoop, and Iceberg.

Albert Quiroga works as a senior solutions architect at Amazon, where he is helping to design and architect one of the largest data lakes in the world. Prior to that, he spent four years working at AWS, where he specialized in big data technologies such as EMR and Athena, and where he became an expert on AWS Glue. Albert has worked with several Fortune 500 companies on some of the largest data lakes in the world and has helped to launch and develop features for several AWS services.

Ishan Gaur has more than 13 years of IT experience in software development and data engineering, building distributed systems and highly scalable ETL pipelines using Apache Spark, Scala, and various ETL tools such as Ab Initio and Datastage. He currently works at AWS as a senior big data cloud engineer and is an SME of AWS Glue. He is responsible for helping customers to build out large, scalable distributed systems and implement them in AWS cloud environments using various big data services, including EMR, Glue, and Athena, as well as other technologies, such as Apache Spark, Hadoop, and Hive.

About the reviewers

Akira Ajisaka is an open source developer who has over 10 years of engineering experience in big data. He contributes to the open source community and is an Apache Software Foundation member and Apache Hadoop PMC member. He has worked for the AWS Glue ETL team since 2022 and is learning a lot about Apache Spark.

Keerthi Chadalavada is a senior software engineer with AWS Glue. She is passionate about building cloud-based, data-intensive applications at scale. Her recent work includes enabling data engineers to build event-driven ETL pipelines that respond in near real time to data events and provide the latest insights to business users. In addition, her work on Glue Blueprints enabled data engineers to build templates for repeatable ETL pipelines and enabled non-data engineers without technical expertise to use these templates to gain faster insights from their data. Keerthi holds a master's degree in computer science from Ohio State University and a bachelor's degree in computer science from Bits Pilani, India.

Table of Contents

3

Data Ingestion

Section 2 – Data Preparation, Management, and Security

4

Data Preparation

5

Data Layouts

6

Data Management

7

Metadata Management

8

Data Security

9

Data Sharing

10

Data Pipeline Management

Section 3 – Tuning, Monitoring, Data Lake Common Scenarios, and Interesting Edge Cases

11

Monitoring

12

Tuning, Debugging, and Troubleshooting

13

Data Analysis

14

Machine Learning Integration

15

Architecting Data Lakes for Real-World Scenarios and Edge Cases

Index

Other Books You May Enjoy

Preface

These days, organizations have gravitated toward data-driven business. Today, data integration across various data sources has become a key driver for businesses. In the cloud, data integration services such as AWS Glue do the undifferentiated heavy lifting based on serverless infrastructure. AWS Glue helps you to integrate data across different sources and build a data lake at scale in a serverless fashion without maintaining infrastructure.

This book shows you how AWS Glue can be used to solve real-world problems, along with teaching you about data processing, data integration, and building data lakes. It allows you to learn how to perform various aspects of data integration techniques such as data ingestion from various sources, data layout optimization, data and metadata management, and data pipeline management. Further, it covers data analysis use cases such as ad hoc queries, visualization, and real-time analysis using AWS Glue. Additional topics such as CI/CD, data quality validation, data sharing, and data security aspects, such as access control, encryption, auditing, and networking, are also covered. Toward the end, the book focuses on providing various monitoring options and the best practices for tuning, debugging, and troubleshooting.

The book takes you through the AWS Glue features such as jobs, the Data Catalog, crawlers, DataBrew, Glue Studio, custom connectors, and so on, in addition to AWS Lake Formation.

By the end of this book, you will be able to integrate data across different sources and build a data platform for scalable analysis using AWS Glue.

Who this book is for

This book is designed for data engineers, ETL developers, and data analysts who want to understand how AWS Glue can help to solve their business problems. Basic knowledge of AWS data services is assumed. Experience with AWS Glue is also preferred but not required. Even without prior knowledge, you can start learning AWS Glue with the book. Most of the features are accompanied by a walkthrough to help you understand the concepts that are explained in each chapter.

What this book covers

Chapter 1, Data Management – Introduction and Concepts, introduces basic concepts associated with data management.

Chapter 2, Introduction to Important AWS Glue Features, introduces some important AWS Glue features.

Chapter 3, Data Ingestion, describes how to ingest data across multiple data stores.

Chapter 4, Data Preparation, describes typical data preparation use cases with both a GUI-based approach and a source code-based approach using AWS Glue.

Chapter 5, Designing Data Layouts, describes how to optimize data layout on Amazon S3 using AWS Glue.

Chapter 6, Data Management, describes how to manage, clean up, and enrich data using AWS Glue.

Chapter 7, Metadata Management, describes how to populate and maintain metadata based on data using AWS Glue.

Chapter 8, Data Security, describes how to secure your data by access control, encryption, auditing, and network security using AWS Glue.

Chapter 9, Data Sharing, describes how to share your data across multiple accounts to democratize your data lake.

Chapter 10, Data Pipeline Management, describes how to build and orchestrate a data-processing pipeline using AWS Glue.

Chapter 11, Monitoring, describes how to monitor a data lake and AWS Glue components.

Chapter 12, Tuning, Debugging, and Troubleshooting, describes the best practices to tune, debug, and troubleshoot typical use cases.

Chapter 13, Data Analysis, describes common options to analyze data using AWS analytics services.

Chapter 14, Machine Learning Integration, describes how to utilize your data for a machine learning workload.

Chapter 15, Architecting Data Lakes for Real-World Scenarios and Edge Cases, describes end-to-end examples of architecting data lakes.

To get the most out of this book

All walkthroughs will require a web browser (Google Chrome, Mozilla Firefox, Microsoft Edge, or Safari) installed on a computer in order to use AWS Management Console, and you'll need an AWS account to access the AWS Console and utilize AWS resources. Next to that, you'll need to install the **AWS Command Line Interface (AWS CLI)** *on a computer to run commands:*

Software/hardware covered in the book	Operating system requirements
An AWS account	Windows, macOS, or Linux
The AWS CLI	
A web browser (Google Chrome, Mozilla Firefox, Microsoft Edge, or Safari)	

Not all the chapters' walkthroughs require an AWS CLI installation. You'll be informed in each chapter when you need further requirements.

If you are using the digital version of this book, we advise you to type the code yourself or access the code from the book's GitHub repository (a link is available in the next section). Doing so will help you avoid any potential errors related to the copying and pasting of code.

Download the example code files

You can download the example code files for this book from GitHub at `https://github.com/PacktPublishing/Serverless-ETL-and-Analytics-with-AWS-Glue`. If there's an update to the code, it will be updated in the GitHub repository.

We also have other code bundles from our rich catalog of books and videos available at `https://github.com/PacktPublishing/`. Check them out!

Download the color images

We also provide a PDF file that has color images of the screenshots and diagrams used in this book. You can download it here: `https://packt.link/fTqGe`.

Conventions used

There are a number of text conventions used throughout this book.

`Code in text`: Indicates code words in text, database table names, folder names, filenames, file extensions, pathnames, dummy URLs, user input, and Twitter handles. Here is an example: "We used the `glueContext.write_dynamic_frame.from_options()` method to write the data to Amazon S3."

A block of code is set as follows:

```
root
 |-- ColumnA: string (nullable = true)
 |-- ColumnB: string (nullable = true)
```

Bold: Indicates a new term, an important word, or words that you see onscreen. For instance, words in menus or dialog boxes appear in **bold**. Here is an example: "This can be done by navigating to **AWS Glue Studio console** | **Connectors** | **Marketplace Connectors** and subscribing to **Cloudwatch Metrics connector for AWS Glue**."

> Tips or Important Notes
> Appear like this.

Get in touch

Feedback from our readers is always welcome.

General feedback: If you have questions about any aspect of this book, email us at customercare@packtpub.com and mention the book title in the subject of your message.

Errata: Although we have taken every care to ensure the accuracy of our content, mistakes do happen. If you have found a mistake in this book, we would be grateful if you would report this to us. Please visit www.packtpub.com/support/errata and fill in the form.

Piracy: If you come across any illegal copies of our works in any form on the internet, we would be grateful if you would provide us with the location address or website name. Please contact us at copyright@packt.com with a link to the material.

If you are interested in becoming an author: If there is a topic that you have expertise in and you are interested in either writing or contributing to a book, please visit authors.packtpub.com.

Share Your Thoughts

Once you've read *Serverless ETL and Analytics with AWS Glue*, we'd love to hear your thoughts! Scan the QR code below to go straight to the Amazon review page for this book and share your feedback.

https://packt.link/r/1-800-56498-8

Your review is important to us and the tech community and will help us make sure we're delivering excellent quality content.

Section 1 – Introduction, Concepts, and the Basics of AWS Glue

In this section, you will learn about the basics of AWS Glue and the general trends in data management. You will be introduced to the important AWS Glue features and ways to ingest data using AWS Glue from heterogeneous sources.

This section includes the following chapters:

- *Chapter 1, Data Management – Introduction and Concepts*
- *Chapter 2, Introduction to Important AWS Glue Features*
- *Chapter 3, Data Ingestion*

1

Data Management – Introduction and Concepts

A vast amount of data is being generated by people, organizations, devices, and software applications, and the volume of data being generated is growing rapidly. The numbers vary significantly, depending on the source, but it is estimated that approximately 60% to 80% of data gathered by organizations is **dark data**. Essentially, data is being collected, processed, and stored for a long time by organizations for compliance reasons, but the data is not used for any other purposes, such as analytics or direct monetization. In most cases, storing and securing this data can be more expensive than the value extracted.

In today's digital economy, organizations are striving to be data-driven by basing their strategic business decisions on intelligence that's been obtained from data gathered from various sources. Until recently, organizations thought of data purely in the context of transactions and locked it away in heavily siloed databases that were built for transaction processing; however, this was not suitable for open-ended analysis. All this changed with advancements in data processing techniques and drops in the costs involved in processing and analyzing data. Organizations are now adopting data-driven approaches for key business decisions.

In this chapter, we will cover the following topics:

- Types of data processing – OLTP and OLAP
- Data warehouses and data marts
- Data lakes
- Data lakehouse
- Data mesh
- Apache Spark on the AWS cloud

- AWS Glue
- Querying data using AWS

The topics in this chapter will introduce us to different data management techniques and different tools and services offered by the AWS cloud. These concepts will help you understand the different design approaches you can take to build effective data integration and management setups that are suitable to your use cases when using AWS Glue.

Types of data processing – OLTP and OLAP

Traditionally, data storage systems have been classified as **Online Transaction Processing (OLTP)** and **Online Analytical Processing (OLAP)**. OLTP systems are responsible for day-to-day business executions. For instance, when you call your phone carrier's customer service to add a new value pack to your phone plan, the customer service agent quickly pulls up the account information for your phone number and adds your desired value pack. The system that's used by the customer service agent is designed to be fast so that the customer wait time can be minimized, which allows the customer service agent to be more efficient and serve customers faster. The system is also designed so that it updates the data quickly so that a large number of concurrent transactions can be processed. This allows the customer service agent to confirm that the value pack has been successfully applied to the account. Other examples include banking and shopping applications.

These faster updates are achieved by using a **normalized data model**. Normalization is the process of structuring the dataset as per a set of *normal-forms* to reduce redundancy and enhance data integrity. The normalized data model ensures that you don't update multiple tables with the same information for a user operation. This is done by reducing the redundancy of the data in these systems. For example, if a customer updates their `preferred_name`, we can make this change in one table; the rest of the dependent tables will use `customer_id` to fetch updated information. So, a typical SQL query for the CRM application that's used by the customer service agent contains the `customer_id = 'xxxxxx'` expression or `data_plan_id = 'xxxxxx'` in the WHERE clause.

These OLTP systems are not designed for obtaining or analyzing trends – for example, a query for gathering the mobile data usage (volume) of all customers over the last 2 years. Such queries involve joining a lot of tables on the OLTP side because of normalizations and usually results in poor performance as the amount of data scales up.

This problem can be solved by using OLAP systems. OLAP systems typically use the **data warehouse** of an organization, where they are utilized for executing complex queries over a large amount of data. They generally store historical datasets.

So, while both OLAP and OLTP have different ways of storing data and are designed for different use cases, the data on which they operate can be the same – the data is just modeled differently. Since both systems work on the same data, the data must be moved from one system to another. OLTP systems support live business transactions, so data generally originates there. This data is then brought into a data warehouse through an **Extract, Transform, Load (ETL)** or **Extract, Load, Transform (ELT)** tool so that it can then be consumed by OLAP systems. The following table explains the differences between OLTP and OLAP:

Online Transaction Processing (OLTP)	Online Analytical Processing (OLAP)
Primarily designed to handle a large volume of short transactional operations.	Primarily designed to handle analytical workloads on large volumes of historical data.
Designed for handling day-to-day operations.	Designed to aid in planning and decision-making processes.
Data is updated quite frequently.	Data updates are infrequent compared to OLTP.
Queries are simple, standardized, and target small amounts of data – one or few records.	Queries are often quite complex, involve aggregations, and target large amounts of data.
Uses a normalized data model and typically contains a lot of tables.	Data is typically denormalized with fewer tables.

Table 1.1 – Differences between OLTP and OLAP

Now that we understand the fundamentals of the OLTP and OLAP models, let's explore different data management systems, such as data warehouses, data marts, data lakes, data lakehouses, and data meshes.

Data warehouses and data marts

In an organization, it is not uncommon for day-to-day operations to be performed and stored in several transactional operating systems. However, when higher-level business decisions are to be made using data gathered from these systems, it would be easier to collate necessary information from these sources and build a centralized repository for datasets to gather actionable intelligence.

A data warehouse is a centralized repository of data that's been gathered from various sources within an organization. The collated data within this repository is analyzed and can be used to make business decisions. A data mart, on the other hand, is a subset of a data warehouse aligned toward a specific business unit within an organization.

The concept of data warehouses was introduced in the late 1980s. Data warehouses are subject-oriented, integrated, time-variant, and non-volatile. This means that data warehouses are designed to be able to make sense of the data in a specific subject rather than ongoing operations, such as sales, marketing, and HR. Data warehouses are also designed to integrate data for several different source systems, such as **Enterprise Resource Planning (ERP)**, **Human Resource Management Systems (HRMSs)**, **Customer Relationship Management (CRM)**, **Financial Management Systems (FMSs)**, and any other operational systems within an organization. The data within a data warehouse is usually structured, but it can be unstructured as well. Data warehouses also allow users to analyze the data at different grains of time, such as `year`, `month`, and `day`. The data in data warehouses is non-volatile and maintains history. So, changes in the source systems result in newer entries in the data warehouses where the new state of the data is used while preserving the old state of the data.

In Inmon's top-down data warehousing approach, data architects and modelers start by looking at the holistic data landscape of an organization and identifying the main subject areas and entities under it. Inmon's data warehouse is normalized and avoids redundancy. This simplifies the data ingestion process but is not optimized for queries. Hence, **data marts** are built on top of data warehouses and users access these data marts for their queries.

While data marts can be based on a star or snowflake schema, the star schema is generally preferred because it results in faster queries due to fewer joins. In 1996, Ralph Kimball introduced the star schema methodology to the data management world. This follows the bottom-up approach and creates data marts based on the business requirements instead of starting with an enterprise data warehouse.

In a data mart, data is stored at multiple levels and the table at the correct level is picked for processing the data. The atomic level by which the facts may be defined is known as the *grain* or *granularity* of the table.

For example, let's consider a retail sales dataset for a retail store chain operating in different countries. A customer could buy several products in a single sale and the same customer could buy higher quantities of the same product within the same sale. We can have a table that contains region information that can be linked to sales and product tables.

So, while selecting a grain, it is beneficial to have the fact table populated with the most atomic grain. This allows us to be as granular as we want with the information we query. If we define the grain at the *sales transaction* level, we can query individual sales transactions and get information such as the amount per sale, payment method, and so on. However, we won't be able to get the product information in a particular sale. To mitigate this, let's say we define the grain at the *product in a sales transaction* level. We can query product-related information along with sales information.

These different levels of pre-computation help us avoid heavy computations at query time. For example, if a user is querying for `sales_amount` at the region level, it might be far easier to select the data from the table that contains the `sales_amount` and `region` columns.

As we can see, data marts are helpful for working with datasets related to a specific context or a business line. However, a centralized data warehouse is beneficial when our analysis needs data to be aggregated from a variety of sources across the organization to extract actionable intelligence from the dataset.

A fresh approach to data warehousing came with the introduction of **data vaults**. This is a hybrid approach that incorporates the best-normalized model and a denormalized star schema. This approach to data modeling can be quite helpful when working with multi-source systems or data sources that have constantly changing relationships. This makes it easier to ingest data from multiple sources. Also, because of the way the data is modeled, data vaults make it easier to audit and track data.

Data transformation is a requirement for the data to be loaded into a data warehouse. This creates entry barriers and lags in delivering value to customers. Generally, organizations have multiple sources of data and they must be imported into a data warehouse to make business decisions or even to know if it adds value. Later, if the user discovers that combining the data from certain sources is not delivering the value that was initially expected, then this results in time and resources being wasted. Also, it is not always possible to forecast the analytical requirements in a world where businesses have to constantly evolve to stay relevant. What happens if a business user needs historical data that isn't available in the data warehouse? Around 2015, data lakes were created to solve these problems.

Data lakes

A **data lake** can be defined as a centralized repository that allows you to store all structured and unstructured data at any scale. With today's hyper scalers providing cheap and durable storage, it is now possible for organizations to store all of their data in the cloud without significant cost implications. Data lakes are broken down into **layers** or **zones**.

In the first layer of the data lake, data is generally stored as-is. This reduces the entry barrier and enables organizations to move all of their data to the "lake" without significantly increasing development or maintenance costs. Because the first layer of the data lake is an as-is copy of the data, organizations can use an automated configuration-based pipeline to create newer sources.

Organizations usually pick a replication tool such as **AWS Data Migration Service** (**AWS DMS**) to bring the data into the data lake. While AWS DMS involves taking care of the replication infrastructure, it is mostly a hands-off mechanism for hydrating the lake. Organizations may also use a push mechanism to FTP to transfer the files to an AWS **Simple Storage Service** (**S3**)-based data lake using AWS Transfer Family.

Data from the first layer is compressed and partitioned, and audited columns are added during data preparation so that they can be used by downstream systems more effectively. Having all the data in the data lake enables data analysts to do the initial discovery to find out the value of combining data from various sources. If the value is discovered, then necessary transformations are applied in an ETL pipeline so that the target is hydrated with newer data periodically or through a streaming arrangement. These automated transformations are then loaded into the final layer of a data lake and used for user consumption.

Data lakehouse

Challenged by the newer demands to derive value from the vast and ever-increasing unstructured data, it became important to come up with a new arrangement that does not try to force unstructured data into the strict models of a data warehouse. The data lakehouse blurs the lines between data lakes and data warehouses by enabling the **atomicity, consistency, isolation, and durability** (**ACID**) properties on the data in the data lake and enabling multiple processes to concurrently read and write data.

With this, transformed data in open formats such as **Apache Parquet** can be consumed for **feature engineering** and **machine learning** (**ML**) workloads and can also be used for analytics.

Data mesh

While cheap, durable storage helped in storing vast volumes of data, this data had to be secured properly. Since data from a vast variety of sources is stored in the lake, it becomes difficult to define the ownership and management of this data. This requirement resulted in a paradigm of serving data as a product and setting the ownership of the product. This thought process led to the creation of the **data mesh**.

Data meshes ensure that data lakes don't become another monolith that the organization's IT teams now have to manage. This decentralization leads to the democratization of data, which fuels innovation without hindering access to the data. Although data is decentralized and offered as a service, the permission model that's applied to create a data lake ensures interoperability to reduce the barriers to accessing data products for users that have the right permissions.

Distributed computing for big data

Before the advent of big data, ELT and ETL tools usually had a server and an orchestrator that was responsible for reading the data from the OLTP systems and populating the data warehouse. Some of these tools used the compute of these intermediate servers, while others used the compute of the target to process the data. Traditionally, these ETL/ELT systems were used to pull data once a day and during off-business hours. This was done to reduce the impact of the data being pulled from the OLTP systems. When a system required higher data processing capabilities, organizations would scale up the ETL/ELT servers.

This arrangement worked fine for a few years but the volume of data kept increasing, and scaling the ETL/ELT systems became cost prohibitive. With the world increasingly becoming more data-centric, the amount of data produced continued to grow. It is estimated that 90% of the data today has been generated in the last 2 years.

Not only has the volume of data increased, but organizations also want to get the data faster for quicker decision-making.

In a connected world, the number of variables that impact a business decision has increased, so there is a need to get data from multiple different sources to make a decision. For example, for a retail company to find out the discount to be applied to a certain product, it can no longer just rely on the cost price of the product and the profit that it expects from the sale. It would be beneficial to know the cost of keeping the product on the shelf before it is sold, along with knowing the approximate time for which the product is expected to stay on the shelf. The retail company may also want to know the price of the same product on competitor websites, along with the price of similar products with better features.

Here, the cost price can be obtained from the company's ERP data. The percentage of expected profit might be a business transformation logic that uses their "secret sauce." The cost of keeping the product on the shelf will be based on the cumulative sum of all the costs of the store. The approximate duration for which the product will be on the shelf might come from an ML model. The price of the same product sold by the competitors can be scraped from their websites and the cost of similar products with better features can be obtained from third-party market research. So, modern decision-making involves making sense of data from a variety of sources.

Big data is a collection of data derived from various sources and is characterized by the volume, velocity, variety, veracity, and value of the data. These are known as the **5 V's of big data**. While we collect the data from a variety of sources at a certain velocity and volume like never before, we also want to make sure that the collected data is accurate and can be trusted. This can be achieved using a series of validation steps based on the data being collected. Finally, once we have the trusted data, we want to be able to derive value from it.

When importing the data into a data lake or a data warehouse, the old arrangements of scaling up do not work, so we must deal with the 5 V's of big data. The solution to these challenges came in the form of distributed computing.

Distributed computing systems distribute the workload of any given query to multiple workers instead of a single worker. The workloads being distributed across multiple worker nodes meant that organizations could now add nodes to increase the computing power rather than vertically scaling the node. The advantage of this approach is that we can process data on multiple nodes in parallel. This allows us to keep up with the high velocity of incoming data where one single node may not be enough.

With the advent of distributed computing in big data processing and analytics, several engines and frameworks were developed to handle different aspects of data processing and analysis. One of the most popular processing and analytics engines is **Apache Spark**.

Apache Spark

Apache Spark is an open source unified analytics engine that was originally developed in 2009 at UC Berkeley. It became a top-level Apache project in February 2014. It has over 1.7K contributors and over 30K star gazers on GitHub. The following is a quote from the Spark documentation (`https://spark.apache.org/docs/latest/index.html`):

"Apache Spark is a unified analytics engine for large-scale data processing. It provides high-level APIs in Java, Scala, Python, and R, and an optimized engine that supports general execution graphs. It also supports a rich set of higher-level tools including Spark SQL for SQL and structured data processing, MLlib for machine learning, GraphX for graph processing, and Structured Streaming for incremental computation and stream processing."

At a high level, a Spark cluster consists of a set of executors running a **Java Virtual Machine (JVM)**. One of these executors runs the Driver program. This driver program is responsible for creating a SparkContext. A SparkContext is the entry point for Spark features. Spark applications are instances of this SparkContext, which connects to a Cluster Manager.

The following diagram shows the workflow that's used by Apache Spark to execute the workload. Here, the user submits the workload using the `spark-submit` command; then the Spark driver coordinates with the Cluster Manager to execute the workload within the executors on the worker nodes:

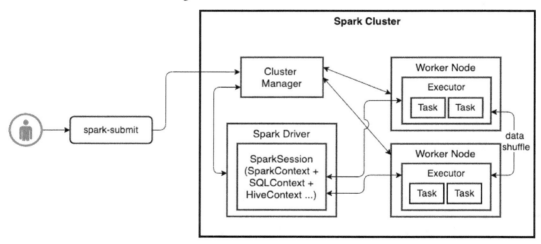

Figure 1.1 – Overview of Apache Spark's workload execution

A Cluster Manager can be Spark's standalone cluster manager, Mesos, Apache Hadoop **Yet Another Resource Negotiator (YARN)**, or Kubernetes. Cluster Managers are responsible for allocating containers to various Spark applications running on the cluster. With YARN, Spark can run in either cluster mode or client mode.

In client mode, the driver program runs on the machine that submitted the Spark Job. In cluster mode, the driver program runs on one of the executors. Executors are responsible for executing the tasks that are sent through `SparkContext` and run in YARN's JVMs containers. When we invoke an action in a Spark application, a Spark Job is created. A list of actions available in Spark can be found in the Apache Spark documentation (`https://spark.apache.org/docs/latest/rdd-programming-guide.html#actions`). To execute a Job, an execution plan must be created based on a **Directed Acyclic Graph (DAG)**.

A DAG scheduler converts the logical execution plan into a physical execution plan. A DAG consists of **stages**. A Spark stage is a set of independent tasks all computing the same function that is needed as part of a Spark Job. Each stage is further divided into **tasks**. All of these tasks can be run in parallel on the CPU cores of the executors. Once Spark acquires the executors, `SparkContext` sends the tasks to the executors to perform.

Spark also has a component called `SparkSQL` which allows users to write SQL queries for data transformation. `SparkSQL` is enabled by the **Catalyst** and **Tungsten** engines.

Catalyst is responsible for creating a physical plan from a logical plan, while Tungsten is responsible for generating the byte code that will be executed on the cluster.

This new architecture of data processing came with challenges. Organizations now had to quickly develop a new skill set to manage clusters of nodes that were used for data processing. Also, what do you do with all these ETL compute nodes when they are not used for processing?

Apache Spark on the AWS cloud

The problem of unused compute resources was solved by the hyperscalers of the world. One of the leading hyperscalers is AWS. AWS has two offerings for managed Spark: Amazon EMR and AWS Glue. With Amazon EMR, customers get higher control of the underlying compute and can run Spark workloads on Amazon EC2 instances, on Amazon **Elastic Kubernetes Service** (**EKS**) clusters, or on-premises using EMR on AWS Outposts. Customers can also work with other open source tools such as Apache Spark, Apache Hive, Apache HBase, Apache Flink, Apache Hudi, and Presto on Amazon EMR.

AWS Glue

On August 14, 2017, AWS released a new service called AWS Glue. AWS Glue is a serverless data integration service. AWS Glue also provides some easy-to-use features that almost eliminate the administrative overhead of infrastructure management and simplify how common data integration tasks can be integrated.

Let's look at some of the notable components of the AWS Glue feature set:

- **AWS Glue DataBrew**: Glue DataBrew is used for data cleansing and enrichment through another GUI. Creating AWS Glue DataBrew Jobs does not require the user to write any source code and the Jobs are created with the help of a GUI.

- **AWS Glue Data Catalog**: AWS Glue Data Catalog is a central catalog of metadata that can be used with other AWS services such as Amazon Athena, Amazon Redshift, and Amazon EMR.

- **AWS Glue Connections**: Glue Connections are catalog objects that help organize and store connection information to various data stores. AWS Glue Connections can also be created for Marketplace AWS Glue Connectors, which allows you to integrate with third-party data stores, such as Apache Hudi, Google Big Query, and Elastic Search.

- **AWS Glue Crawlers**: Crawlers can be used to crawl existing data and populate an AWS Glue Data Catalog with metadata.

- **AWS Glue ETL Jobs**: Glue ETL Jobs enables users to extract source data from various data stores, process it, and write output to a data target based on the logic defined in the ETL script. Users can take advantage of Apache Spark-based ETL Jobs to handle their workload in a distributed fashion. Glue also offers Python shell Jobs for ETL workloads; these don't need distributed processing.

- **AWS Glue Interactive Sessions**: Interactive sessions are managed interactive environments that can be used to develop and test AWS Glue ETL scripts.

- **AWS Glue Schema Registry**: AWS Glue Schema Registry allows users to centrally control data stream schemas and has integrations with Apache Kafka, Amazon Kinesis, and AWS Lambda.

- **AWS Glue Triggers**: AWS Glue Triggers are data catalog objects that allow us to either manually or automatically start executing one or more AWS Glue Crawlers or AWS Glue ETL Jobs.

- **AWS Glue Workflows**: Glue Workflows can be used to orchestrate the execution of a set of AWS Glue Jobs and AWS Glue Crawlers using AWS Glue Triggers.

- **AWS Glue Blueprints**: Blueprints are useful for creating parameterized workflows that can be created and shared for similar use cases.

- **AWS Glue Elastic Views**: Glue Elastic Views helps users replicate the data from one store to another using familiar SQL syntax.

This book will focus on learning about AWS Glue, diving deep into the features listed here, and learning about how these features help solve the data problems of the modern world. We will also learn about the fundamental concepts of AWS LakeFormation, which are important for securely managing and administering the data assets of an organization.

Querying data using AWS

At the beginning of this chapter, we focused on various ways to collect and organize the data from various systems to enable various downstream workloads, such as feature engineering, data exploration, and analytics. While data lakes and data meshes have reduced the entry barrier to democratize data, you may still need to access data from various purpose-built stores.

Today's applications are built around the microservice architecture, which allows teams to split vertically based on their functionality and scale independently. Organizations may have their two pizza teams working on different microservices. Each of these teams is independent and can pick its own purpose-built data stores to support its application.

In an ideal world, data from all of these purpose-built stores should flow into the data lake, but this might not always be the case. In a world where the speed of decision-making is paramount, data analysts may want to access the data and combine it even before the data starts hydrating the data lake.

This requirement led to the need for modern tools to support querying data across multiple different sources. In the AWS ecosystem, both Amazon Athena and Amazon Redshift allow you to query data across multiple data stores.

While using Amazon Athena to query S3 data cataloged in AWS Glue Catalog is quite common, Amazon Athena can also be used to query data from Amazon CloudWatch Logs, Amazon DynamoDB, Amazon DocumentDB, Amazon RDS, and JDBC-compliant relational data sources such MySQL and PostgreSQL under the Apache 2.0 license using AWS Lambda-based data source connectors. Athena Query Federation SDK can be used to write a customer connector too. These connectors return data in Apache Arrow format. Amazon Athena uses these connectors and manages parallelism, along with predicate pushdown.

Similarly, Amazon Redshift also supports querying Amazon S3 data through Amazon Redshift Spectrum. Redshift also supports querying data in Amazon RDS for PostgreSQL, Amazon Aurora PostgreSQL-Compatible Edition, Amazon RDS for MySQL, and Amazon Aurora MySQL-Compatible Edition through its Query Federation feature. Amazon Redshift offloads part of the computations to the target data stores and uses its parallel processing capabilities for the query's operation.

To handle the undifferentiated heavy lifting, AWS Glue introduced a new feature called AWS Glue Elastic Views. It allows users to use familiar SQL. It combines and materializes the data from various sources into the target. Since AWS Glue Elastic Views is serverless, users do not have to worry about managing the underlying infrastructure or keeping the target hydrated.

Summary

In this chapter, we discussed data collection practices that are used by organizations and the issue of dark data. We also discussed different storage and processing techniques, such as OLTP and OLAP, and how organizations are using a combination of these two techniques to extract value from the data gathered. We briefly discussed the evolution of data management strategies such as data warehousing, data lakes, the data lakehouse, and data meshes and the role played by ETL and ELT processes in ingesting data into OLAP systems for analysis.

Then, we introduced the Apache Spark framework and talked about how Spark executes workloads by dividing them into different Spark Jobs, stages, and tasks. After this, we discussed different services in the AWS cloud that can be used to execute Spark workloads. We introduced AWS Glue and the different features available in Glue that make it a full-fledged data integration platform and not just a managed ETL service.

In the next chapter, we will discuss the different microservices that are available in AWS Glue and how they work. We will also focus on some Glue-specific features/enhancements that make AWS Glue an ideal service for your data integration workloads.

2
Introduction to Important AWS Glue Features

In the previous chapter, we talked about the evolution of different data management strategies, such as data warehousing, data lakes, the data lakehouse, and data meshes, and the key differences between each. We introduced the Apache Spark framework, briefly discussed the Spark workload execution mechanism, learned how Spark workloads can be fulfilled on the AWS cloud, and introduced AWS Glue and its components.

In this chapter, we will discuss the different components of AWS Glue so that we know how AWS Glue can be used to perform different data integration tasks.

Upon completing this chapter, you will be able to define data integration and explain how AWS Glue can be used for this. You will also be able to explain the fundamental concepts related to different features of AWS Glue, such as AWS Glue Data Catalog, AWS Glue connections, AWS Glue crawlers, AWS Glue Schema Registry, AWS Glue jobs, AWS Glue development endpoints, AWS Glue interactive sessions, and AWS Glue triggers.

In this chapter, we will cover the following topics:

- Data integration
- Integrating data with AWS Glue
- Features of AWS Glue

Now, let's dive into the concepts of data integration and AWS Glue. We will discuss the key components and features of AWS Glue that make it a powerful data integration tool.

Data integration

Data integration is a complex operation that involves several tasks – **data discovery, ingestion, preparation, transformation**, and **replication**. Data integration is the very first step in deriving insights from data so that data can be shared across the organization for collaboration and faster decision-making.

The data integration process is often iterative. Upon completing a particular iteration, we can query and visualize the data and make data-driven business decisions. For this purpose, we can use AWS services such as Amazon Athena, Amazon Redshift, and Amazon QuickSight, as well as some other third-party services. The process is often repeated until the right quality data is obtained. We can set up a job as part of our data integration workflow to profile the data obtained against a specific set of rules to ensure that it meets our requirements. For instance, AWS Glue DataBrew offers built-in capabilities to define data quality rules and allows us to profile data based on our requirements. We will be discussing AWS Glue DataBrew Profile jobs in detail in *Chapter 4, Data Preparation*. Once the right quality data is obtained, it can be used for analysis, **machine learning** (**ML**), or building data applications.

Since data integration helps drive the business forward, it is a critical business process. This also means there is less room for error as this directly impacts the quality of the data that's obtained, which, in turn, impacts the decision-making process.

Now, let's briefly explore how data integration can be simplified using AWS Glue.

Integrating data with AWS Glue

AWS Glue was initially introduced as a **serverless ETL service** that allows users to crawl, catalog, transform, and ingest data into AWS for analytics. However, over the years, it has evolved into a **fully-managed serverless data integration service**.

AWS Glue simplifies the process of data integration, which, as discussed earlier, usually involves discovering, preparing, extracting, and combining data for analysis from different data stores. These tasks are often handled by multiple individuals/teams with a diverse set of skills in an organization.

As mentioned in the previous section, data integration is an iterative process that involves several steps. Let's take a look at how AWS Glue can be used to perform some of these tasks.

Data discovery

AWS Glue Data Catalog can be used to discover and search data across all our datasets. Data Catalog enables us to store table metadata for our datasets and makes it easy to query these datasets from several applications and services. AWS Glue Data Catalog can not only be used by AWS services such as AWS Glue, AWS EMR, Amazon Athena, and Amazon Redshift Spectrum, but also by on-premise or third-party product implementations that support the Hive metastore using the open source AWS Glue Data Catalog Client for Apache Hive Metastore (`https://github.com/awslabs/aws-glue-data-catalog-client-for-apache-hive-metastore`).

AWS Glue Crawlers enable us to populate the Data Catalog with metadata for our datasets by crawling the data stores based on the user-defined configuration.

AWS Glue Schema Registry allows us to manage and enforce schemas for data streams. This helps us enhance data quality and safeguard against unexpected schema drifts that can impact the quality of our data significantly.

Data ingestion

AWS Glue makes it easy to ingest data from several standard data stores, such as HDFS, Amazon S3, JDBC, and AWS Glue. It allows data to get ingested from SaaS and custom data stores via custom and marketplace connectors.

Data preparation

AWS Glue enables us to de-duplicate and cleanse data with built-in ML capabilities using its **FindMatches** feature. With FindMatches, we can label sets of records as either *matching* or *not matching* and the system will learn the criteria and build an ETL job that we can use to find duplicate records. We will discuss FindMatches in detail in *Chapter 14, Machine Learning Integration*.

AWS Glue also enables us to interactively develop, test, and debug our ETL code using AWS Glue development endpoints, AWS Glue interactive sessions, and AWS Glue Jupyter Notebooks. Apart from notebook environments, we can also use our favorite IDE to develop and test ETL code using AWS Glue development endpoints or AWS Glue local development libraries.

AWS Glue DataBrew provides an interactive visual interface for cleaning and normalizing data without writing code. This is especially beneficial to novice users who do not have Apache Spark and Python/Scala programming skills. AWS Glue DataBrew comes pre-packed with over 250 transformations that can be used to transform data as per our requirements.

Using AWS Glue Studio, we can develop highly scalable Apache Spark ETL jobs using the visual interface without having in-depth knowledge of Apache Spark.

Data replication

The Elastic Views feature of AWS Glue enables us to create views of data stored in different AWS data stores and materialize them in a target data store of our choice. We can create materialized views by using PartiQL to write queries.

At the time of writing, AWS Glue Elastic Views currently supports Amazon DynamoDB as a source. We can materialize these views in several target data stores, such as Amazon Redshift, Amazon OpenSearch Service, and Amazon S3.

Once materialized views have been created, they can be shared with other users for use in their applications. AWS Glue Elastic Views continuously monitors changes in our dataset and updates the target data stores automatically.

In this section, we mentioned several AWS Glue features and how they aid in different data integration tasks. In the next section, we will explore the different features of AWS Glue and understand how they can help implement our data integration workload.

Features of AWS Glue

AWS Glue has different features that appear disjointed, but in reality, they are interdependent. Often, users have to use a combination of these features to achieve their goals.

The following are the key features of AWS Glue:

- AWS Glue Data Catalog
- AWS Glue Connections
- AWS Glue Crawlers and Classifiers
- AWS Glue Schema Registry
- AWS Glue Jobs
- AWS Glue Notebooks and interactive sessions
- AWS Glue Triggers
- AWS Glue Workflows
- AWS Glue Blueprints
- AWS Glue ML
- AWS Glue Studio
- AWS Glue DataBrew
- AWS Glue Elastic Views

Now that we know the different features and services involved in executing an AWS Glue workload, let's discuss the fundamental concepts related to some of these features.

AWS Glue Data Catalog

A Data Catalog can be defined as an inventory of data assets in an organization that helps data professionals find and understand relevant datasets to extract business value. A Data Catalog acts as metadata storage (or a *metastore*) that contains metadata stored by disparate systems. This can be used to keep track of data in data silos. Typically, the user is expected to provide information about data formats, locations, and serialization deserialization mechanisms, along with the query. Metastores make it easy for us to capture these pieces of information during table creation and can be reused every time the table is used. Metastores also enable us to discover and explore relevant data in the data repository using metastore service APIs. The most popular metastore product that's used widely in the industry is Apache Hive Metastore.

AWS Glue Data Catalog is a persistent metastore for data assets. The dataset can be stored anywhere – AWS, on-premise, or in a third-party provider – and Data Catalog can still be used. AWS Glue Data Catalog allows users to store, annotate, and share metadata in AWS. The concept is similar to Apache Hive Metastore; however, the key difference is that AWS Glue Data Catalog is serverless and there is no additional administrative overhead in managing the infrastructure.

Traditional Hive metastores use **relational database management systems (RDBMSs)** for metadata storage – for example, MySQL, PostgreSQL, Derby, Oracle, and MSSQL. The problem with using RDBMS for Hive metastores is that relational database servers need to be deployed and managed. If the metastore is to be used for production workloads, then we need to factor **high availability (HA)** and redundancy into the design. This will increase the complexity of the solution architecture and the cost associated with the infrastructure and how it's managed. AWS Glue Data Catalog, on the other hand, is fully managed and doesn't have any administrative overhead (deployment and infrastructure management).

Each AWS account has one Glue Data Catalog per AWS region and is identified by a combination of `catalog_id` and `aws_region`. The value of `catalog_id` is the 12-digit AWS account number. The value of `catalog_id` remains the same for each catalog in every AWS region. For instance, to access the Data Catalog in the North Virginia AWS region, `aws_region` must be set to `'us-east-1'` and the value of the `catalog_id` parameter must be the 12-digit AWS account number – for example, `123456789012`.

AWS Glue Data Catalog is comprised of the following components:

- Databases
- Tables
- Partitions

Now, let's dive into each of these catalog item types in more detail.

Databases

A database is a logical collection of metadata tables in AWS Glue. When a table is created, it must be created under a specific database. A table cannot be present in more than one database.

Tables

A table in a Glue Data Catalog is a resource that holds the metadata for any given dataset. The following diagram shows the metadata of a table stored in the Data Catalog:

Figure 2.1 – Metadata of a table stored in a Data Catalog

All tables contain information such as the name, input format, output format, location, and schema of the dataset, as well as table properties (stored as key-value pairs – primarily used to store table statistics, the compression format, and the data format) and **Serializer-Deserializer (SerDe)** information such as SerDe name, the serialization library, and SerDe class parameters.

The SerDe library information in the table's metadata informs the query processing engine of which class to use to translate data between the table view and the low-level input/output format. Similarly, `InputFormat` and `OutputFormat` specify the classes that describe the original data structure so that the query processing engine can map the data to its table view. At a high level, the process would look something like this:

- **Read operation: Input data | InputFormat | Deserializer | Rows**

- **Write operation: Rows | Serializer | OutputFormat | Output data**

> **Table Versions**
>
> It is important to note that AWS Glue supports versioning catalog tables. By default, a new version of the table is created when the table is updated. However, we can use the `skipArchive` option in the AWS Glue `UpdateTable` API to prevent AWS Glue from creating an archived version of the table. Once the table is deleted, all the versions of the table will be removed as well.

Partitions

Tables are organized into partitions. Partitioning is an optimization technique by which a table is further divided into related parts based on the values of a particular column(s). A table can have a combination of multiple partition keys to identify a particular partition (also known as `partition_spec`).

For instance, a table for `sales_data` can be partitioned using the `country`, `category`, `year`, and `month` columns.

The following is an example query for this:

```
SELECT *
FROM sales_data
WHERE country='US' AND category='books' AND year='2021' AND
month='10'
```

Over time, as data grows, the number of partitions that can be added to a table can grow significantly based on the partition keys defined in the table. Fetching metadata for all these partitions can introduce a huge amount of latency. To address this issue, Glue allows users to add indexes for partition keys (refer to https://docs.aws.amazon.com/glue/latest/dg/partition-indexes.html) and when the `GetPartitions` API is called by the query processing engine with a particular query expression, the API will try to return a subset of partitions instead of all partitions. By default, if partition indexes are not defined on a table, the `GetPartitions` API will return all the partitions and perform filtering on the returned API response.

Now, let's consider an example database setup, as shown in the following diagram. If partition indices idx_1, idx_2, and idx_3 are not defined, all the partitions in the table are returned when the GetPartitions API is called on the table_1 or table_2 table in the catalog_database database. However, if the partition indices are defined, only the partitions for a specific table with indices that match the values passed in the query will be returned. This reduces the effort involved by the query engine in fetching partition metadata:

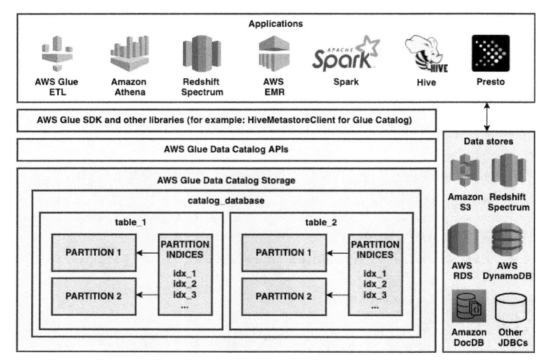

Figure 2.2 – Structure of AWS Glue Data Catalog

Limitations of Using Partition Indexes

Once a partition index has been added to a table in Glue Data Catalog, index keys' data types will be validated for all new partitions added to this table. It is important to make sure the values for the columns listed as partition indexes adhere to the data type defined in the schema. If this is not the case, the partition won't be created. Once a table has been created, the names, data types, and the order of the keys registered as part of the partition index cannot be modified.

Now that we understand the fundamentals of AWS Glue Data Catalog, in the next section, we'll explore AWS Glue connections and understand how they enable communication with VPC/on-premise data stores.

Glue connections

AWS Glue connections are resources stored in AWS Glue Data Catalog that contain connection information for a given data store. Typically, an AWS Glue connection contains information such as login credentials, **connection strings** (**URIs**), and VPC configuration (VPC subnet and security group information), which are required by different AWS Glue resources to connect to the data store. The contents of an AWS Glue connection differ from one connection type to another.

Aws Glue Connection is a feature available in AWS Glue that is not present in traditional Hive Metastores. Connections enable AWS Glue workloads (crawlers, ETL Jobs, development endpoints, and interactive sessions) to access data stores that are typically not exposed to the public internet – for example, RDS database servers and on-premise data stores.

Glue users can define connections that can be used to connect to data sources or targets. At the time of writing, there are eight types of Glue connections, each of which is designed to establish a connection with a specific type of data store: **JDBC**, **Amazon RDS**, **Amazon Redshift**, **Amazon DocumentDB**, **MongoDB**, **Kafka**, **Network**, and **Custom/Marketplace** connections.

The parameters required for each connection type are different based on the type of data store the connection will be used for. For instance, the JDBC connection type requires SSL configuration, JDBC URI, login credentials, and VPC configuration.

The Network connection type is useful when users wish to route the traffic via an Amazon S3 VPC endpoint and do not want their Amazon S3 traffic to traverse the public internet. This pattern is usually used by organizations for security and privacy reasons. The Network connection type is also useful when users wish to establish connectivity to a custom data store (for example, an on-premise Elasticsearch cluster) within the ETL job and not define connection parameters in a Glue connection.

When a Glue connection is attached to any Glue compute resource (Jobs, Crawlers, development endpoints, and interactive sessions), behind the scenes, Glue creates EC2 **Elastic Network Interfaces** (**ENIs**) with the VPC configuration (subnet and security groups) specified by the user. These ENIs are then attached to compute resources on the server side. This mechanism is used by AWS Glue to communicate with VPC/on-premise data stores.

> **Elastic Network Interfaces (ENIs)**
> An ENI is essentially a virtual network interface that facilitates networking capabilities for compute resources on AWS.

Let's use the following diagram to understand how Glue uses ENIs to communicate with VPC/on-premise data stores:

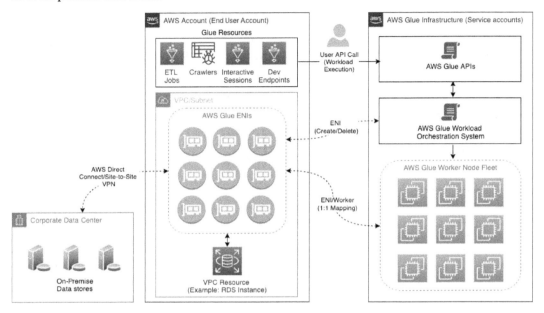

Figure 2.3 – VPC-based data store access from AWS Glue using ENIs

Here, when a user makes an API call to execute the AWS Glue workload, the request is submitted to the AWS Glue workload orchestration system, which will calculate the amount of compute resources required and allocates workers from the worker node fleet.

If the workload being executed requires VPC connectivity, ENIs are created in the end user AWS account and are attached to worker nodes. There is a 1:1 mapping between the worker nodes and the ENIs; the worker nodes use these ENIs to communicate with the data stores. These data stores can be present in an AWS account or they could be present in the end user's corporate data center.

ENIs that are created during workload execution are automatically cleared by AWS Glue (this can take up to 10 to 15 minutes). AWS Glue uses the same IAM role that's used for workload execution to delete ENIs once the workload has finished executing. If the IAM role is not available during ENI deletion (for instance, if the IAM role was deleted immediately after workload execution), the ENIs will stay active indefinitely until they are manually deleted by the user.

> **Note**
>
> It is important to make sure that the subnet being used by the Glue connection has enough IP addresses available as each Glue resource creates multiple ENIs (each of which consumes one IP address) based on the compute capacity required for workload execution.
>
> At the time of writing, a Glue resource can only use one subnet. If multiple connections with different subnets are attached, the subnet settings from the first connection will be used by default. However, if the first connection is unhealthy for any reason – for instance, if the availability zone is down – then the next connection is used.

In the next section, we will explore Glue Crawlers and classifiers and how they aid in data discovery.

AWS Glue crawlers

A Crawler is a component of AWS Glue that helps crawl the data in different types of data stores, infers the schema, and populates AWS Glue Data Catalog with the metadata for the dataset that was crawled.

Crawlers can crawl a wide variety of data stores – Amazon S3, Amazon Redshift, Amazon RDS, JDBC, Amazon DynamoDB, and DocumentDB/MongoDB to name a few. This is a powerful tool that's available for data discovery in AWS Glue.

> **Glue Connections for Crawlers**
>
> For a crawler to crawl a VPC resource or on-premise data stores such as Amazon Redshift, JDBC data stores (including Amazon RDS data stores), and Amazon DocumentDB (MongoDB compatible), a Glue connection is required.
>
> Crawlers are capable of crawling S3 buckets without using Glue connections. However, a Network connection type is required if you must keep S3 request traffic off the public internet.
>
> For a crawler with a Glue connection, it is recommended to have at least 16 IP addresses available in the subnet. When a connection is attached to a Glue resource, multiple ENIs are created to run the workload.

Now that we know what data stores are supported by AWS Glue crawlers, let's explore how they work. Take a look at the following diagram:

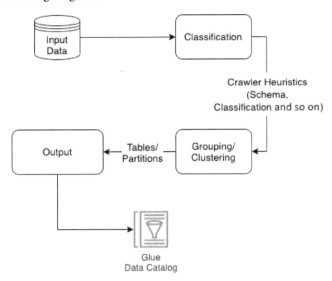

Figure 2.4 – Workflow of a Glue crawler

The workflow of a crawl can be divided into three stages:

1. **Classification**: In this stage, the crawler traverses the input data store and uses classifiers (built-in/custom) to classify the source data. When a crawler is created, users can choose one or more custom classifiers that will be used by the crawler during classification to identify the format of the data to infer the schema. Input data is evaluated against the list of classifiers in the same order; the `certainty=1.0` value (100% certainty) is returned for the first classifier to successfully recognize the data store. This will be used for schema inference. If none of the custom classifiers are successful in recognizing the data store, the crawler will move on to evaluate the data store against a list of built-in classifiers (`https://docs.aws.amazon.com/glue/latest/dg/add-classifier.html#classifier-built-in`). Finally, the `certainty` score decides how the data store is classified. If none of the classifiers return `certainty=1.0`, the output of the classifier with the highest `certainty` value will be used by Glue for table creation. If no classifier returned a `certainty` value that was higher than `0.0`, Glue creates a table with the UNKNKOWN classification. The crawler will use the selected classifier to infer the schema of the dataset.

2. **Clustering/Grouping**: The output from the classification stage is used by the crawler and the data is grouped based on crawler heuristics (schema, classification, and other properties). Table or partition objects are created based on clustered data using Glue crawler's internal logic wherein schema similarity, `compressionType`, directory structure, and other factors are considered.

3. **Output**: In this stage, the table or partition objects that were created in the clustering stage will be written to Glue Data Catalog using Glue API calls. If the table(s) already exists and this is the first run of the crawler, a new table with a hash string suffix will be created. However, if the crawler is running on an existing catalog table or if this is the crawler's subsequent run, updates to catalog table(s) will be handled according to the crawler's `SchemaChangePolicy` settings. (In some edge cases, the `SchemaChangePolicy` property will be ignored and new tables and partitions might be created. This depends on the data source type defined in the crawler.) The tables created by the crawler are placed in the database that's been nominated. If no database has been set up in the crawler settings, the tables will be placed in the `default` database.

> **Note**
>
> At the time of writing, the maximum runtime for any crawler is 24 hours. After 24 hours, the crawler's run is automatically stopped with the `CANCELLED` status.
>
> Users are allowed to specify a table prefix in the crawler's settings. The length of this prefix cannot exceed 64 characters.
>
> The maximum length of the name of the table cannot exceed 128 characters. The crawler automatically truncates the names generated to fit this limit.
>
> If the name of the table that's generated is the same as the name of an existing table, the Glue crawler automatically adds a hash string suffix to ensure that the table name is unique.

For Amazon S3 data store crawls, the crawler will read all the files in the path specified by default. The crawler will classify each of the files available in the S3 path and persist the metadata to the crawler's service side storage (not to be confused with AWS Glue Data Catalog). Metadata gets reused and the new files are crawled during the subsequent crawler runs and the metadata stored on the service side is updated as necessary.

> **Note**
>
> When a new version of an existing file is uploaded to Amazon S3 after a crawl, a subsequent crawl will consider this a new file. Then, the new file will be included in the new crawl.

For plain text file formats (`CSV`, `TSV`, `JSON`), it is not feasible to crawl the entire file for larger files to evaluate the schema. Therefore, the crawler will read the initial 1 to 10 MB of data of each file, depending on the file format, and ensure that at least one record is read (if the record's size is greater than 1 MB). The schema is inferred based on the data read into the buffer.

For the JDBC, Amazon DynamoDB, and Amazon DocumentDB (with MongoDB compatibility) data stores, the stages of the crawler workflow are the same, but the logic that's used for classification and clustering is different for each data store type. The classification of the table(s) is decided based on the data store type/database engine.

For JDBC data stores, Glue connects to the database server, and the schema is obtained for the tables that match the `include` path value in the crawler settings. Similar logic is used for DocumentDB/MongoDB data stores and the schema of MongoDB collections is inferred.

In the next section, we'll explore some of the key features of AWS Glue crawlers.

Key features of Glue crawlers

AWS Glue crawlers have several features and configuration options that make it easy to discover data and populate the Data Catalog. In the following sub-sections, we will look at some of the features of AWS Glue crawlers that help optimize the data discovery process.

Data sampling – DynamoDB and DocumentDB/MongoDB

By default, Glue performs a full scan of the DynamoDB table and MongoDB collection to infer the schema. This operation can be time-consuming when the table is not a high throughput table. To address this issue, we can enable the **Data sampling** feature. When sampling is enabled, Glue will scan a subset of the data rather than perform a full scan.

Data sampling – Amazon S3

By default, Glue will read all the files in the Amazon S3 data store. The **Data sampling** feature is available for Amazon S3 data stores as well. This will reduce crawler runtime significantly. Users can specify the number of objects (in a value range of 1 to 249) in each leaf directory to be crawled. This feature is helpful when the users have prior knowledge of data formats and the schemas in the directories do not change.

Amazon S3 data store – incremental crawl

In Amazon S3, crawlers are used to scan new data and register new partitions in Glue Data Catalog. This can be further optimized by enabling the **Incremental Crawls** feature (`https://docs.aws.amazon.com/glue/latest/dg/incremental-crawls.html`). This feature is best suited for datasets that have stable schemas. When this feature is enabled, only new directories that have been added to the dataset are crawled. This feature can be enabled in the AWS Glue console by selecting the **Crawl new folders only** checkbox.

Amazon S3 data store – table-level specification

While discussing the Clustering/Grouping stage in the crawler workflow, we talked about how crawlers use internal logic based on data store properties (schema similarity, compression, and directory structure) to decide whether a directory that's stored in Amazon S3 is a partition or a table. In some use cases, two or more tables can have a similar schema, which causes the crawler to mark these tables as partitions of the same tables instead of creating separate tables. Using the `TableLevelConfiguration` option in the Grouping policy, we can inform the crawler of where the tables are located and how we want the partitions to be created. Let's consider an example.

Imagine that we have the following directory structure in an Amazon S3 bucket:

```
s3://myBucket/prefix/data/year=2021/month=10/day=08/hour=12/
file1.parquet
s3://myBucket/prefix/data/year=2021/month=11/day=10/hour=12/
file2.parquet
```

All the Parquet files in the S3 location have the same schema. If we point the crawler to `s3://myBucket/prefix/data/` and run the crawler, it will create a single table and four partition keys – `year`, `month`, `day`, and `hour`. However, consider a scenario where we want to create separate tables for each month. Typically, the solution is to add multiple `include_path` for the crawler to crawl – for example, `s3://myBucket/prefix/data/year=2021/month=10/` and `s3://myBucket/prefix/data/year=2021/month=11/`. Now, if there are hundreds of such paths and we want to create a table for all of them, it would not be feasible to add all the paths to the crawler configuration.

The same outcome can be achieved by using the **Table level** feature. We can set the `Table Level` parameter to 5 in crawler output settings. This will instruct the crawler to create the tables at level 5 from root (which corresponds to `month` in the directory structure specified previously). Now, the crawler will create two tables called `month_10` and `month_11`.

In this section, we discussed some of the key features of Glue crawlers that can be enabled to enhance the performance or precision of the crawler. Please refer to the AWS Glue documentation for an exhaustive list of available crawler features.

Custom classifiers

While discussing the different stages in the crawler workflow, we mentioned it is possible to add custom classifiers to Glue crawlers. Classifiers are responsible for determining the file's classification string (for example, `parquet`) and the schema of the file. When built-in classifiers are not capable of crawling the dataset or the table that's been created requires customization, users can define custom classifiers and crawlers that will use the logic defined to create schema based on the type of classifier.

> **Note**
> If a custom classifier definition gets changed after a crawl, any data that was previously crawled will not be reclassified as the crawler keeps track of metadata for previously crawled data. If a mistake was made during classifier configuration, just fixing the classifier configuration will not help. The only way to reclassify already classified data is to delete and recreate the crawler with the updated classifier attached.

At the time of writing, users can define the following types of custom classifiers:

- **Grok classifiers**: Grok patterns are named sets of regular expressions that can match one line of data at a time. When the dataset matches the grok pattern specified, the structure of the dataset is determined and the data is tokenized and mapped to fields defined in the pattern specified. The `GrokSerDe` serialization library is used for tables created in Glue Data Catalog.

- **XML classifiers**: XML classifiers allow users to define the tag in the XML files that contains the records. For instance, let's consider the following XML sample:

```
<?xml version="1.0"?>
<catalog>
   <book id="bk101">
      <author>Gambardella, Matthew</author>
      <title>XML Developer's Guide</title>
   </book>
   <book id="bk102">
      <author>Ralls, Kim</author>
      <title>Midnight Rain</title>
   </book>
</catalog>
```

In this case, using `book` as the XML row tag will create a table containing two columns – `author` and `title`.

> **Note**
> It is important to note that an element that holds the record cannot be self-closing. For example, `<book id="bk102"/>` will not parse. Empty elements should have a separate starting and closing tag; for example, `<book id="bk102"> </book>`.

- **JSON Classifiers**: Using JSON classifiers, users can specify the JSON path where individual records are present. This classifier uses `JsonPath` expressions as input and accesses the items in the JSON based on the path specified. The syntax for `JsonPath` can be found in `https://docs.aws.amazon.com/glue/latest/dg/custom-classifier.html#classifier-values-json`.

Let's consider the following sample JSON dataset:

```
{
  "book": [
    {
      "category": "reference",
```

```
        "author": "Nigel Rees",
        "title": "Sayings of the Century",
      },
      {
        "category": "fiction",
        "author": "Herman Melville",
        "title": "Moby Dick",
      }
    ]
  }
```

To extract individual books as records, we can use the `$.book[*]` JSON path.

- **CSV Classifiers**: CSV classifiers allow users to specify different options to crawl delimited files. Users can specify custom delimiters, quote symbols, options about the header, and validations (this allows files with a single column – trim whitespace before column identification).

In the next section, we will discuss the **AWS Glue Schema Registry** (**GSR**) and how we can handle evolving schemas to stream data stores centrally.

AWS Glue Schema Registry

With organizations' growing need for real-time analytics, streaming data processing is becoming more and more important in an enterprise data architecture. Organizations collect real-time data from a wide variety of sources, including IoT sensors, user applications, application/security logs, and geospatial services. Collecting real-time data gives organizations visibility into aspects of their business and customer activity and enables them to respond to emerging situations. For example, sensors in industrial equipment send data to streaming applications. The application monitors the data that's been sent by the sensors and detects any potential faults in the machinery.

Over time, as organizations grow, more data sources (for example, additional sensors or trackers) can be used to enrich the data streams with additional information that's vital to the business. This creates a problem for all the downstream applications that already consume these data streams as they must be upgraded to handle these schema changes. Schema registries can be used to address the issues caused by schema evolution and allow streaming data producers and consumers to discover and manage schema changes, as well as adapt to these changes based on user settings.

GSR is a feature available in AWS Glue that allows users to discover, control, and evolve schema for streaming data stores centrally. Glue Schema registries support integrations with a wide variety of streaming data stores such as Apache Kafka, Amazon Kinesis Data Streams, Amazon **Managed Streaming for Apache Kafka** (**MSK**), Amazon Kinesis Data Analytics for Apache Flink, and AWS Lambda by allowing users to enforce and manage schemas.

AWS Glue Schema Registry is fully managed, serverless, and available for users free of cost. At the time of writing, GSR supports the AVRO, JSON, and **protocol buffer** (**protobuf**) data formats for schemas. JSON schema validation is supported by the Everit library (`https://github.com/everit-org/json-schema`).

> **Note**
> The AWS Glue Schema Registry currently supports the Java programming language. Java version 8 (or above) is required for both producers and consumers.

Schema registries use serialization and deserialization processes to help stream data that producers and consumers enforce a schema on records. If a schema is not available in the schema registry, it must be registered for use (auto-registration of the schema can be enabled for any new schema to be auto-registered).

Upon registering a schema in the schema registry, a schema version identifier will be issued to the serializer. If the schema is already available in the GSR and the serializer is using a newer version of the schema, the GSR will check the compatibility rule to make sure that the new version is compatible. The schema will be registered as a new version in the GSR.

When a producer has its schema registered, the GSR serializer validates the schema of the record with where the schema is registered. If there is a mismatch, an exception will be returned. Producers typically cache the schema versions and match the schema against the versions available in the cache. If there is no version available in the cache that matches the schema of the record, GSR will be queried for this data using the `GetSchemaVersion` API.

If the schema is validated using a version in the GSR, the schema version ID and definition will be cached locally by the producer. If the record's schema is compliant with the schema registered, the record is decorated with the schema version ID and then serialized (based on the data format selected), compressed, and delivered to the destination.

Once a serialized record has been received, the deserializer uses the version ID available in the payload to validate the schema. If the deserializer has not encountered this schema version ID before, the GSR is queried for this and the schema version is cached in local storage.

If the schema version IDs in the GSR/cache match the version in the serialized record, the deserializer decompresses and deserializes the data and the record is handed off to the consumer application. However, if the schema version ID doesn't match the version IDs available in cache or the GSR, the consumer application can log this event and move on to other records or halt the process based on user configuration.

SerDe libraries can be added to both producer and consumer applications by adding the `software.amazon.glue:schema-registry-serde` Maven dependency (`https://mvnrepository.com/artifact/software.amazon.glue/schema-registry-serde`). Refer to `https://docs.aws.amazon.com/glue/latest/dg/schema-registry-integrations.html` for example producer and consumer implementations.

In the next section, we will explore one of the key components of AWS Glue: ETL jobs.

AWS Glue ETL jobs

ETL is one of the main components of data integration. Designing an ETL pipeline to ingest and transform data can be time-consuming as data grows over time. Setting up, managing, and scaling the infrastructure takes up most of the effort in a typical on-premise data engineering project. Glue ETL almost eliminates the effort involved in setting up infrastructure as it is fully managed and serverless. All the effort involved in setting up hosts, configuration management, and patching is handled behind the scenes by the Glue ETL engine so that the user can focus on developing ETL scripts and managing the necessary dependencies. Of course, Glue ETL is not a silver bullet that eliminates all the challenges involved in running an ETL workload, but with the right design and strategy, it can be a great fit for almost all organizations.

At the time of writing, Glue allows users to create three different types of ETL jobs – Spark ETL, Spark Streaming, and Python shell jobs. The key differences between these job types are in the libraries/packages that are injected into the environment during job orchestration on the service side and billing practices.

During job creation, users can use the AWS Glue wizard to generate an ETL script for Spark and Spark Streaming ETL jobs by choosing the source, destination, column mapping, and connection information. However, for Python shell jobs, the user will have to provide a script. At the time of writing, Glue ETL supports Scala 2 and PySpark (Java and R jobs are currently not supported) for Spark and Spark Streaming jobs and Python 3 for Python shell jobs.

When Glue ETL was introduced, Python 2 support was available in Glue ETL v0.9 and 1.0. However, since Python 2 was sunsetted by the open source community, ETL job environments that used Python 2 were phased out. This is specified in the policy available in the Glue EOS milestones documentation (`https://docs.aws.amazon.com/glue/latest/dg/glue-version-support-policy.html#glue-version-support-policy-milestones`).

> **Note**
>
> AWS Glue allows multiple connections to be attached to ETL jobs. However, it is important to note that a Glue job can use only one subnet for VPC jobs. If multiple connections are attached to a job, only the first connection is attached to the ETL job.

There are some advanced features that users can select during job creation, such as job bookmarks, continuous logging, Spark UI, and capacity settings (the number of workers and worker type). Glue allows users to inject several job parameters (including Spark configuration parameters) so that they can alter the default Spark behavior.

Glue ETL introduces quite a lot of advanced Spark extensions/APIs and transformations to make it easy to achieve complex ETL operations. Let's look at some of the important extensions/features that are unique to Glue ETL.

GlueContext

The GlueContext class wraps Apache Spark's SparkContext object and allows you to interact with the Spark platform. GlueContext also serves as an entry point to several Glue features – DynamicFrame APIs, job metrics, continuous logging, job bookmarks, and more. The GlueContext class provides methods to create DataSource and DataSink variables, which is essential in reading/writing Glue DynamicFrames. GlueContext is also helpful in setting the number of output partitions (the default is 20) in a DynamicFrame when the number of output partitions is below the minimum threshold (the default is 10).

GlueContext can be initialized using the following code snippet:

```
sc = SparkContext()
glueContext = GlueContext(sc)
```

Once the GlueContext class has been initialized, we can use the object created to extract the SparkSession object:

```
spark = glueContext.spark_session
```

DynamicFrame

DynamicFrame is a key functionality of Glue that enables users to perform ETL operations efficiently. As defined in the AWS Glue documentation (https://docs.aws.amazon.com/glue/latest/dg/glue-etl-scala-apis-glue-dynamicframe-class.html), a DynamicFrame is a distributed collection of *self-describing* DynamicRecord objects (comparable to a Row in Spark DataFrame, but DynamicRecords do not require them to adhere to a set schema). Since the records are self-describing, DynamicFrames do not require a schema to be created and can be used to read/transform data with inconsistent schemas. SparkSQL performs two passes over the dataset to read data since a Spark DataFrame expects a well-defined schema for data ingestion – the first one to infer the schema from the data source and the second to load the data. Even though SparkSQL supports schema inference, it is still limited in its capabilities. Glue infers the schema for a given dataset at runtime when required and does not pre-compute the schema. Any schema inconsistencies that are detected are encoded as choice (or union) data types that can be later resolved to make the dataset compatible with targets that require a fixed schema.

DynamicFrames can be created using different APIs, depending on the use case. The following syntax can be used to create a DynamicFrame using a Glue Data Catalog table in PySpark (documentation on this can be found at `https://docs.aws.amazon.com/glue/latest/dg/aws-glue-api-crawler-pyspark-extensions-dynamic-frame-reader.html#aws-glue-api-crawler-pyspark-extensions-dynamic-frame-reader-from_catalog`):

```
datasource0 = glueContext.create_dynamic_frame.from_
catalog(name_space='my_database', table_name='my_table',
transformation_ctx='datasource0')
```

This statement will create a DynamicFrame object called `datasource0` for the `my_table` table in the `my_database` database. This statement will use the `GlueContext` object, which uses the Glue SDK, to connect to Glue Data Catalog and fetch the data store classification and properties to create the object. Additionally, users can pass additional options into this statement by using the `additional_options` parameter and a pushdown predicate filter expression to apply filters to the dataset while it is being read using the `push_down_predicate` parameter.

In the preceding source code example, we used the `from_catalog` method to create `datasource0`. Similarly, DynamicFrames can be created using the following methods:

- `from_options`: This method allows users to create DynamicFrames by manually specifying the connection type, options, and format. This method provides users with the flexibility to customize options for a data store.

- `from_rdd`: This method allows users to create DynamicFrames using Spark **Resilient Distributed Datasets (RDDs)**.

The `DynamicFrame` class provides several transformations that are unique to Glue and also allows conversion to and from Spark DataFrames. This makes it incredibly easy to integrate the existing source code and take advantage of the operations that are available in Spark DataFrames but not yet available in Glue DynamicFrames. Users can convert a DynamicFrame into a Spark DataFrame using the following syntax:

```
df = datasource0.toDF()
```

Here, `datasource0` is the DynamicFrame and `df` is the Spark DataFrame that was returned.

Similarly, a Spark DataFrame can be converted into a Glue DynamicFrame using the following code snippet:

```
from awsglue.dynamicframe import DynamicFrame
dyf = DynamicFrame.fromDF(dataframe=df, glue_ctx=glueContext,
name="dyf")
```

Both Spark DataFrames and Glue DynamicFrames are high-level Spark APIs that interact with Spark RDDs. That being said, the structure of a DynamicFrame is significantly different from that of a DataFrame.

While a DynamicFrame provides a flexible set of APIs to access and transform datasets, there are some areas where DataFrames outshine DynamicFrames. For instance, since DynamicFrames are based on raw RDDs and not Spark DataFrames, it does not take advantage of Spark's catalyst optimizer. This is the reason why some aggregation operations (such as joins) perform better with Spark DataFrames than Glue DynamicFrames. In such cases, we can convert it into Spark DataFrame to take advantage of the performance boost offered by the catalyst optimizer. Also, some functions/classes are only available for Spark DataFrames, such as Spark MLlib and SparkSQL functions.

It is important to note that converting a Glue DynamicFrame into a Spark DataFrame requires a full Map stage in Spark. This should only be used when necessary. DynamicFrame to DataFrame conversion blocks Spark from optimizing workloads based on upstream code and dramatically reduces efficiency.

Job bookmarks

Bookmarking is a key feature available in Glue ETL that allows users to keep track of data that was processed and written. During the next job run, only new data will be processed. This is an extremely useful option that helps in processing large datasets that are constantly growing. While specifying the syntax for DynamicFrame creation from the Data Catalog table earlier, the `transformation_ctx` parameter (`https://docs.aws.amazon.com/glue/latest/dg/monitor-continuations.html#monitor-continuations-implement-context`) was mentioned in the code snippet. This parameter is used as the identifier for the job bookmark's state, which is persisted across job runs. Job bookmarks are supported for S3 and JDBC-based data stores. At the time of writing, the `JSON`, `CSV`, `Apache Avro`, `XML`, `Parquet`, and `ORC` file formats are supported with S3 data stores. For an Amazon S3 data source, job bookmarks keep track of the *last modified timestamp* of the objects processed. This information is then persisted in the bookmark storage on the service side. During the next jobRun, the information that was collected by the bookmark in the previous jobRun will be used to filter out already processed objects; then, new objects are processed.

> **Note:**
> A new version of an already existing object is still considered a new object and will be processed in the new jobRun.
>
> At the time of writing, Glue DynamicFrames only support Spark `SaveMode.Append` mode for writes. So, if a new version of an object was added to the data store, there is a possibility of data duplication in the target data store. This must be handled by the user with custom logic in the ETL script.

For JDBC data stores, job bookmarks use key(s) specified by the user and the order of the keys to track the data being processed. If no keys are specified by the user, Glue will use the primary key of the JDBC table. It is important to note that Glue will not accept the primary key, which is not sequential (there shouldn't be gaps in the values). In such cases, we just have to specify the column manually as the key in the `jobBookmarkKeys` parameter in `additional_options` (`connection_options` for the `from_options` API). This will force Glue to use the key for bookmarking.

> **Note – Bookmarking with JDBC Data Stores**
>
> If more than one key is specified, Glue will combine the keys to form a single composite key. However, if a key is not specified, Glue will use the primary key of the JDBC table as the key (only if the key is increasing/decreasing sequentially). If keys are specified by the user, gaps are allowed for these keys. However, the keys have to be sorted – either increasing or decreasing.

GlueParquet

Parquet is one of the most popular file formats used for data analytics workloads. We already know that DynamicFrames contain self-describing dynamic records with flexible schema requirements – the same principle can be applied while writing parquet datasets. By setting the output format as `glueparquet`, users can take advantage of the custom-built parquet writer, which computes the schema dynamically during write operations.

This writer computes a schema for the dataset that's available in memory. Performing a pass over the dataset in memory is computationally cheaper compared to performing a pass over the data in disk or Amazon S3. A buffer is created for each column that's encountered in this pass and data is inserted into these buffers. If the writer comes across a new column, a new buffer is initialized and data is written into it. When the file is to be written to the target, the buffers for all the columns are aggregated and flushed. This approach helps avoid schema computation during a parquet write to the target.

This writer can be used by setting `format="glueparquet"` or `format=parquet` along with the `format_option` parameter, where `useGlueParquetWriter` is set to `true`. The data that's written to the target data store is still in parquet format, however, the writer uses different logic to write data to the target.

It is important to note that the `GlueParquet` writer only supports schema evolution – that is, adding/removing columns – and does not support changing data types for existing columns. The `glueparquet` format can only be used for write operations. To read the data written by this writer, we still have to use `format=parquet`.

Now that we understand the fundamentals of AWS Glue ETL Jobs, we will explore Glue development endpoints, which can be used by end users to develop ETL scripts for ETL Jobs.

Glue development endpoints

When Glue ETL was introduced, the orchestration service on the service side provisioned Spark clusters on-demand and configured them. This approach introduced a significantly high cold-start of about 10 to 15 minutes (with a timeout of 25 minutes). However, this all changed with the introduction of Glue v2.0, which used a different infrastructure provisioning mechanism. This cut down the cold-start from 10 to 15 minutes to 10 to 30 seconds (with a timeout of 5 minutes).

Glue ETL is a heavily customized environment with a lot of proprietary classes and libraries pre-packaged and ready for use. Developing Glue ETL scripts proved to be a challenge as Glue ETL was not initially designed for instant feedback. One mistake in the ETL script during development can take up to 10 to 15 minutes for the job to start running – only then will the user be able to see the mistake. This can be a bit frustrating and lead to poor developer experience.

Glue development endpoints were introduced to address this pain point. This feature allows users to create an environment for Glue ETL development wherein the developer/data engineer can use Notebook environments (Jupyter/Zeppelin), **read-eval-print loop** (**REPL**) shells, or IDEs to develop ETL scripts and test them instantly using the endpoint.

Glue development endpoints are essentially long-running Spark clusters that run on the service side with all the pre-packaged libraries and dependencies available in the ETL environment ready for use. Apache Livy and Zeppelin Daemon are also installed in a development environment, which enables users to use Jupyter and Zeppelin notebook environments for ETL script development.

While Glue development endpoints provided a mechanism for users to develop and test Glue ETL scripts, it required users to create and manage development endpoints and notebook servers. Glue interactive sessions made this process easier by allowing users to use their own notebook environments.

In the next section, we'll explore interactive sessions in more detail.

AWS Glue interactive sessions

Glue interactive sessions introduced the optimizations that are used for Glue ETL v2.0 infrastructure provisioning to development environments. This can be used by users via custom-built Jupyter kernels. Glue interactive sessions are not long-running Spark clusters and can be instantaneously created or torn down (using the `%delete_session` magic command). The cold-start duration is significantly less (approximately 7 to 30 seconds) compared to development endpoints (10 to 20 minutes).

Interactive sessions make it easier for users to access the session from Jupyter notebook environments hosted anywhere (the notebook server can be running locally on a user workstation as well) with minimal configuration. The session is created on-demand when the user starts the session in the notebook using the `%new_session` magic command and can be configured to auto-terminate when there is no user activity for a set period (with the `%idle_timeout` magic variable).

To set up Glue interactive sessions, all we need is a Jupyter environment with Python 3.6 or above with Glue kernels installed and connectivity to AWS Glue APIs. We can follow the steps available at `https://docs.aws.amazon.com/glue/latest/dg/interactive-sessions.html` to set up an interactive sessions environment.

The configuration for the interactive session (similar to the ETL job configuration) can be done using the magic variables that are available in Glue kernels. An exhaustive list of the magic variables that are available in Glue kernels can be found at `https://docs.aws.amazon.com/glue/latest/dg/interactive-sessions-magics.html`.

As we can see, with minimal setup, we can start developing ETL scripts from any Jupyter environment, so long as Glue kernels are installed and connectivity to Glue APIs is available.

In the next section, we will explore Glue triggers, which allow us to orchestrate complex Glue workloads since we can execute Glue jobs or crawlers on-demand, based on a schedule or the outcome of a condition.

Triggers

Triggers are Glue Data Catalog objects that can be used to start (manually or automatically) one or more crawlers or ETL jobs. Triggers allow users to chain crawlers and ETL jobs that depend on each other.

There are three types of triggers:

- **On-demand triggers**: These triggers allow users to start one or more crawlers or ETL jobs by activating the trigger. This can be done manually or via an event-driven API call.

- **Scheduled triggers**: These time-based triggers are fired based on a specified `cron` expression.

- **Conditional triggers**: Conditional triggers fire when the previous job(s)/crawler(s) satisfy the conditions specified. Conditional triggers watch the status of the jobs/crawlers specified – success, failed, timeout, and so on. If the list of conditions specified is satisfied, the trigger is fired.

> **Note**
>
> A **scheduled/conditional** trigger must be in the ACTIVATED state (and not in the CREATED/DEACTIVATED state) for the trigger to start firing based on a schedule or a specific condition. This is the first thing that the user can check if a **scheduled/conditional** trigger is not firing as expected.
>
> When multiple glue resources are chained using triggers, the dependent job/crawler is started, provided that the previous job/crawler was started by a trigger.
>
> If we are designing a chain of dependent jobs/crawlers, it is important to make sure that all the jobs and crawlers in the chain are descendants of the same scheduled/on-demand trigger.

Triggers can be part of a Glue workflow or they can be independent. We can design a chain of dependent jobs and crawlers. However, Glue workflows are preferable while designing complex multi-job ETL operations. We will discuss Glue workflows and blueprints in detail later in this book.

Summary

In this chapter, we introduced different AWS Glue microservices, including Glue Data Catalog, crawlers, classifiers, connections, ETL jobs, development endpoints, the schema registry, and triggers. We also discussed the key features of each of those different microservices to understand how they aid in different stages of data integration.

Then, we explored the structure of Glue Data Catalog, Glue connections, and the mechanisms used by crawlers and classifiers for data discovery. We also talked about the different classes/APIs that are available in AWS Glue ETL that help with data preparation and transformation. After this, we briefly explored development endpoints and interactive sessions, which make it easy for data engineers/developers to test and write ETL jobs. Then, we explored AWS Glue Triggers and understood how they help us orchestrate complex ETL workflows by allowing Glue users to chain crawlers and ETL jobs based on specific conditions or a schedule.

In the next chapter, we will discuss some of the key features of AWS Glue ETL jobs in detail and explore how they can be used to prepare and ingest data from different types of data stores.

3
Data Ingestion

In the previous chapter, we discussed the fundamental concepts and inner workings of the various features/microservices that are available in AWS Glue, such as Glue Data Catalog, connections, crawlers, and classifiers, the schema registry, Glue ETL jobs, development endpoints, interactive sessions, and triggers. We also explored how AWS Glue crawlers aid in **data discovery** by crawling different types of data stores – Amazon S3, JDBC (Amazon RDS or on-premises databases), and DynamoDB/ MongoDB/DocumentDB infer the schema and populate AWS Glue Data Catalog. While discussing Glue ETL in the previous chapter, we introduced a few of the important extensions/features of Spark ETL, including `GlueContext`, `DynamicFrame`, `JobBookmark`, and `GlueParquet`. In this chapter, we will see them in action by looking at some examples.

In this chapter, we will be discussing some of the components of AWS Glue mentioned in the previous paragraph – specifically **Glue ETL jobs, the schema registry**, and **Glue custom/Marketplace connectors** – in further detail and exploring data ingestion use cases, such as **ingesting data** from file/object stores, JDBC-compatible data stores, streaming data sources, and SaaS data stores, to demonstrate the capabilities of Glue. We know that AWS Glue supports three different types of ETL jobs – **Spark**, **Spark Streaming**, and a **Python Shell**. Each of these job types is designed to handle a specific type of workload and the environment in which the workload is executed varies, depending on the type of ETL job. For instance, Python Shell jobs allow users to execute Python scripts as a shell in AWS Glue. These jobs run on a single host on the server side. Spark/Spark Streaming ETL, on the other hand, allows you to execute PySpark/Scala-based ETL jobs in a distributed environment and allows users to take advantage of Spark libraries to execute ETL workloads.

In the upcoming sections, we will explore how Glue ETL can be used to ingest data from different data stores, including file/object stores, JDBC data stores, Spark Streaming data sources, and SaaS data stores.

By completing this chapter, you will be able to articulate and explain the features of AWS Glue ETL that help with ingesting data from file/object stores, HDFS, JDBC, Spark Streaming, and SaaS data stores and compose ETL scripts for them. You will also be able to explain the mechanism used by Glue job bookmarks to perform incremental data ingestion from Amazon S3 object stores and JDBC data stores. You will be able to create and use JDBC/custom/Marketplace connectors to ingest data from custom JDBC and SaaS data stores.

In this chapter, we will cover the following topics:

- Data ingestion from file/object stores
- Data ingestion from JDBC data stores
- Data ingestion from streaming data sources
- Data ingestion from SaaS data stores

Now, let's explore how we can ingest data from different types of data stores using AWS Glue and the salient features of AWS Glue that make it easy to ingest data from data stores.

Technical requirements

To get started with this chapter, you will need a workstation that's running Linux, macOS, or Windows with at least 7 GB of storage and 4 GB of RAM. While the code snippets can be run directly on AWS Glue (an AWS account is required to access AWS Glue), you can still run most of the code snippets in this chapter on your workstation directly. The code snippets in this chapter are available in this book's GitHub repository at `https://github.com/PacktPublishing/Serverless-ETL-and-Analytics-with-AWS-Glue/tree/main/Chapter03`.

There are several options available for setting up the Glue development environment on your workstation. Please refer to the AWS Glue documentation at `https://docs.aws.amazon.com/glue/latest/dg/aws-glue-programming-etl-libraries.html` for instructions regarding each of those options.

Now, let's explore how we can ingest data from different types of data stores one by one.

Data ingestion from file/object stores

This is one of the most common use cases for Glue ETL, where the source data is already available in file storage or cloud-based object stores. Here, depending on the type of job being executed, the methods or libraries used to access the data store differ.

There are several file/object storage services available today – Amazon S3, HDFS, Azure Storage, Google Cloud Storage, IBM Cloud Object Storage, FTP, SFTP, and HTTP(s) to name a few. In this section, we will focus on two of the most popular file/object stores that are used with AWS Glue – Amazon S3 and HDFS.

Data ingestion from Amazon S3

Data ingestion from Amazon S3 is by far the most commonly used design pattern for ETL in AWS Glue. Most organizations already have some mechanism to move data to Amazon S3, typically by using the AWS CLI/SDKs directly, AWS Transfer Family (`https://aws.amazon.com/aws-transfer-family/`), or some other third-party tools.

If we are using Python Shell jobs, the user can take advantage of several Python packages that allow them to connect to the desired file storage. If the user wishes to read an object from Amazon S3, they can use the Amazon S3 Boto3 client to get and read objects using Python packages/functions (for example, native Python functions and pandas), depending on the file format.

The following code snippet can be used with an AWS Glue Python Shell ETL job to read a CSV from an Amazon S3 bucket, transform the file from CSV into JSON, and write the output to another Amazon S3 location (the source code for this is available in this book's GitHub repository at `https://github.com/PacktPublishing/Serverless-ETL-and-Analytics-with-AWS-Glue/tree/main/Chapter03`):

```python
import boto3, io, pandas as pd
client = boto3.client('s3')
# nyc-tlc - https://registry.opendata.aws/
src_bucket = 'nyc-tlc' # SOURCE_S3_BUCKET_NAME
target_bucket = 'TARGET_S3_BUCKET_NAME'
src_object = client.get_object(
    Bucket=src_bucket,
    Key='trip data/yellow_tripdata_2021-07.csv'
)
# Read CSV and Transform to JSON
df = pd.read_csv(src_object['Body'])
jsonBuffer = io.StringIO()
df.to_json(jsonBuffer, orient='records')
# Write JSON to target location
client.put_object(
    Bucket=target_bucket,
    Key='target_prefix/data.json',
    Body=jsonBuffer.getvalue()
)
```

Here, the source data is a CSV file stored in an Amazon S3 location. The preceding script is downloading the data using the `get_object()` method, which is available in the AWS Python SDK (`boto3`), reading and transforming the CSV file using the `pandas` library, and writing to a different Amazon S3 location using the `put_object()` method.

The same source code can be executed in several ways in AWS – any Amazon EC2 instance with Python installed, an AWS Lambda function, or within a Docker container using Amazon ECR/AWS Batch, to name a few. Out of all the approaches listed, AWS Lambda and AWS Glue Python Shell jobs are the only ones that are serverless. Now, the question is, "*Why should we use AWS Glue Python Shell jobs over AWS Lambda?*"

While AWS Lambda and AWS Glue Python Shell jobs are both capable of running Python scripts, Python Shell jobs are designed for ETL workloads and can be orchestrated easier with other Glue components such as crawlers, Spark jobs, and Glue triggers using AWS Glue workflows. AWS Lambda functions can use a maximum of 512 MB of storage space (the /tmp directory), up to 10,240 MB of memory, and up to 6 vCPUs – the functions can run for up to a maximum of 15 minutes.

Glue Python Shell jobs, on the other hand, use the concept of **Data Processing Units** (**DPUs**) for capacity allocation, where one DPU provides four vCPUs and 16 GB of memory. Users can use either 0.0625 DPU or a 1 DPU capacity for Python Shell jobs. Essentially, a Python Shell job can use up to four vCPUs and 16 GB of memory and the user can configure the timeout value for Python Shell jobs (the default is 48 hours). At the time of writing, Glue Python Shell jobs are allocated 20 GB of disk space by default, though this may change in the future.

Now, let's consider the same ETL operation we performed in the previous script but using AWS Glue Spark ETL instead.

Let's consider the following code snippets:

- Using AWS Glue DynamicFrame (code snippet 1): The following code snippet shows how to read data from Amazon S3 and write the transformed data to another Amazon S3 location:

```
dy_frame = glueContext.create_dynamic_frame.from_options(
    connection_type="s3",
    connection_options = {"paths": ["s3://nyc-tlc/trip
data/yellow_tripdata_2021-07.csv"]},
    format="csv",
    format_options = {"withHeader": True}
)
datasink = glueContext.write_dynamic_frame.from_options(
    frame = dy_frame, connection_type = "s3",
    connection_options = {
    "path": "s3://TARGET_BUCKET_NAME/target_prefix/"
    },
    format = "json"
)
```

- Without using AWS Glue DynamicFrame (code snippet 2): The following code snippet implements a similar workflow to the previous one, but this time, we will not be using AWS Glue DynamicFrames:

```
df = spark.read.option("header","true").csv("s3://
nyc-tlc/trip data/yellow_tripdata_2021-07.csv")
df.write.json("s3://TARGET_BUCKET_NAME/target_prefix/")
```

Now, both of these code snippets are essentially performing the same operations. This begs the question, *"What is the advantage of using Glue* `DynamicFrame` *over a Spark DataFrame?"*

As discussed in the previous chapter, DynamicFrames are structurally different from Spark DataFrames since they have several optimizations enabled under the hood.

AWS Glue Spark ETL and EMRFS

In the preceding examples, as you may have noticed, we specified Amazon S3 paths in the `s3://BUCKET_NAME/prefix` format to read or write data. Notice that the `s3://` protocol string was used instead of `s3a://` or `s3n://`, which you may have seen examples of in the Spark documentation or blog articles online.

Under the hood, AWS Glue ETL (Spark) uses an EMRFS (`https://docs.aws.amazon.com/emr/latest/ReleaseGuide/emr-fs.html`) driver by default to read from Amazon S3 data stores when the path begins with the `s3://` URI scheme (class: `com.amazon.ws.emr.hadoop.fs.EmrFileSystem`), regardless of whether Apache Spark DataFrames or AWS Glue DynamicFrames are used. The EMRFS driver was originally developed for Amazon EMR and has been since adopted by AWS Glue for Amazon S3 reads and writes from Glue Spark ETL.

While users can still use `s3a://` (class: `org.apache.hadoop.fs.s3a.S3AfileSystem`) and `s3n://` (class: `org.apache.hadoop.fs.s3native.NativeS3FileSystem`) to read from Amazon S3 data stores, it is strongly discouraged as different classes would be used to read the data store and bypass configuration properties, as well as making optimizations that have been set up on the server side.

In addition, it is important to note that `NativeS3FileSystem` (`s3n://`) has reached **End-of-Life** (**EoL**) and must not be used.

Schema flexibility

Since DynamicRecords are self-describing, a schema is computed on the fly and there is no need to perform an additional pass over the source data.

Advanced options for managing schema conflicts

DynamicFrames make it easier to handle schema conflicts by introducing `ChoiceType` whenever a schema conflict is encountered instead of defaulting to the most compatible data type (usually, this is `StringType`). For instance, if one of the columns has integer/long values and string values, Spark infers it as `StringType` by default. However, Glue creates `ChoiceType` and allows the user to resolve the conflict.

Let's consider an example where the `Provider Id` column has `StringType` and numeric (`LongType`) values. When the data is read using a Spark DataFrame, the column will be inferred as `StringType` by Spark:

```
root
 |-- ColumnA: string (nullable = true)
 |-- ColumnB: string (nullable = true)
```

Now, when the same dataset is read using AWS Glue DynamicFrames, the column is represented with `ChoiceType` and lets the user decide how to resolve the type conflict:

```
root
 |-- ColumnA: string
 |-- ColumnB: choice
 |      |-- long
 |      |-- string
```

When a column is recognized as `ChoiceType`, the user can resolve the conflict by using the `ResolveChoice` class in Glue. There are four different options for the user to choose from: `cast`, `make_cols`, `make_struct`, and `project`. Let's take a look:

- `cast`: The user can cast the column to long using the following statement:

  ```
  new_dyf = dyf.resolveChoice(specs = [('ColumnB
  ','cast:long')])
  ```

  The column will use `LongType` as the data type. For the string values that could not be cast to `LongType`, Glue inserts `null` values.

- `make_struct`: We can convert this into a struct using `make_struct`, which will produce a struct column in the DynamicFrames, with each containing both `StringType` and `LongType` values.

- `make_cols`: This option can be used by the user to create separate columns for each of the data types detected. In this instance, two new columns will be produced: `ColumnB_long` and `ColumnB_string`.

- `project`: This option can be used when the user is only concerned about retaining values of a specific type. In this case, if the `project:long` action is used, this will result in a DynamicFrame where the values that are not `long` are dropped.

Now that we know how to manage schema conflicts in Glue ETL, let's explore other features of Glue ETL that make it easy for us to transform and ingest data.

AWS Glue-specific ETL transformations and extensions

Several transformations and ETL actions are unique to Glue DynamicFrames – Unbox, SplitFields, ResolveChoice, and Relationalize to name a few. Please refer to the AWS Glue documentation at https://docs.aws.amazon.com/glue/latest/dg/aws-glue-programming-python.html for an exhaustive list of transformations and extensions supported by DynamicFrames.

Job bookmarks

To take advantage of job bookmarks – a key feature of Glue ETL – it is necessary to use Glue DynamicFrames.

Grouping

We have all come across or heard of a classic problem in big data processing – *reading a large number of small files*. Spark launches a separate task for each data partition for each stage; if the file size is less than the block size, Spark will launch one task per file. Consider a scenario where there are billions of such files/objects in the data store – this will lead to a huge number of tasks being created, which will cause unnecessary delays due to scheduling logic (any given executor can run a finite number of tasks in parallel, depending on the number of CPU cores available). Using the **Grouping** feature in Glue ETL (https://docs.aws.amazon.com/glue/latest/dg/grouping-input-files.html), users can group input files to combine multiple files into a single task. This can be done by specifying the target size of groups in bytes with groupSize. Glue ETL automatically enables this feature if the number of input files is higher than 50,000.

For example, in the following code snippet, we are reading JSON data from Amazon S3 while performing grouping. This allows us to control the task size rather than letting Spark control the task size based on the number of input files:

```
dy_frame = glueContext.create_dynamic_frame.from_options(
    connection_type="s3",
    connection_options = {
        'paths': ["s3://s3path/"],
        'recurse':True,
        'groupFiles': 'inPartition',
        'groupSize': '1048576'
    }, format="json")
```

> **Note**
> groupFiles is supported for DynamicFrames that have been created using the csv, ion, grokLog, json, and xml formats. This option is not supported for Avro, Parquet, or ORC data formats.

Optimizing Amazon S3 reads using S3ListImplementation

When a DynamicFrame, DataFrame, or RDD is created in Spark, Spark creates a list of files in the Spark driver memory to be included in the object. Only when the list is completely in memory is the object created in Spark. This becomes a problem when we are dealing with a huge number of files – the Spark driver will run out of memory and the ETL job will fail.

Glue provides a mechanism to handle this issue: S3ListImplementation. This allows DynamicFrame to lazily load the file listing. S3ListImplementation works by calling the Amazon S3 ListObjectsV2 API and fetching the list of objects in batches of 1,000. Job Bookmark and Pushdown predicate filters are applied to each batch and the next batch is fetched using the pagination token that's returned in each API response. This process repeats until the job traverses all the files in the Amazon S3 path supplied.

The following code snippet demonstrates how we can enable the S3ListImplementation feature in Glue DynamicFrames to read data from an Amazon S3 data store:

```
dyf = glueContext.create_dynamic_frame.from_catalog(
    database = "db_name",
    table_name = "million_files_table",
    transformation_ctx = " dyf",
    additional_options = {
        "useS3ListImplementation": True
    }
)
```

As you may have observed, this feature is only beneficial when some form of filtering is enabled – Job Bookmark or Pushdown predicates. If job bookmarks are not enabled or if the list of files is still bigger than what the driver can handle, S3ListImplementation will not help and the job will fail due to the driver running out of memory.

In such cases, the best option is to perform **workload partitioning using Bounded Execution** or to *push down partitions further and batch your ETL job*, as we will see in the next section.

Workload partitioning with Bounded Execution for Amazon S3 data stores

The Bounded Execution feature was introduced to allow users to mitigate issues originating from inefficient Spark scripts, data abnormalities, and in-memory execution of large-scale transformations.

Workload partitioning allows users to run ETL jobs on unprocessed data with an upper bound on the *data size* or *the number of files* that can be processed within a job run. For instance, if there are 4,000 files, the user can set the upper bound for the number of files to 1,000, which will limit the number of files that are processed during this JobRun. We can use four separate JobRuns to process the entire dataset instead of processing the entire dataset within a single JobRun.

> **Note**
>
> It is important to use this feature in conjunction with job bookmarks to avoid reprocessing the same dataset over and over again.

This feature can be used when the jobs are failing due to driver or executor memory issues, which can occur due to data skew (a hot partition issue), too many objects being listed, or large data shuffles.

- **By the number of files**: Bounded Execution can be implemented in AWS Glue ETL to limit the data that's read in an ETL job run to a specific number of files. The following code snippet demonstrates how we can implement this:

```
dyf_4000 = glueContext.create_dynamic_frame.from_catalog(
    database = "db_name",
    tableName = "four_thousand_file_table",
    transformation_ctx = "dyf_4000",
    additional_options = {"boundedFiles": "1000"}
)
```

- **By the volume of data**: Bounded Execution can also be implemented to limit the volume of data that's ingested per job run instead of limiting the run to a specified number of files. The following code snippet demonstrates how this can be implemented:

```
dyf_volume = glueContext.create_dynamic_frame.from_
catalog(
    database = "db_name",
    tableName = "four_thousand_file_table",
    transformation_ctx = "dyf_volume",
    # Volume in bytes
    additional_options = {"boundedSize": "1000000000"}
)
```

Data ingestion from HDFS data stores

While it is true that several features/optimizations in AWS Glue Spark ETL are designed for data ingestion from Amazon S3 data stores, it is still possible to ingest data from HDFS data stores (or any data store supported by Apache Spark). Data can be read from a Hadoop cluster hosted in on-premises data centers or by a third-party provider.

The following code snippet demonstrates data ingestion from a HDFS location in an Amazon EMR cluster:

```
df = spark.read.parquet("hdfs://EMR_MASTER:8020/parquet/")
df.write.mode("overwrite").parquet("s3://TARGET/prefix/")
```

Data ingestion from JDBC data stores

For many organizations hydrating data lakes by ingesting the data from OLTP, data stores are the primary use case for using ETL tools/frameworks. Typically, these ETL jobs are run periodically to keep the data lake up to date. As discussed in *Chapter 1, Data Management - Introduction and Concepts*, there are quite a few options available in AWS to achieve this outcome. The most popular ones are **AWS DMS** and **AWS Glue**.

Users can set up AWS DMS replication instances to capture ongoing changes from the source data store. At the time of writing, this feature supports Microsoft SQL Server, PostgreSQL, Oracle, and MySQL databases. Please refer to the AWS DMS documentation at https://docs.aws.amazon.com/dms/latest/userguide/CHAP_Task.CDC.html for more information on this feature.

Another option is to use AWS Glue Spark ETL to read JDBC data stores and move the data to Amazon S3 or other target data stores supported by Apache Spark. With this option, users do not need to set up replication tasks or instances and AWS Glue Spark ETL supports advanced transformations. AWS Glue leverages Apache Spark's capability of handling JDBC operations and adds quite a few optimizations under the hood for JDBC read/write operations. We will be unpacking a few of the key optimizations available using examples shortly.

Let's consider a simple ETL operation where the job is moving data from a JDBC-compatible data store (we will be using a MySQL database for our example here) to the Amazon S3 target location in Parquet format. In the following code snippet, we are connecting to a MySQL 5.7 host that is using the world_x sample dataset. This is available in the MySQL documentation at https://dev.mysql.com/doc/world-x-setup/en/:

```
mysql_options = {
    "url": "jdbc:mysql://DB_HOST:3306/world_x",
    "dbtable": "city",
    "user": "admin",
    "password": "password"
}
dyf = glueContext.create_dynamic_frame.from_options(
    connection_type="mysql",
    connection_options=mysql_options
)
```

```
sink = glueContext.write_dynamic_frame.from_options(
    frame = dyf,
    connection_type = "s3",
    connection_options = {"path": "s3://TARGET/prefix/"},
    format = "parquet"
)
```

A similar outcome can be achieved in AWS Glue Spark ETL using Apache Spark DataFrames instead of AWS Glue DynamicFrames:

```
mysql_options = {
    "url": "jdbc:mysql://DB_HOST:3306/world_x",
    "dbtable": "city",
    "user": "admin",
    "password": "password"
}
df = spark.read \
        .format("jdbc") \
        .option("url", mysql_options["url"]) \
        .option("dbtable", mysql_options["dbtable"]) \
        .option("user", mysql_options["user"]) \
        .option("password", mysql_options["password"]) \
        .load()
df.write.parquet("s3://TARGET/prefix")
```

Even though both code snippets do the same thing, the second code snippet is missing quite a few optimizations that were used by the first code snippet under the hood.

Both code snippets offer similar performance for smaller datasets. However, when the second code snippet is run on a larger dataset, it will run into executor **out-of-memory** (**OOM**) issues. This is because Apache Spark sets the default value of the fetchsize JDBC option to 0. On the other hand, AWS Glue DynamicFrames use a fetchsize value of 1000 rows by default.

The fetchsize parameter informs the JDBC driver of the number of rows to read in one round trip. Since the default value for this parameter is 0, the entire table will be read in one round trip. This is not a problem for smaller tables that can easily fit into executor on-heap memory space, but the same cannot be said for larger tables where the volume of data is larger than the executor heap memory allocation; this will lead to **executor OOM errors**, causing the ETL job to fail.

It is important to note that with both of these approaches, Apache Spark connects to the JDBC data store over a single connection by default. Considering Spark ETL jobs are executed in a distributed environment, a single executor is active while Spark is reading data from the JDBC data store; the rest of the executors are idle. To distribute the workload across all the available executors, we can parallelize JDBC reads by specifying a few additional parameters.

Let's explore how to implement parallel JDBC reads from Glue Spark ETL using DynamicFrames. Upon checking the schema for the table we used in the preceding example (city), we can see that the table has a primary key ID (integer) and four other string columns – Name, CountryCode, District, and Info.

Now, in AWS Glue ETL, we can parallelize JDBC reads using hashexpression – an integer column or a WHERE condition that yields an integer value or a hashfield. This is a column (of any data type) in the table using which we can partition the dataset. Here, it is preferable to use a key that has an even distribution of values. For example, we can use the month column in a transactions table to partition the data. However, if one of the months has an extremely high number of transactions, then this introduces a data skew and affects performance.

In our sample dataset, since we have a primary key with integer values, we can use this column as our hashexpression and specify the number of partitions desired (hashpartitions). The Glue ETL libraries will launch parallel SELECT queries based on the hashpartitions value that is set. The following code snippet demonstrates how we can implement input partitioning on the same dataset:

```
mysql_options = {
    "url": "jdbc:mysql://DB_HOST:3306/world_x",
    "dbtable": "city",
    "user": "admin",
    "password": "password",
    "hashexpression": "ID",
    "hashpartitions": '10'}
dyf = glueContext.create_dynamic_frame.from_options(
    connection_type="mysql",
    connection_options=mysql_options
)
sink = glueContext.write_dynamic_frame.from_options(
    frame = dyf,
    connection_type = "s3",
    connection_options = {"path": "s3://TARGET/prefix/"},
    format = "parquet"
)
```

Once we execute the preceding code snippet, AWS Glue ETL will launch 10 SELECT queries in parallel, each of which will query a different partition of data while using hashexpression to split the data. We can enable MySQL general_log (refer to the instructions outlined in the knowledge center article at https://aws.amazon.com/premiumsupport/knowledge-center/rds-mysql-logs/ for AWS RDS) and check the query history to see this in action:

```
mysql> SELECT argument FROM mysql.general_log WHERE argument LIKE '%city%' AND
command_type = 'Prepare' ORDER BY argument;
+-----------------------------------------------------------------------------+
| argument                                                                    |
+-----------------------------------------------------------------------------+
| SELECT * FROM (select * from city WHERE ID % 10 = 0) as city                |
| SELECT * FROM (select * from city WHERE ID % 10 = 0) as city WHERE 1=0      |
| SELECT * FROM (select * from city WHERE ID % 10 = 1) as city                |
| SELECT * FROM (select * from city WHERE ID % 10 = 1) as city WHERE 1=0      |
| SELECT * FROM (select * from city WHERE ID % 10 = 2) as city                |
| SELECT * FROM (select * from city WHERE ID % 10 = 2) as city WHERE 1=0      |
| SELECT * FROM (select * from city WHERE ID % 10 = 3) as city                |
| SELECT * FROM (select * from city WHERE ID % 10 = 3) as city WHERE 1=0      |
| SELECT * FROM (select * from city WHERE ID % 10 = 4) as city                |
| SELECT * FROM (select * from city WHERE ID % 10 = 4) as city WHERE 1=0      |
| SELECT * FROM (select * from city WHERE ID % 10 = 5) as city                |
| SELECT * FROM (select * from city WHERE ID % 10 = 5) as city WHERE 1=0      |
| SELECT * FROM (select * from city WHERE ID % 10 = 6) as city                |
| SELECT * FROM (select * from city WHERE ID % 10 = 6) as city WHERE 1=0      |
| SELECT * FROM (select * from city WHERE ID % 10 = 7) as city                |
| SELECT * FROM (select * from city WHERE ID % 10 = 7) as city WHERE 1=0      |
| SELECT * FROM (select * from city WHERE ID % 10 = 8) as city                |
| SELECT * FROM (select * from city WHERE ID % 10 = 8) as city WHERE 1=0      |
| SELECT * FROM (select * from city WHERE ID % 10 = 9) as city                |
| SELECT * FROM (select * from city WHERE ID % 10 = 9) as city WHERE 1=0      |
+-----------------------------------------------------------------------------+
20 rows in set (0.20 sec)
```

Figure 3.1 – Queries generated by AWS Glue ETL when hashpartitions is specified

Based on the query log shown in the preceding screenshot, we can see that the SELECT query looks similar to the following template:

```
SELECT * FROM (select * from table_name WHERE hashexpression %
hashpartitions = partition_num) as table_name
```

AWS Glue ETL also executes the same query with an additional condition, WHERE 1=0, before executing the actual query. This query returns no results; however, it returns the schema for the partition.

Similarly, if we use hashfield instead of hashexpression, the query follows a similar pattern but instead, the modulo operator will be used on the hash generated based on the hashfield column value.

The following is an example query (for partition #0):

```
SELECT * FROM (select * from city WHERE
CONV(SUBSTRING(MD5(CONCAT('',CountryCode)), -8, 8), 16, 10) %
10 = 0) as city
```

Based on the queries we've executed, we can see that the pattern looks similar to the following template:

```
SELECT * FROM (select * from table_name WHERE
CONV(SUBSTRING(MD5(CONCAT('',hashfield)), -8, 8), 16, 10) %
hashpartitions = partition_num) as table_name
```

Now, it is important to note that the sample query or the template mentioned previously is for a MySQL database engine. If a different database engine is being queried, the hashing syntax or functions used will be completely different. That being said, the overall logic will be similar to the example mentioned previously.

JDBC reads from Spark DataFrames can be optimized similarly using the `partitionColumn`, `lowerBound`, `upperBound`, and `numPartitions` parameters – refer to the Apache Spark documentation at `https://spark.apache.org/docs/3.1.1/sql-data-sources-jdbc.html` for more information on these parameters. Apache Spark will use these parameters to create partitions using `partitionColumn` to parallelize JDBC reads. It is important to note that `lowerBound` and `upperBound` are just used to decide the partition's stride; data will not be filtered based on these values.

Now, the difference between the approach used by AWS Glue's `hashexpression`- or `hashfield`-based partitioning and Apache Spark's built-in approach is that Apache Spark can split the data using the built-in approach without generating hashes. The built-in approach to split the data is more efficient during SQL query runtime compared to AWS Glue's approach as it avoids performing multiple full scans of the source table. However, Apache Spark's approach is vulnerable to data skews, which may lead to other performance issues during the ETL job's runtime. So, it is important to examine the dataset and consider the use case before choosing one approach over the other to partition the dataset.

So far, we've explored JDBC reads from AWS Glue DynamicFrames using code snippets where we created Python dictionaries to define connection properties such as the JDBC URL, username, password, database name, and table name. However, it is not recommended to hardcode credentials directly into an ETL script. This is not a problem when a catalog table is being used to connect to the JDBC data store as JDBC credentials are stored in an AWS Glue connection and can be encrypted using an AWS KMS key. However, if the `create_dynamic_frame.from_options()` method is being used to read from the JDBC data store, we can leverage AWS Glue's integration with AWS Secrets Manager to keep JDBC user credentials away from the ETL script. We can store the `username` and `password` properties in AWS Secrets Manager in the following format:

```
{
    "username": "admin",
    "password": "password"
}
```

Once the credentials have been stored in AWS Secrets Manager, we can grant permissions to the AWS IAM role that's used by the Glue ETL job to read these credentials (refer to the AWS Secrets Manager documentation for sample policies: `https://docs.aws.amazon.com/secretsmanager/latest/userguide/auth-and-access_examples.html#auth-and-access_examples_read`) and make the following change to the connection options in the Python dictionary:

```
mysql_options = {
    "url": "jdbc:mysql://DB_HOST:3306/world_x",
    "dbtable": "city",
    "secretId": "glue_sec/mysqltestdb" # secret ARN or Name
}
```

Now, AWS Glue will automatically fetch the username/password combination from AWS Secrets Manager when connecting to the JDBC data store.

It is also possible for users to pass a custom JDBC driver for JDBC data stores that are supported by AWS Glue by passing the `customJdbcDriverS3Path` and `customJdbcDriverClassName` parameters. This option is helpful when users wish to use the advanced version of JDBC compared to the one available in the AWS Glue environment by default:

```
mysql_options = {
    "url": "jdbc:mysql://DB_HOST:3306/world_x",
    "dbtable": "city",
    "customJdbcDriverS3Path":"s3://bucket/pre/mysql8.jar",
    "customJdbcDriverClassName":"com.mysql.cj.jdbc.Driver",
    "secretId": "glue_sec/mysqltestdb" # secret ARN or Name
}
```

For a list of JDBC data stores supported and the JDBC driver versions available in AWS Glue ETL, please refer to the AWS Glue documentation at `https://docs.aws.amazon.com/glue/latest/dg/migrating-version-30.html#migrating-version-30-appendix-jdbc-driver`.

We can also build a custom JDBC connector if the database server engine is not natively supported by AWS Glue ETL. In the next section, we will explore how this can be achieved using AWS Glue Studio.

AWS Glue custom JDBC connectors

So far, we have focused on reads/writes for JDBC data store types directly supported by AWS Glue – Microsoft SQL Server (`"connectionType": "sqlserver"`), MySQL (`"connectionType": "mysql"`), Oracle DB (`"connectionType": "oracle"`), PostgreSQL (`"connectionType": "postgresql"`), and Amazon Redshift (`"connectionType": "redshift"`).

However, there are plenty of other JDBC data store types that were not mentioned previously. AWS Glue added a feature to AWS Glue Studio that allows users to define custom JDBC connectors. This would use either a `custom.jdbc` or `marketplace.jdbc` connection type, depending on the connector definition in AWS Glue Studio.

Users can create custom JDBC connectors using the **Create custom connector** option in AWS Glue Studio by uploading the JDBC JAR file to S3 and specifying the JDBC class name and base URL.

For instance, if we have to run an ETL workload that reads from MySQL Database v8.0 and use advanced parameters that are only supported by MySQL Connector J/8.0, we can create a custom connector with different configuration properties. Here, we are using a comma-separated list JDBC URL format supported by MySQL Connector J/8.0 and we are defining placeholders (`${varName}`) instead of specifying actual values. The advantage of this approach is that we can reuse connectors to create multiple connections.

The following are the parameters we used to create a MySQL Connector J/8.0 custom JDBC connector in the AWS Glue Studio management console:

- **Connector S3 URL**: `s3://bucket/pre/mysql-connector-java-8.0.23.jar`
- **Name**: `mysql-8-connector`
- **Connector type**: `JDBC`
- **Class name**: `com.mysql.cj.jdbc.Driver`
- **JDBC URL Base**: `jdbc:mysql://(host=${host},port=${port}, user=${username},password=${password})/${dbname}`
- **URL parameter delimiter**: `&`

Once the connector has been set up, we can set up a secret in AWS Secrets Manager with the following key-value pairs:

```
{
    "username": "admin",
    "password": "password",
    "engine": "mysql",
    "host": "database.hostname.internal",
    "port": "3306",
    "dbname": "world_x"
}
```

Once the secret has been set up, we can create a connection (let's assume the name of the connection is `mysql-8-connection-rds`) in Glue Studio using the connector and select the secret and network options (VPC, subnet, and security group).

We can use the following AWS Glue DynamicFrame code snippet to read data using the custom JDBC connection that was just created:

```
dyf = glueContext.create_dynamic_frame.from_options(
    connection_type="custom.jdbc",
    connection_options={
        "dbTable": "city",
        "connectionName": "mysql-8-connection-rds",
    }
)
dyf.toDF().show(truncate=False)
```

The preceding code snippet will read the dataset from the MySQL 8 database using a custom JDBC connector and print the top 20 rows to logs.

Now that we know how we can ingest data from JDBC data stores, in the next section, we will learn how to ingest data from streaming data sources such as Apache Kafka and AWS Kinesis.

Data ingestion from streaming data sources

We explored fundamental concepts regarding data ingestion from streaming data sources in the previous chapter when we discussed AWS **Glue Schema Registry** (**GSR**). In this section, we will learn how to implement data ingestion from streaming data sources such as Amazon Kinesis and Apache Kafka using AWS Glue Spark ETL.

Stream processing can be defined as the act of continuously incorporating new data to compute a result wherein the input data is *unbounded* and *has no predetermined beginning or end*. Apache Spark has two components for stream processing: **Spark Streaming** and **Structured Streaming**.

According to the Apache Spark documentation (https://spark.apache.org/docs/3.1.1/streaming-programming-guide.html), "*Spark Streaming is an extension of the core Spark API that enables scalable, high-throughput, fault-tolerant stream processing of live data streams.*"

Spark Streaming introduces a high-level abstraction layer called a **discretized stream** (also known as a **Dstream**), which represents a continuous stream of data and exposes a programming model to operate on the underlying data in the stream.

Structured Streaming, on the other hand, is a stream processing engine built on the Spark SQL engine. Structured Streaming is known to be both scalable and fault-tolerant and as an added benefit, we can express operations on streaming data in the same way we do so for batch data. This extends the `Dataset` and `Dataframe` APIs with streaming capabilities and uses a declarative model to acquire data from a stream or set of streams.

Stream processing in AWS Glue ETL uses Apache Spark's Structured Streaming. Streaming ETL in AWS Glue allows users to hydrate their data lakes or data warehouses by ingesting streaming data while allowing users to take advantage of Glue DynamicFrames. The same set of advanced ETL transforms is available in AWS Glue ETL for batch data processing. Glue ETL supports streaming data ingestion from Apache Kafka and Amazon Kinesis. Data is read in *micro-batches* with a specified window size (100 seconds by default).

Unlike Spark batch jobs, Structured Streaming jobs require a schema for the data. We can use a schema stored in Glue Data Catalog as the source for the schema so that it can be integrated with the Glue Schema Registry.

The following code snippets will show us how to ingest streaming data from an Apache Kafka stream. The ETL source code is similar for the AWS Kinesis streaming data source. The key difference is in the setup that's involved in creating a Glue Data Catalog table and the parameters that will be passed.

Now, the read statement looks almost similar to batch data reads when using a Glue Data Catalog table. However, the difference here is that we are creating a DataFrame instead of a DynamicFrame:

```
df_kafka = glueContext.create_data_frame.from_catalog(
    database = "default",
    table_name = "kafka_stream",
    transformation_ctx = "datasource0",
    additional_options = {
        "startingOffsets": "earliest",
        "inferSchema": "true"
    }
)
```

The preceding code snippet looks similar for a Kinesis data stream. However, the only difference would be the parameters that are passed in `additional_options` – we can pass Amazon Kinesis connection properties (refer to `https://docs.aws.amazon.com/glue/latest/dg/aws-glue-programming-etl-connect.html#aws-glue-programming-etl-connect-kinesis` for a list of AWS Kinesis properties that can be used) instead of Apache Kafka connection properties (refer to `https://docs.aws.amazon.com/glue/latest/dg/aws-glue-programming-etl-connect.html#aws-glue-programming-etl-connect-kafka` for a list of Apache Kafka connection properties that can be used).

The next step is to define a method that will be executed on each micro-batch during stream processing:

```
def processBatch(data_frame, batchId):
    if (data_frame.count() > 0):
        datasource0 = DynamicFrame.fromDF(
```

```
            data_frame,
            glueContext,
            "from_data_frame"
        )
        now = datetime.datetime.now()
        path_datasink1 = "s3://bucket/destination/" + "/
ingest_year=" + "{:0>4}".format(str(now.year)) + "/ingest_
month=" + "{:0>2}".format(str(now.month)) + "/ingest_day=" +
"{:0>2}".format(str(now.day)) + "/ingest_hour=" + "{:0>2}".
format(str(now.hour)) + "/"
        datasink1 = glueContext.write_dynamic_frame.from_
options(
            frame = datasource0,
            connection_type = "s3",
            connection_options = {
                "path": path_datasink1
            },
            format = "parquet",
            transformation_ctx = "datasink1"
        )
```

In the preceding code snippet, we converted the DataFrame into a DynamicFrame and built the target S3 path to write the micro-batch that's being processed by obtaining the year, month, day, and hour values using the `datetime` Python library. We used the `glueContext.write_dynamic_frame.from_options()` method to write the data to Amazon S3.

Since the preceding code snippet defines a method for writing data to Amazon S3, this method has to be called on each micro-batch. This is where Glue ETL's `forEachBatch()` method comes into the picture. Using this method, we can call the `processBatch()` method on each micro-batch and specify Structured Streaming-related options such as `windowSize` and `checkpointLocation`:

```
glueContext.forEachBatch(
    frame = data_frame_datasource0,
    batch_function = processBatch,
    options = {
        "windowSize": "100 seconds",
        "checkpointLocation": "s3://bucket/checkpoint_loc/"
    }
)
```

If our use case requires the data to be transformed in some way, we can implement the DynamicFrame transformations that are available in Glue ETL in the `processBatch()` method after converting the DataFrame into a DynamicFrame.

AWS Glue Schema Registry

We introduced **AWS GSR** in the previous chapter. In this chapter, we will explore how the schema registry works in detail.

AWS GSR is fully managed and serverless and available to users free of cost. At the time of writing, GSR supports the AVRO and JSON data formats for the schema. JSON Schema validation is supported via the Everit library, which is available at `https://github.com/everit-org/json-schema`.

> **Note**
> AWS GSR currently supports the Java programming language. Producers and consumers need to be running Java 8 or above.

Schema registries use *serialization* and *deserialization* processes to help streaming data producers and consumers enforce schemas on records.

If a schema is not available in the schema registry, it must be registered for use (auto-registration of the schema can be enabled for any new schema). Upon registering a schema in the schema registry, a schema **version identifier (version ID)** will be issued to the serializer.

If the schema is already available in GSR and the serializer is using a newer version of the schema, GSR will check the compatibility rule to make sure that the new version is compatible. If it is, the schema will be registered as a new version in GSR.

When a producer has its schema registered, the GSR serializer validates the schema of the record with the schema that's been registered. If there is a mismatch, an exception will be returned. Producers typically cache the schema versions and match the schema against the versions available in the cache. If there is no version available in the cache that matches the schema of the record, GSR will be queried for the same using the `GetSchemaVersion` API. If the schema is validated using a version in GSR, the schema version ID and definition will be cached locally by the producer. If the record's schema is compliant with the schema that's been registered, the record is decorated with the schema version ID and then serialized (based on the data format selected), compressed, and delivered to the destination.

Once a serialized record has been received, the deserializer uses the version ID available in the payload to validate the schema. If the deserializer has not encountered this schema version ID before, GSR is queried for it and the schema version is cached in local storage.

If the schema version IDs in GSR/the cache match with the version in the serialized record, the deserializer decompresses and deserializes the data and the record is handed off to the consumer application. However, if the schema version ID doesn't match the version IDs available in the cache or GSR, the consumer application can log this event and move on to other records or halt the process based on the user's configuration.

SerDe libraries can be added to both producer and consumer applications by adding the `software.amazon.glue:schema-registry-serde` Maven dependency (refer to `https://mvnrepository.com/artifact/software.amazon.glue/schema-registry-serde` for more information). Please refer to the AWS GSR documentation at `https://docs.aws.amazon.com/glue/latest/dg/schema-registry-integrations.html` for example producer and consumer implementations.

In this section, we explored how to ingest data from streaming data sources and understood the mechanism that's used by AWS GSR to centrally manage evolving schemas.

In the next section, we will learn how to ingest data from SaaS data stores.

Data ingestion from SaaS data stores

So far, we have explored ways to ingest data from file/object stores, JDBC, and streaming data sources using AWS Glue ETL. Apart from these methods, organizations can take advantage of Marketplace connectors or create their own connectors to ingest data from a data store that is not directly supported by AWS Glue ETL. This feature was added to AWS Glue as part of the Glue Studio release in December 2020.

For example, with this new capability, we can take advantage of connectors for Salesforce, SAP, and Snowflake. If a connector is not readily available in AWS Marketplace, we can build custom connectors so that we can integrate custom-built Spark connectors and Athena Federated Query connectors into our ETL jobs.

Connectors for popular data stores such as Snowflake, SAP, Salesforce, Apache Hudi, Google BigQuery, Delta Lake, Elasticsearch, and CloudWatch Logs are readily available on AWS Marketplace. Depending on the publisher of a given connector, there might be a subscription fee for connector usage. At the time of writing, all the connectors that have been published by AWS on Marketplace are available for use at no additional cost.

If a connector is not available for a data store, users can build a connector and use it in their ETL workload. While exploring methods to ingest data from JDBC data stores, we unpacked the process of creating custom JDBC connectors. In this section, we will explore how to use a Marketplace connector to ingest data from a SaaS product. We will be ingesting data from AWS CloudWatch Logs for our example. However, before we can proceed, we will have to set up a connector by subscribing to the CloudWatch connector on AWS Marketplace.

This can be done by navigating to the **AWS Glue Studio console** | **Connectors** | **Marketplace Connectors** and subscribing to **Cloudwatch Metrics connector for AWS Glue**. For a detailed set of instructions for subscribing to Marketplace connectors, please refer to the AWS Glue Studio documentation at `https://docs.aws.amazon.com/glue/latest/ug/connectors-chapter.html#subscribe-marketplace-connectors`. Once the subscription process is complete, a connection will be created in AWS Glue Studio with the name specified during the setup process.

We can use the following code snippet to read metrics data from AWS CloudWatch metrics:

```
dyf = glueContext.create_dynamic_frame.from_options(
    connection_type="marketplace.athena",
    connection_options={
        "schemaName": "default",
        "tableName": "metrics",
        "connectionName": "CloudWatchMetricsConnector",
    }
)
```

Once the metrics data has been read into a DynamicFrame, we can either transform the data or write the data straight to the target. In this use case, we'll write the data to an Amazon S3 location in Parquet format and set up a Glue Data Catalog table for the target dataset so that it can immediately be queried from Amazon Redshift Spectrum or Amazon Athena:

```
target = glueContext.getSink(
    path="s3://bucket/target/",
    connection_type="s3",
    updateBehavior="UPDATE_IN_DATABASE",
    partitionKeys=[],
    compression="snappy",
    enableUpdateCatalog=True
)
target.setCatalogInfo(
    catalogDatabase="default",
    catalogTableName="cw_metrics"
)
target.setFormat("glueparquet")
target.writeFrame(dyf)
```

The preceding code snippet will write the data to the Amazon S3 target location and create a table called `cw_metrics` in the `default` database in AWS Glue Data Catalog.

In this section, we ingested metrics data from AWS CloudWatch for AWS resources in a specific AWS Region. Users can ingest data from a data store for which a connector is not readily available by building a custom connector using the Apache Spark DataSource API or the Amazon Athena DataSource API; detailed instructions and examples are available in `aws-samples/aws-glue-samples` in this book's GitHub repository at `https://github.com/aws-samples/aws-glue-samples/tree/master/GlueCustomConnectors/development`.

Summary

In this chapter, we discussed the methods and different optimization features that can be used in AWS Glue ETL to ingest data from file/object stores, JDBC-compatible data stores, and streaming data stores. We also explored serialization and deserialization, which are used by AWS GSR to handle evolving schemas. Then, we introduced Glue Studio Marketplace connectors, using which we can ingest data from SaaS. Finally, we briefly discussed how users can build custom JDBC/Spark/Athena Federated Query connectors to ingest data from data stores that are not directly supported by AWS Glue and when there is no connector readily available in AWS Marketplace.

In the next chapter, we will be discussing data preparation strategies. We'll explore different factors that can be considered while choosing the right service/tool. We will also discuss the different available options: visual data preparation versus source code-/SQL-based data preparation and the different transformation classes that are available in AWS Glue ETL to help with preparing data.

Section 2 – Data Preparation, Management, and Security

In this section, you will learn about using the right tool (such as Glue Studio, Glue DataBrew, Lambda, and EMR) for the right purpose. You will also learn about good data layout practices along with data sharing, data security, metadata management, and various ways of orchestration. You will explore the common data transformation tasks customers have.

This section includes the following chapters:

- *Chapter 4, Data Preparation*
- *Chapter 5, Designing Data Layouts*
- *Chapter 6, Data Management*
- *Chapter 7, Metadata Management*
- *Chapter 8, Data Security*
- *Chapter 9, Data Sharing*
- *Chapter 10, Data Pipeline Management*

4
Data Preparation

In the previous chapter, we explored fundamental concepts surrounding data ingestion and how we can leverage AWS Glue to ingest data from various sources, such as file/object stores, JDBC data stores, streaming data sources, and SaaS data stores. We also discussed different features of AWS Glue ETL, such as schema flexibility, schema conflict resolution, advanced ETL transformations and extensions, incremental data ingestion using job bookmarks, grouping, and workload partitioning using bounded execution in detail with practical examples. Doing so allowed us to understand how each of these features can be used to ingest data from data stores in specific use cases.

In this chapter, we will be introducing the fundamental concepts related to **data preparation**, different strategies that can help choose the right service/tool for a specific use case, visual data preparation, and programmatic data preparation using AWS Glue.

Upon completing this chapter, you will be able to explain how to perform data preparation operations in AWS Glue using a visual interface and source code. You will also be able to articulate different features of AWS Glue DataBrew, AWS Glue Studio, and AWS Glue ETL. You will also be able to write simple ETL scripts in AWS Glue ETL to prepare the data using some of the most popular transformations and extensions. Finally, you will be able to articulate the importance of planning and the different factors that must be taken into consideration while choosing a tool/service to implement a data preparation workflow.

In this chapter, we will cover the following topics:

- Introduction to data preparation
- Data preparation using AWS Glue
- Selecting the right service/tool

Now, let's dive into the fundamental concepts of data preparation and understand how data preparation can be done using AWS Glue and the different services/tools we can utilize to perform data preparation tasks quite easily.

Technical requirements

Please refer to the *Technical requirements* section in *Chapter 3, Data Ingestion*, as they are the same for this chapter as well.

In the upcoming sections, we will be discussing the fundamental concepts of data preparation, the importance of data preparation, and how we can prepare data using different tools/services in AWS Glue.

Introduction to data preparation

Data preparation can be defined as the process of sanitizing and normalizing the dataset using a combination of transformations to prepare the data for downstream consumers. In a typical data integration workflow, prepared data is consumed by analytics applications, visualization tools, and machine learning pipelines. It is not uncommon for the prepared data to be ingested by other data processing pipelines, depending on the requirements of the consuming entity.

When we consider a typical data integration workflow, quite often, data preparation is one of the more challenging and time-consuming tasks. It is important to ensure the data is prepared correctly according to the requirements as this impacts the subsequent steps in the data integration workflow significantly.

The complexity of the data preparation process depends on several factors, such as the schema of the source data, schema drift, the volume of data, the transformations to be applied to obtain the data in the required schema, and the data format, to name a few. It is important to account for these factors while planning and designing the data preparation steps of the workflow to ensure the quality of the output data and to avoid a **garbage in, garbage out** (**GIGO**) situation.

Now that we know the fundamental concepts and the importance of the data preparation steps in a data integration workflow, let's explore how we can leverage AWS Glue to perform data preparation tasks.

Data preparation using AWS Glue

It is normal for data to grow continuously over time in terms of volume and complexity, considering the huge number of applications and devices generating data in a typical organization. With this ever-growing data, a tremendous amount of resources are required to ingest and prepare this data – both in terms of manpower and compute resources.

AWS Glue makes it easy for individuals with varying levels of skill to collaborate on data preparation tasks. For instance, novice users with no programming skills can take advantage of **AWS Glue DataBrew** (`https://aws.amazon.com/glue/features/databrew/`), a visual data preparation tool that allows data engineers/analysts/scientists to interact with and prepare the data using a variety of pre-built transformations and filtering mechanisms without writing any code.

While AWS Glue DataBrew is a great tool for preparing data using a **graphical user interface** (**GUI**), there are some use cases where the built-in transformations may not be flexible enough or the user may prefer a programmatic approach to prepare data over using the GUI-based approach. In such cases, AWS Glue enables users to prepare data using **AWS Glue ETL**. Users can leverage AWS Glue Studio – AWS Glue's new graphical interface – to author, execute, and monitor ETL workloads. Although Glue Studio offers a GUI, users may still require programmatic knowledge of AWS Glue's transformation extensions and APIs to implement data preparation workloads, especially when implementing custom transformations using SQL or source code.

Now that we know about the different data preparation options that are available in AWS Glue, let's dive deep into each of them while looking at practical examples to understand them.

Visual data preparation using AWS Glue DataBrew

AWS Glue makes it possible to prepare data using a visual interface through AWS Glue DataBrew. As mentioned previously, AWS Glue DataBrew is a visual data preparation tool wherein users can leverage over 250 pre-built transformations to filter, shape, and refine data according to their requirements. AWS Glue DataBrew makes it easy to gather insights from raw data, regardless of the level of technical skill that the individuals interacting with the data have. More importantly, since DataBrew is serverless, users can explore and reshape terabytes of data without creating expensive long-running clusters, thus eliminating any administrative overhead involved in managing infrastructure.

Getting started with AWS Glue DataBrew is quite simple. To use DataBrew, you can create a project and connect it to a data store to obtain raw data. AWS Glue DataBrew can ingest raw data from Amazon S3, Amazon Redshift, JDBC data stores (including on-premise database servers), and AWS Data Exchange. We can also ingest data from external data stores such as Snowflake. You can even upload a file directly from the AWS Glue DataBrew console and specify an Amazon S3 location to store this uploaded file. At the time of writing, AWS Glue DataBrew supports the CSV, TSV, JSON, JSONL, ORC, Parquet, and XLSX file formats.

AWS Glue DataBrew can also ingest data from a wide range of external **Software-as-a-Service (SaaS)** providers via Amazon AppFlow. There are several external SaaS providers supported via Amazon AppFlow, including Amplitude, Datadog, Google Analytics, Dynatrace, Marketo, Salesforce, ServiceNow, Slack, and Zendesk, to name a few. This feature enables users to prepare the data by applying the necessary transformations while interacting with the data on a visual interface. This data can be further integrated with datasets from other data stores or SaaS applications. This helps the users take a holistic approach to analyzing and gathering insights from their datasets, which have been spread across different data stores or SaaS platforms. The following screenshot outlines the grid-like visual interface and different options available in the AWS Glue DataBrew project workspace:

Figure 4.1 – AWS Glue DataBrew project workspace

Once a project has been created and a dataset has been attached to the project, you can specify the AWS IAM role that can be used by this project to interact with other AWS services and a sampling strategy. This includes specifying the number of rows the visual editor has to load and whether these rows can be chosen at random or whether they have to be from the beginning or the end of the dataset.

After creating the project, AWS Glue DataBrew loads the project workspace and you will see your data in a grid-like interface (*Figure 4.1*). You can explore the data with ease using the project workspace and you will also be able to gain insights into each column using the statistics populated under the column name in the interface. Detailed statistics can be viewed for individual columns by clicking on the column name. By doing this, AWS Glue DataBrew generates insights based on the sample data that's been loaded into the project workspace and displays them in the **Column details** panel on the right-hand side of the workspace.

The following screenshot shows the list of recommendations that were generated for the `human_rights` column in the sample dataset after it was loaded into the AWS Glue DataBrew project workspace:

Figure 4.2 – AWS Glue DataBrew – recommended transformations

Based on the data type and the sample data that's loaded into the workspace, AWS Glue DataBrew also generates a list of recommended transformations that can be applied. For instance, if the values for a specific column in the dataset are missing, the list of recommended transformations includes different strategies to handle missing values, such as deleting rows with missing values, filling with an empty value, filling with the last valid value, filling with the most frequent value, and filling with a custom value. To apply one of these transforms, all we have to do is click **Apply as step** next to the transform.

As we make changes by applying different transformations, AWS Glue DataBrew captures the sequence of transformations that have been applied and builds a *recipe*. You can click on the column name and select a transformation from the top ribbon to apply a transformation for that column. Once you are happy with the recipe that's been generated, you can publish this recipe and it will be saved in AWS DataBrew (*Figure 4.2a*). This recipe can be downloaded as a JSON file and can be reused by importing the file as a new recipe in DataBrew. This is useful when you want to share the recipe with DataBrew users in other AWS accounts:

Figure 4.3 – Options to create, publish, and import/export recipes

In the preceding screenshot, several options are highlighted. Option *1* allows us to toggle the sidebar, which displays the current version of the recipe. The same recipe can be published using option *2*. A recipe can be exported or imported using option *3*. Finally, option *4* allows us to create a job from the recipe. Now that we know how to build, export, and import a recipe, let's explore different types of jobs in AWS Glue DataBrew.

Recipe jobs

A recipe can be used to create a **recipe job** in AWS Glue DataBrew, which will allow you to run the steps on your dataset (refer to option *4* in *Figure 4.3*). The job can be set up to run on-demand or at regular intervals by specifying a schedule. At the time of writing, AWS Glue DataBrew allows you to write transformed data to Amazon S3, Amazon Redshift, and JDBC data. Additional settings can be specified for the job, depending on the type of output destination data store.

For instance, if you are writing the data to an Amazon S3 location, you can specify options such as output format, compression codec, and output encryption using AWS KMS. The list of available options changes with the type of output data store selected. Other configuration items can be set for the job run, such as **Maximum number of units** (maximum number of DataBrew nodes that can be used), **Number of retries**, **Job timeout** (in minutes), and **CloudWatch logs for the job run**.

Profile jobs

In the previous section, you learned how to define a recipe and create a job from this recipe. Wouldn't it be great if most of the heavy lifting involved in understanding the data is handled by AWS Glue DataBrew so that we can plan the transformations better? AWS Glue DataBrew has another type of job called a **profile job** that addresses this exact issue. A profile job can be defined to evaluate the dataset and generate statistics and a summary that will help us understand the data better. This will, in turn, help us decide the type of transformations required to prepare the data.

A profile job run generates a data profile in AWS Glue DataBrew that contains a summary of the dataset and statistics for each column and any advanced summaries selected by the user. Profile jobs allow users to generate a correlations summary of different numeric columns available. It also allows you to profile the dataset based on advanced rules such as **personally identifiable information** (**PII**) detection. The dataset is evaluated using pre-built rules that analyze the column names and the values to flag any potential PII data in a given dataset. This is extremely helpful to make sure the dataset complies with data governance policies set forth by the organization or an external governing body.

The following screenshot shows what a sample data profile looks like:

Figure 4.4 – Data profile overview

In the lower half of the preceding screenshot, we can see that AWS Glue DataBrew has generated different summaries of the dataset based on the data types of the columns in the dataset. For instance, we can see a value distribution chart and the minimum, maximum, mean, median, mode, standard deviation, and other statistics for numeric columns. The summary also captures any missing data. We can use these pieces of information and design appropriate transformations in the recipe job to filter and reshape data based on our requirements.

Now that we know how profile jobs can be used to generate statistics and summaries for a given dataset, let's learn how to enrich these summaries with information based on user-defined rules.

Controlling data quality using DQ Rules

AWS Glue DataBrew allows us to define a ruleset that governs the quality of the dataset based on specified rules. The dataset is evaluated against the user-defined rules and violations are flagged in the data profile generated by the profile job run. This allows us to enrich the data profile with additional information based on the custom rules defined.

Upon creating a dataset in AWS Glue DataBrew, you can navigate to the **DQ Rules** option in the navigation panel and define a new ruleset for the dataset that's been created.

A **data quality** (**DQ**) ruleset is a collection of rules that defines the data quality for the dataset. This is achieved by comparing different data metrics with expected values. Once a ruleset has been defined, we can associate this ruleset with a profile job. After the job run, we will be able to see additional information under the **Data quality rules** tab in the generated data profile. This view includes the list of user-defined rules that were evaluated and a summary of whether all the columns adhered to these rules.

The following screenshot shows that the sample dataset was evaluated against two user-defined rules. The dataset passed the checks for one rule (**Check Dataset For Duplicate Rows**) and a few columns failed the checks for the other rule (**Check All Columns For Missing Values**):

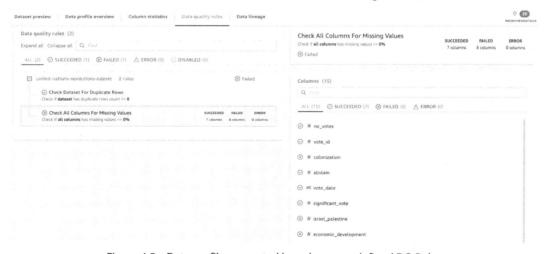

Figure 4.5 – Data profile generated based on user-defined DQ Rules

Using the insights generated by profile jobs, you can plan the data preparation steps according to your requirements and write the output to your destination data store. For instance, now that we know there are missing values in some of the columns, we can define transformations to handle those missing values – for example, populate with the last valid value, populate it with an empty string, or use a custom value.

Similarly, data masking transformations such as redaction, substitution, and hash functions can be applied to columns flagged as PII. We can even encrypt the data using probabilistic (using an AWS KMS key) or deterministic encryption (using a secret in AWS Secrets Manager) and decrypt the data when necessary.

There are over 250 transformations available in AWS Glue DataBrew for cleaning, reshaping, and preparing data based on the requirements and new transformations are being added to DataBrew frequently. A complete list of all the recipe steps and functions can be found in the AWS Glue DataBrew documentation at `https://docs.aws.amazon.com/databrew/latest/dg/recipe-actions-reference.html`.

Usage patterns for services/tools differ from one organization to another. An organization can choose to use AWS Glue DataBrew as its tool of choice for all data preparation workloads. However, if an organization prefers to use SQL or ETL scripts for their data preparation workload, AWS Glue DataBrew can be used for prototyping a data preparation pipeline. Then, data engineers can use the recipe in DataBrew as a reference to the authoring Glue ETL job. This allows other individuals within an organization who do not have Spark/Glue ETL programming skills to actively collaborate in data preparation workflows. Using this approach will reduce the effort and time taken by engineers to explore the data and design the data preparation steps from scratch.

Now that we know how we can leverage AWS Glue DataBrew for data preparation using a visual interface, let's learn how to prepare data using a source code-based approach in AWS Glue.

Source code-based approach to data preparation using AWS Glue

While AWS Glue DataBrew offers a visual interface-based approach to tackle data preparation tasks in a data integration workflow, AWS Glue offers AWS Glue ETL and AWS Glue Studio as source code/SQL-based approaches for the same. AWS Glue ETL and AWS Glue Studio require us to have some level of Glue/Spark programming knowledge to implement ETL jobs, which aids in data preparation as we get a much higher level of flexibility compared to AWS Glue DataBrew. With AWS Glue DataBrew, we can use pre-built transformations to prepare data. Since there are no such restrictions in AWS Glue ETL and AWS Glue Studio, we can design and develop custom transformations based on our requirements using existing Glue/Spark ETL APIs and extensions.

AWS Glue ETL and AWS Glue Studio

In *Chapter 2, Introduction to Important AWS Glue Features*, and *Chapter 3, Data Ingestion*, we briefly discussed some of the features of AWS Glue ETL and how they aid in data ingestion. In this section, we will explore different features of AWS Glue ETL and AWS Glue Studio and how these can be leveraged to prepare data.

Based on our discussion in *Chapter 2, Introduction to Important AWS Glue Features*, we know that a `DynamicRecord` is a data structure in AWS Glue in which individual rows/records in the dataset are processed and that a `DynamicFrame` is a distributed collection of `DynamicRecord` objects. To use Glue ETL transformations, the dataset must be represented as a Glue `DynamicFrame`, not an Apache Spark `DataFrame`. We can author ETL scripts using several methods on AWS Glue Studio, Interactive Sessions, or even locally on our development workstation using our preferred IDE or text editor since AWS Glue runtime libraries are publicly available. You can refer to the AWS Glue documentation at `https://docs.aws.amazon.com/glue/latest/dg/aws-glue-programming-etl-libraries.html` to explore different ETL job development options.

AWS Glue Studio is a new visual interface that makes it easy to author, run, and monitor AWS Glue ETL Jobs. AWS Glue Studio enables us to design and develop ETL jobs using a visual editor (*Figure 4.6*), implement complex operations such as PII detection and redaction, provide interactive ETL script development using Jupyter notebooks, set up custom/marketplace connectors to connect to SaaS/custom data stores, and easily monitor ETL job runs using a unified monitoring dashboard:

Figure 4.6 – Visual job editor in AWS Glue Studio

In the next section, we'll learn how to clean and prepare data using some of the transformations and extensions available in AWS Glue ETL.

Data transformation using AWS Glue ETL

Data preparation can be done in AWS Glue ETL by making use of built-in extensions and transformations. A complete list of extensions and transformations, syntax, and usage instructions can be found in the AWS Glue ETL documentation:

- AWS Glue Scala ETL jobs: `https://docs.aws.amazon.com/glue/latest/dg/glue-etl-scala-apis.html`

- AWS Glue PySpark ETL jobs: `https://docs.aws.amazon.com/glue/latest/dg/aws-glue-programming-python.html`

In this section, we will explore some of the most commonly used transformations in AWS Glue ETL.

ApplyMapping

The `ApplyMapping` transformation allows us to specify a declarative mapping of columns to a specified DynamicFrame. This transformation takes a DynamicFrame and a list of tuples, each consisting of the column name and data type mapping in the source and target DynamicFrames. This transformation is helpful when we want to rename columns or restructure a nested schema or change the data type of a column. It is important to specify a mapping for all the columns that are to be present in the target DynamicFrame. If a mapping is not defined for a column, that column will be dropped in the target DynamicFrame.

For example, let's assume there's a dataset with the following nested schema:

```
root
|-- email: string
|-- employee: struct
|    |-- employee_id: int
|    |-- employee_name: string
```

We can rename the `email` column `employee_email` and move the column under the `employee` struct using the following `ApplyMapping` transformation:

```
mappingList = [("email", "string", "employee.employee_email",
"string"), ("employee.employee_id", "int", "employee.employee_
id", "int"), ("employee.employee_name", "string", "employee.
employee_name", "string")]
applyMapping0 = ApplyMapping.apply(frame=datasource0,
mappings=mappingList)
```

In the preceding snippet, `mappingList` is the list of mapping tuples being passed to the `ApplyMapping` transform. We can also see the mapping tuple that maps the `email` column to `employee.employee_email`. This mapping is essentially renaming the column `employee_email` and moving the column under the `employee` struct. Now, when we print the schema of the `applyMapping0` DynamicFrame, we will see the following:

```
>>> applyMapping0.printSchema()
root
|-- employee: struct
|       |-- employee_email: string
|       |-- employee_id: int
|       |-- employee_name: string
```

As you can see, by using the `ApplyMapping` transformation, we were able to achieve two things:

- Rename the column `employee_email`.
- Reshape the schema of the dataset to move the `email` column under the `employee` struct.

Now, let's look at another commonly used transformation: `Relationalize`.

Relationalize

The `Relationalize` transform helps us reshape a nested schema of the dataset by flattening it. Any array columns that are present are pivoted out. This transformation is extremely helpful when we are working with a dataset that has a nested schema structure and we want to write the output to a relational database.

Let's see this transformation in action. Let's assume there is a dataset with the following schema. You will be able to find the source code and sample dataset for this example in this book's GitHub repository at `https://github.com/PacktPublishing/Serverless-ETL-and-Analytics-with-AWS-Glue/tree/main/Chapter04`:

```
>>> datasource1.printSchema()
root
|-- company: string
|-- employees: array
|       |-- element: struct
|       |       |-- email: string
|       |       |-- name: string
```

Now, let's apply the `Relationalize` transformation to flatten this schema. In return, we will get a `DynamicFrameCollection` populated with DynamicFrames. Any array columns present in the dataset are pivoted out to a separate DynamicFrame:

```
relationalize0 = Relationalize.apply(frame=datasource1,
staging_path='/tmp/glue_relationalize', name='company')
```

To list the keys for the different DynamicFrames that have been generated, we can use the `keys()` method on the returned `DynamicFrameCollection`. In the preceding example, that would be `relationalize0`:

```
>>> relationalize0.keys()
dict_keys(['company', 'company_employees'])
```

Now, since two DynamicFrames in `DynamicFrameCollection` were returned, it would be easier to interact with them separately if we extract them from `DynamicFrameCollection`. We could `select()` each of those DynamicFrames and use the `show()` method to see their contents. Alternatively, we can use the `SelectFromCollection` transformation to select individual DynamicFrames:

```
>>> company_Frame = relationalize0.select('company')
>>> company_Frame.toDF().show()
+----------+---------+
|   company|employees|
+----------+---------+
|DummyCorp1|        1|
|DummyCorp2|        2|
|DummyCorp3|        3|
+----------+---------+
>>> emp_Frame = relationalize0.select('company_employees')
>>> emp_Frame.toDF().show()
+---+-----+-------------------+------------------+
| id|index|employees.val.email|employees.val.name|
+---+-----+-------------------+------------------+
|  1|    0|   foo@company1.com|              foo1|
|  1|    1|   bar@company1.com|              bar1|
|  2|    0|   foo@company2.com|              foo2|
|  2|    1|   bar@company2.com|              bar2|
|  3|    0|   foo@company3.com|              foo3|
```

```
|   3|    1|    bar@company3.com|                    bar3|
+---+-----+--------------------+-------------------+
```

As you may recall, the `Relationalize` transform has pivoted the `employees` column and created a new DynamicFrame with additional columns: `id` and `index`. The `id` column acts similarly to a foreign key for the `employees` column in the `company` DynamicFrame.

However, for us to be able to write the flattened data to a relational database, we need the data to be present in one DynamicFrame. To bring both of these DynamicFrames together, we can use the `Join` transform. Let's look at the `Join` transform and see how it works.

Join

The `Join` transform, as its name suggests, joins two DynamicFrames. A `Join` transform in AWS Glue performs an equality join. If you are interested in performing other types of `Join` (for example, broadcast joins) in Glue ETL, you will have to convert the DynamicFrame into a Spark DataFrame.

Let's continue with our example and join the two DynamicFrames that were created by `Relationalize` while using `employees` and `id` as the keys:

```
join0 = Join.apply(frame1 = company_Frame, frame2 = emp_Frame,
keys1 = 'employees', keys2 = 'id')
```

Let's use the `show()` function to see the joined data:

```
>>> join0.toDF().show(truncate=False)
```

This will result in the following output:

employees.val.name	company	employees	index	employees.val.email	id
foo2	DummyCorp2	2	0	foo@company2.com	2
bar2	DummyCorp2	2	1	bar@company2.com	2
foo3	DummyCorp3	3	0	foo@company3.com	3
bar3	DummyCorp3	3	1	bar@company3.com	3
foo1	DummyCorp1	1	0	foo@company1.com	1
bar1	DummyCorp1	1	1	bar@company1.com	1

Figure 4.7 – Output demonstrating a Join transformation

As you can see, the `email` and `name` array columns have been renamed `employees.val.email` and `employees.val.name`, respectively. This is the result of pivoting the array in the `Relationalize` transformation. This can be corrected using the `RenameField` transformation before joining the DynamicFrames.

Now, let's look at the `RenameField` transformation to see how we can rename columns.

RenameField

The `RenameField` transformation allows us to rename columns. This transformation takes three parameters as input – a DynamicFrame where the column needs to be renamed, the name of the column to be renamed, and the new name for the column.

In our example, we saw that after the array was pivoted by the `Relationalize` transform, the `email` and `name` array columns were renamed `employees.val.email` and `employees.val.name`, respectively. To rename the columns so that they have their original names, we can use the following code snippet:

```
renameField0 = RenameField.apply(frame = join0, old_name =
"`employees.val.email`", new_name = "email")
renameField1 = RenameField.apply(frame = renameField0, old_name
= "`employees.val.name`", new_name = "name")
```

You may have noticed the wrapping backquotes (`) for the old column names in the preceding code snippet. This is because we have a dot (.) character in the name of the column itself and here, the dot character does not represent a nested structure. To suppress the default behavior of the dot character, we have wrapped the column names in backquotes.

We can confirm that the columns have been successfully renamed by printing the schema of the `renameField1` DynamicFrame. Now that we have a flattened schema structure and the columns have been renamed according to our requirements using transformations such as `Relationalize`, `Join`, and `RenameField`, we can safely write the resultant DynamicFrame to a table in a relational database.

Now, let's look at some of the other transformations available in AWS Glue ETL.

Unbox

The `Unbox` transformation is helpful when a column in a dataset contains data in another format. Let's assume that we are working with a dataset that's been exported from a table in a relational database and that one of the columns has a JSON object stored as a string.

If we continue to use `string` data types for this JSON object, we won't be able to analyze the data present in this column as easily as the downstream application may not know how to parse it. Even if it does, the queries would be extremely complex. Since the purpose of the data preparation step is to clean and reshape the data, it is much better to address this within the data preparation workflow.

Let's assume that our dataset has the following schema:

```
root
|-- location: string
|-- companies_json: string
```

When we use the show() method on the DynamicFrame, we will see that there is a JSON string in the companies_json column:

```
+-----------+--------------------+
|   location|       companies_json|
+-----------+--------------------+
|Seattle, WA|{"jsonrecords":[{...|
+-----------+--------------------+
```

Now, let's see how the Unbox transform can help us unpack this JSON object and merge the schema of the JSON object with the DynamicFrame schema:

```
>>> unbox0 = Unbox.apply(frame = datasource2, path =
"companies_json", format = "json")
>>> unbox0.printSchema()
root
|-- location: string
|-- companies_json: struct
|    |-- jsonrecords: array
|    |    |-- element: struct
|    |    |    |-- company: string
|    |    |    |-- employees: array
|    |    |    |    |-- element: struct
|    |    |    |    |    |-- email: string
|    |    |    |    |    |-- name: string
```

As we can see, the schema from the JSON object was merged into DynamicFrame's schema. Now, we can use other transformations to further transform the data or output the DynamicFrame as-is.

Now, there might be situations where you run into an issue when applying a transformation in Glue ETL and you may notice that some or all the records in a DynamicFrame have gone missing. This may happen if there was an error when parsing the records. How do we find out if this has happened? Well, AWS Glue ETL has a transformation that captures the nested error records called ErrorsAsDynamicFrame. Let's take a look at how this works.

ErrorsAsDynamicFrame

This transformation takes a DynamicFrame as input and returns the nested error records that have been encountered up until the creation of the input DynamicFrame. In the Unbox transform example, we used a JSON string nested within a record to demonstrate the capabilities of Unbox. Let's introduce a syntax error into the JSON string of one of the records by removing a curly brace or a comma that will interfere with the normal functioning of the JSON parser.

The following source code can be found in this book's GitHub repository at https://github.com/PacktPublishing/Serverless-ETL-and-Analytics-with-AWS-Glue/tree/main/Chapter04:

```
# Refer GitHub repository for Sample code-gen function
createSampleDynamicFrameForErrorsAsDynamicFrame()
>>> datasource2 =
createSampleDynamicFrameForErrorsAsDynamicFrame()
>>> datasource2.toDF().show()
+-----------+--------------------+
|   location|      companies_json|
+-----------+--------------------+
|Seattle, WA|{"jsonrecords":[{...|
|Sydney, NSW|{"jsonrecords":[{...|
+-----------+--------------------+
>>> unbox0 = Unbox.apply(frame = datasource2, path =
"companies_json", format = "json")
>>> unbox0.toDF().show()
+-----------+--------------------+
|   location|      companies_json|
+-----------+--------------------+
|Seattle, WA|{ [{DummyCorp1, [{...|
+-----------+--------------------+
>>> ErrorsAsDynamicFrame.apply(unbox0).count()
1
>>> ErrorsAsDynamicFrame.apply(unbox0).toDF().show()
+-------------------+
|              error|
+-------------------+
|{{  File "/tmp/66...|
+-------------------+
```

As we can see, the valid JSON record that was in the DynamicFrame was parsed correctly by the parser. However, the invalid record was not parsed and we can see that the `ErrorsAsDynamicFrame` class has captured the errors. As part of the ETL script, we can have validation steps using this class to ensure there were no errors when transforming data.

You may have noticed by now that each of the AWS Glue ETL transforms have two parameters available. These parameters specify the error threshold for each transformation:

- `stageThreshold` specifies the maximum number of errors that can occur in a given transformation for which the job needs to fail.

- `totalThreshold` specifies the maximum number of errors up to and including the current transformation.

We can leverage these parameters to manage error handling behavior in AWS Glue ETL.

There are several other transformations available in AWS Glue ETL that make it easy to reshape and clean data based on our requirements. It would not be practical to discuss each of the transformations available in AWS Glue ETL here as the service has been constantly evolving since it was released and new transformations and extensions are being added by AWS. You can find an exhaustive list of transformations, syntax, and examples in the AWS Glue documentation, as mentioned at the beginning of this section.

Now that we are familiar with AWS Glue DataBrew, AWS Glue ETL, and AWS Glue Studio, it is important to know which tool/service to choose for your workload.

Selecting the right service/tool

In the previous sections, we looked at the different features, transformations, and extensions/APIs that are available in AWS Glue DataBrew, AWS Glue Studio, and AWS Glue ETL for preparing data. With all the choices available and the varying sets of features in each of these tools, how do we pick a tool/service for our use case? There is no hard and fast rule in selecting a tool/service and the choice depends on several factors that need to be considered based on the use case.

As discussed earlier in this chapter, AWS Glue DataBrew empowers data analysts and data scientists to prepare data without writing source code. AWS Glue ETL, on the other hand, has a higher learning curve and requires Python/Scala programming knowledge and a fundamental understanding of Apache Spark. So, if the individuals preparing the data are not skilled in AWS Glue/Spark ETL programming, they can use AWS Glue DataBrew.

One of the important factors to consider while choosing a tool/service is whether the data preparation tasks being planned can be implemented using the tool/service. While AWS Glue DataBrew has a library of over 250 pre-built transformations, they may still not cover some of the transformations required to implement your data preparation workflow or it might be too complex to implement your workflow using built-in transformations in DataBrew. In such cases, we can simplify the workflow by writing an ETL job in AWS Glue ETL since we have the flexibility to write custom transformations. We can leverage built-in AWS Glue ETL transformations or we can custom-design our transformations using Apache Spark APIs.

Another factor that can influence this decision is whether the data preparation workflow that's being implemented is a one-off operation or something that needs to be accomplished quite frequently. If the data preparation tasks are simple and infrequent, it would not justify the effort involved in writing source code manually. In such cases, we can use AWS Glue DataBrew or AWS Glue Studio's visual job editor to set up an ETL job to accomplish our tasks. However, if the tasks are complex, require a higher level of flexibility, and are going to be performed regularly, AWS Glue ETL would be a better choice as we can customize the ETL job based on our requirements.

To summarize, it is important to consider the use case and construct a plan based on the requirements. Some of the key considerations that could factor into the decision-making process are as follows:

- Features offered by a specific tool and whether our tasks can be accomplished using built-in transforms
- The skill sets of individuals within the team
- The complexity of the workflow that is being implemented
- The frequency of data preparation operations

So, it is important to consider the use case at hand, plan your data preparation workflow, and then choose a tool/service to implement your workflow. Otherwise, you could end up wasting a lot of time and effort in designing your workflow using a specific tool/service that was not fit for your use case to begin with.

Summary

In this chapter, we discussed the fundamental concepts and importance of data preparation within a data integration workflow. We explored how we can prepare data in AWS Glue using both visual interfaces and source code.

We explored different features of AWS Glue DataBrew and saw how we can implement profile jobs to profile the data and gather insights about the dataset being processed, as well as how to use a DQ Ruleset to enrich the data profile, use PII detection and redaction, and perform column encryption using deterministic and probabilistic encryption. We also discussed how we can apply transformations, build a recipe using those transformations, create a job using that recipe, and run the job.

Then, we discussed source code-based ETL development using AWS Glue ETL jobs and the different features of AWS Glue Studio before exploring some of the popular transformations and extensions available in AWS Glue ETL. We saw how these transformations can be used in specific use cases while covering source code examples and how we can detect and handle errors during data preparation in AWS Glue ETL.

We talked about different factors that need to be considered while choosing a service/tool in AWS Glue and the importance of considering the use case and planning while designing our data preparation workflow.

In the next chapter, we will discuss the importance of data layouts and how we can design data layouts to optimize analytics workloads. We will be exploring some of the concepts that factor into performance and resource consumption during query execution, such as data formats, compression, bucketing, partitioning, and compactions.

5
Data Layouts

Data analysis is a common practice to make data-driven decisions to accelerate business and grow your company, organization, teams, and more. In a typical analysis process, queries that process and aggregate records in your datasets will be run for your data to understand their business trends. The queries are commonly run from **Business Intelligence** (**BI**) dashboard tools, web applications, automated tools, and more. Then, you will be able to get the results you need such as user subscriptions, marketing reports, sales trends, and more.

For their analytic queries, it's important to consider analytic query performance because they need to timely utilize the analysis data and to quickly make a business decision for their business growth. To accelerate the query performance to quickly obtain the analysis data, you need to care about your dashboard tools, computation engine that processes the large amount of your data, data layout design of your data and its data storage, and more. The combination of these resources affects your analytic query performance so that it's important to understand them.

This chapter focuses on how we design data layouts to optimize your analytic workloads. In particular, to design the data layouts that can maximize your query performance, we need to consider the three important parts such as key techniques for our data to optimize query performance, how we manage our files, and how we optimize our Amazon S3 storage.

By focusing on these three parts, in this chapter, we will learn useful and general techniques to accelerate your analytic workloads, and important functionalities to optimize the workloads that can be achieved using AWS Glue and Lake Formation.

In this chapter, we will cover the following topics:

- Why do we need to pay attention to data layout?
- Key techniques to optimally storing data
- Optimizing the number of files and each file size
- Optimizing your storage by working with Amazon S3

Technical requirements

For this chapter, if you wish to follow some of the walk-throughs, you will require the following:

- Access to GitHub, S3, and the AWS console (specifically AWS Glue, AWS Lake Formation, and Amazon S3)

- A computer with the Chrome, Firefox, Safari, or Microsoft Edge browser installed and the **AWS Command-Line Interface (AWS CLI)**:

  - Regarding the AWS CLI, you can use not only the AWS CLI but also AWS CLI version 2. In this chapter, the AWS CLI (not version 2) is used. You can set up the AWS CLI (and version 2) from `https://docs.aws.amazon.com/cli/latest/userguide/cli-chap-getting-started.html`.

- An AWS account and an accompanying IAM user (or IAM role) with sufficient privileges to complete this chapter's activities. We recommend using a minimally scoped IAM policy to avoid unnecessary usage and making operational mistakes. You can get the IAM policy for this chapter from the relevant GitHub repository, which is shown at `https://github.com/PacktPublishing/Serverless-ETL-and-Analytics-with-AWS-Glue/blob/main/Chapter05/data.json`. This IAM policy includes the following access:

  - Permissions to create a list of IAM roles and policies for creating a service role for an AWS Glue ETL job

  - Permissions to read, list, and write access to an Amazon S3 bucket

  - Permissions to read and write access to Glue Data Catalog databases, tables, and partitions

  - Permissions to read and write access to Glue Studio

- An S3 bucket for reading and writing data with AWS Glue. If you haven't created one yet, you can do so from the AWS console (`https://s3.console.aws.amazon.com/s3/home`) | **Create bucket**. You can also create a bucket by running the `aws s3api create-bucket --bucket <your_bucket_name> --region us-east-1` AWS CLI command.

Why do we need to pay attention to data layout?

As we discussed earlier, it's important to maximize query performance for your analytic workloads because they need to quickly understand for their situation for quick decisions based on the query results. To achieve the most optimal analytics workloads, one of the most important phases is data extraction process that a computation engine retrieves your data from the data location (Relational database, Distributed storage and so on) and reads records. It's because many operations on our analytic workloads are reading data and processing them into what we want based on our running queries. These days, many computation engines that process data are effectively optimized their computation by their community, company and more. However, the data extraction process, especially retrieving and reading data from an external location highly depends on our data layout such as the file number, file format and so on, network speed, and more. Therefore, to achieve optimal data extraction, we should carefully design our **data layout** to optimize our query performance more.

When considering the data layout, you should mainly focus on the following three parts:

- **Key techniques to optimally storing data**: This is the first part. When you store your data, you should pay attention to what file format and compression type you use, and whether you use partitioning and/or bucketing. Because these techniques are import to optimize your query performance. We'll go through the details about the techniques in *Key techniques to optimally storing data* section. Paying attention to how you store data can optimize the processing of your data with a processor engine that actually runs analytic queries such as saving process time to compute data schema by choosing a file format. This has a schema, avoiding processing unnecessary files by filtering your data in advance, and more.

- **Optimizing the number of files and each file size**: This is the second part. It's possible to save processing time by keeping the number of files as small as possible and by keeping each file size the number which is a computation engine's chunk size such as 64MB, 128MB and so on. This is because we can potentially avoid spending time of handling each file by the computation engine.

- **Optimizing data storage based on data access**: This is the last one. Your data size should be incremental and grow continuously, such as continuous web access logs, data sent by IoT devices, and more. Generally, the larger the data size in your storage, the higher the cost of the storage usage you need to pay. Therefore, often you need to archive part of the data and keep other parts based on the access to the data to decrease the storage cost and reduce unnecessary data access for your analytic workloads.

To achieve data retrieval as quickly as possible, and then enhance your analytic workloads, in the next section, we will focus on learning about the previously mentioned points to introduce a good data layout. In particular, this chapter will show how you can meet these requirements with **AWS Glue**, **AWS Lake Formation**, and **Amazon S3**.

Key techniques to optimally storing data

As mentioned earlier, the data extraction process is one of the most important phases to consider when optimizing your analytic workloads. In the usual process of data retrieval, users such as data analysts, business intelligence engineers, and data engineers run queries to a distributed analytics engine such as Apache Spark and Trino. Then, the distributed analytics engine gets information about the data, such as each file location and metadata. Usually, this kind of data is stored in distributed storage such as Amazon S3, HDFS, and more. After getting all the information about the data, the computing engine actually accesses and reads the data that you specify in the queries. Finally, it returns query results to the users.

To make the data retrieval process faster for further analysis, it's important to consider how you store data. In particular, you can optimize workloads for analysis by storing data in the most suitable condition for your analysis. For example, when running analytic queries, if there were a lot of files in your storage, running queries would take more time than if there are a smaller number of files. This is mainly because a distributed analytics engine would need time to get the information about each file, such as each file location and metadata. Based on the information, the computing engine retrieves the data from storage before processing it. In such cases, it's possible to improve the time of the data retrieval process by gathering the files within a smaller number of files and decreasing each file size by compressing it to match the size that the computing engine can process (usually, this size is based on your computing engine's memory capacity). Usually, this processing can be achieved by using computing engines.

To optimally store your data, you should pay attention to file formats, compression types, the splitability of files, and partitioning or bucketing as they can affect the workloads of your analytic queries. We will learn more about this in the following sections.

Selecting a file format

Generally, data can be categorized into **unstructured**, **semi-structured**, and **structured** formats based on whether the data has a specific schema and types. If the data has specific key-value pairs but doesn't have any typed schema, the data can be classified into a semi-structured format such as JSON, CSV, or XML. If the data has specific columns and types, it can be classified into a structured format such as Apache Parquet, Apache ORC, Apache Avro, and more. Otherwise, the data can be generally thought of as an unstructured format such as images and log files.

Selecting a file format affects your query performance. Structured format data with a schema like a relational database table enables a data processing engine to avoid computing the data schema and to extract only necessary data (e.g. values in the columns you want to process) based on user defined queries. In particular, it's recommended that you use the formats that have columnar data structures such as **Apache Parquet** and **Apache ORC** because these formats provide a lot of merits for the analysis. For example, *Apache Spark* that is used on AWS Glue can optimize querying Parquet files by narrowing down access to records based on Parquet format structure. We'll see the merits next, and see how to convert your data to these columnar formats.

Storing your data in columnar formats for effective analytic workloads

As we've seen so far, Apache Parquet and Apache ORC are file formats that have table-like schemas and columnar storage. These formats can effectively provide data processing for your analytic queries based on their columnar format features such as metadata columns, filtering columns and the relevant records, effective compression and encoding schemas, and more.

Actual data in Parquet files consists of row groups, which include arrays of columns. Parquet defines the size of a chunk of the data for each column to store records, which includes columns and pages as **Block size**. By default, this size is defined as 128 MB. Also, ORC has a chunk size to store records called **Stripe size**, which is defined as 64 MB by default. Each chunk in ORC includes index data, row data, and, strip footer. If you store data with a large block or strip size, a processor can execute effective column-based manipulations; however, this is possible to cause multiple I/O operations due to multiple blocks in your storage. On the other hand, if you store data with a small block or strip size, this too needs multiple accesses to each file and possibly reduces its efficiency. Therefore, when you store your data with the Parquet or ORC format, you should store data with the block or stripe size or set a larger block or stripe size based on your data if your data has a lot of columns.

Configuration of Parquet block or ORC stripe size in Glue Spark jobs

You can configure the block or strip size by specifying each relevant parameter to the option method for Spark DataFrameWriter as follows:

```
dataframe.write.option('parquet.block.size', 1024 * 1024)
# 1024 * 1024 bytes = 1MB block size

dataframe.write.option('orc.stripe.size', 1024 * 1024)
# 1MB strip size
```

You can also effectively narrow down your data for Parquet and ORC formats when filtering or querying values in particular columns. Many computation engines such as Apache Spark, Apache Hive and Trino/Presto support a narrow-down feature called **predicate pushdown** or **filter pushdown**. Each block in Parquet and ORC files has statistics of the chunk such as the value range of minimum and maximum. This statistical information is used for your running query to determine which part is necessary to read. If you sort the column value that you use for filtering before processing the data, this can improve your analytic query performance based on its mechanism.

Converting your data to Apache Parquet or Apache ORC formats with AWS Glue

You can convert your data files with a Glue ETL Spark job. Using AWS Glue Studio, you can create the Glue job and it automatically generates the format conversion script. Regarding how to use the Glue Studio, please refer to *AWS Glue ETL and AWS Glue Studio* section in *Chapter 4, Data Preparation*. The following example shows the steps to generate the format conversion script from JSON to Apache Parquet with snappy compression. Follow these steps:

1. Download the sample sales data (`data.json`) on your local machine from `https://github.com/PacktPublishing/Serverless-ETL-and-Analytics-with-AWS-Glue/blob/main/Chapter05/data.json`. Once downloading is completed, upload the file to your Amazon S3 bucket using the command; `aws s3 cp data.json s3://<your-bucket-and-path>/` or from the S3 console (`https://s3.console.aws.amazon.com/s3/buckets`)

2. Access **Jobs** on Glue Studio console (`https://us-east-1.console.aws.amazon.com/gluestudio/home#/jobs`).

3. Choose **Visual with a source and target** and **Create** on the top of right page.

4. In **Data source – S3 bucket** node, set your S3 bucket and path, and choose **Infer schema**.

5. In **Data target – S3 bucket** node, set `Parquet` to **Format**, `Snappy` to **Compression Type** and your S3 bucket and path. You can generate the following diagram and script as the following screenshot.

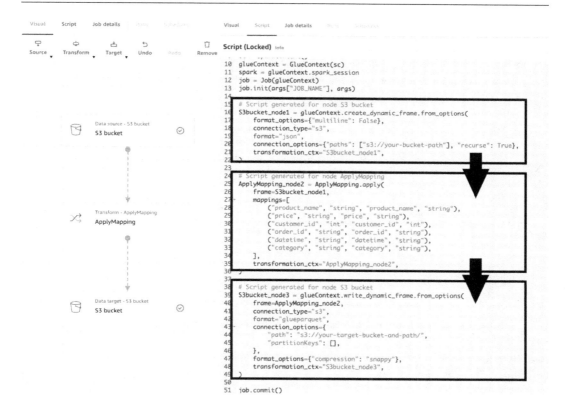

Figure 5.1 – Format conversion Glue job diagram and script on Glue Studio console

6. To run this file format conversion job, choose **Job details** tab and complete all information such as Job name, IAM Role, Job type and so on.

7. After completing all information, choose **Save** and **Run**.

After running the Glue job completed, you can see parquet files with snappy compression in the target S3 bucket and path.

Next, we'll look at several data compression types that can decrease your data size.

Compressing your data

Reducing file size by compression enables you to save data network transfer cost, save query process time, reduce usage of data storage, save the storage cost and so on. For these merits, you should store data with compression. Note that you pay attention to whether the compression type is splittable or not, compression or decompression speed and each compressed file size, which possibly affect your query performance. We will see the file splittability in the *Splittable or Unsplittable files* section and see the file size management in the *Managing number of files and each file size* section.

The following table shows, in Spark, compression formats that are commonly used for Apache Parquet such as `gzip`, `lz4`, `snappy`, and `zstd`, along with their compression ratios and compression/decompression speeds. Each compression ratio and (de)compression speed is measured by running actual data processing jobs, in seconds. Additionally, each of them is normalized by each no compression result and `gzip` compression result, respectively:

Compression types	Compression ratio (normalized by a file size of parquet without compression)	Normalized compression + decompression speed rate (normalized by the `parquet.gzip` compression + decompression speed in seconds)
No compression (the `Parquet` file format)	1.00	n/a
`gzip`	0.61	1.00
`lz4`	0.77	0.64
`snappy`	0.78	0.67
`zstd`	0.60	0.68

Table 5.1 – Comparison of compression ratio and speed between compression types

Each value in the table was measured by running a Glue Spark job. The following list shows what environment the Spark job ran on:

- The test data is all tables in TPC-DS dataset with scale 1000 whose size and file format are 412.3 GB in Apache Parquet files without compression. Refer to Further Reading section about the TPC-DS.

- TPC-DS Glue custom connector (https://aws.amazon.com/marketplace/pp/prodview-xtty6azr4xgey) was used to generate TPC-DS dataset.

- Used analytic engine: Glue 3.0 (Spark 3.1.1).

- Compression speed and ratio were measured by running the Glue job script in the book's GitHub repository (https://github.com/PacktPublishing/Serverless-ETL-and-Analytics-with-AWS-Glue/blob/main/Chapter05/MeasureCompressionSpeedAndRatio.scala).

As shown in *Table 5.1*, compressing the data with gzip, lz4, snappy, and zstd can reduce the file size compared to the case without compression. In addition to reducing file size by the compression technique, compression/decompression speed can affect your processing job. In particular, a data processing job, including gzip compression, is expected to be slower than a job using the other compression types such as lz4, snappy, and zstd, based on *Table 5.1*. Therefore, when compressing your data with a processing job to optimize the data in your storage, you should consider not only the compression ratio but also the compression speed to get compressed data as quickly as possible.

> **Note**
>
> Generally, the higher the compression ratio of an algorithm you specify, the more computation overhead is necessary to compress and decompress data.

So far, we've seen how the compression works for your data and workloads. But how can we actually run the compression job for our data? We can compress our data with AWS Glue. Using Glue Studio, we can generate the compression Glue job script as we've seen in *Converting your data to Apache Parquet or Apache ORC formats with AWS Glue* section. Specifically, we just choose a compression type for the **Data target – S3 bucket** node in *Step 5* of the example in the previous section. The compression type you can choose depends on your file format type. For example, if you set Parquet as the format, you can choose Snappy, LZO, GZIP or Uncompressed. The following example script shows the partial code that is generated by Glue Studio and that writes the Parquet files with GZIP compression.

```
S3bucket_node3 = glueContext.write_dynamic_frame.from_options(
    frame=ApplyMapping_node2,
    connection_type="s3",
    format="glueparquet",
    connection_options={
        "path": "s3://your-target-bucket-and-path/",
        "partitionKeys": [],
    },
    format_options={"compression": "gzip"},
    transformation_ctx="S3bucket_node3",
)
```

You can also compress your data with Spark DataFrame. If you use Spark DataFrame for compression, you need to directly edit your Glue job script on Glue Studio. The following example shows the part of the job script that writes Parquet files with `zstd` compression.

```
COMPRESSION_CODEC = 'zstd'
dataframe.write\
        .option('compression', COMPRESSION_CODEC)\
        .parquet(DST_S3_PATH)
```

The whole script is available at `https://github.com/PacktPublishing/Serverless-ETL-and-Analytics-with-AWS-Glue/blob/main/Chapter05/compression_by_dataframe.py`.

> **Note**
>
> Glue DynamicFrame currently doesn't support `zstd` for reading and writing. You should use Spark DataFrame to compress/decompress data to/from `zstd`.

Next, we'll look at *file splittability*, which is determined by file format and compression type.

Splittable or unsplittable files

When you run analytics queries and process data, it's helpful to know whether the files from your data source are *splittable* or not. *A file is splittable* means whether a processor such as AWS Glue can get the contents of a file by separating it based on the chunk size of the processor when the processor reads the file. When a file is not splittable, a processor cannot separate a file and needs to get the whole file.

Why do we need to think about whether a file is splittable? Well, usually, it affects your data retrieval. Let's assume that your data files are not splittable and each file has a big size that is greater than the size of your memory or storage. A file is not splittable; therefore, a processor cannot separate it as a chunk and needs to read the whole file. However, a processor cannot process a file because each file size is more than the memory and storage size or processor. In particular, with Apache Spark, processing a large size of an unsplittable file might cause an out-of-memory error because Spark processes the data in memory. In other words, you should control each file size appropriately for your processor if your data source has unsplittable files.

Whether it's splittable or not depends on what the file format is and/or how the file has been compressed. The following table shows popular file formats and compression types, and whether they're splittable or unsplittable. Please check the files in your data source if you use a data processor such as AWS Glue, Amazon EMR, or Amazon Athena, which processes and writes the data in your storage:

	Splittable	Unsplittable
File format	• Single-lined CSV • JSON Lines • XML • Apache Parquet • Apache ORC • Apache Parquet or Apache ORC with *row*-level `snappy` or `gzip` compression • Apache Avro	• Multi-lined CSV • Multi-lined JSON
Compression type	• `bzip2` • `zstd`	• `gzip` • `LZO` • `snappy`

Table 5.2 – The splittability of file formats and compression types

From *Table 5.2*, for example, if your data files are in XML format without compression, they're *splittable*. As another example, if your data files are in JSON format with `gzip` compression, they're *unsplittable*.

Partitioning

Partitioning is a technique to store your data separately into different folders based on specified partition keys. Each partition key is related to your data and actually acts as a column. For example, if you have your data in your Amazon S3 bucket as `s3://bucket-name/category=drink/data.json`, the partition key can be recognized as `category`, and its value is `drink`.

By partitioning your data, you can reduce data scan size by querying only the required data. Specifically, a computation engine (such as Spark, Presto and so on.) only reads the data in specified partition keys and values in your query. In the example above, if you specify `drink` for the `category` partition key, the engine only reads the data under the `drink` folder by listing partition values for the key. This can reduce data scan size and improve query performance.

You can define a column as a partition key at table creation. The partition keys and values are registered in your table that is stored in *Apache Hive metastore*. The Hive Metastore is a service to store table metadata and their relevant information, in a database backend such as a relational database. More details about the Hive Metastore is discussed in the *AWS Glue Data Catalog* section in *Chapter 2, Introduction to Important AWS Glue Features*. The computing engine retrieves the list of partition values from the metastore based on your query with a specific range of partition values for keys, and then it reads the data in specified partitions. Therefore, partitioning enables a computation engine to filter partitions and avoid processing unnecessary partitions.

When you partition your data, you should use **Hive style partitioning** such as `/path/to/<partition_key_1>=<value1>/<partition_key_2>=<value2>` compared to non key-value style such as `/path/to/value1/value2`. Using Hive style partitioning, partition keys can be processed as table columns and the values are filtered by `WHERE` clause in SQL-like query such as `WHERE category = 'drink'`. Also, you can automatically register partition values in your Hive Metastore for the key by `MSCK REPAIR TABLE <your-table-name>` Hive query that can be run by not only Glue Spark jobs but also Athena. For more details about Hive query for Hive style partitioned tables, please refer to `https://docs.aws.amazon.com/athena/latest/ug/partitions.html`.

The Glue Data Catalog, which we saw in *Chapter 2, Introduction to Important AWS Glue Features*, can be used as an external Hive metastore. You can register partition keys and values in your table in the Glue Data Catalog. For example, you can register `category` as a partition key and `drink` as its value in a Glue Data Catalog table on your S3 bucket structure such as *s3://bucket-name/category=drink/<data files>*. We look at how to register partition keys and values in the Glue Data Catalog in *Registering partition values in a Glue Data Catalog table* section below. By specifying a range of partitions, you can reduce the data scan size in your Glue ETL Spark job because the job only reads the data in the specified partitions. This possibly improves the Glue job performance.

Example - partitioning by AWS Glue ETL Spark job

In this example, we partition the data (`data.json`) in the S3 bucket with Hive style partitioning by a Glue Spark job. Specifically, we partition the S3 bucket as the following structure based on the `data.json` records. In the following folder structure, the partition key is `category`, and the values are `drink`, `grocery` and `kitchen`.

```
s3://bucket-name/
            ├── category=drink/<data files>
            ├── category=grocery/<data files>
            ├── category=kitchen/<data files>
```

To write your data with Hive style partitioning by a Glue job, you can mainly use `partitionKeys` option for Glue DynamicFrame or `partitionBy` method for Spark DataFrame.

As we've seen in *Converting your data to Apache Parquet or Apache ORC formats with AWS Glue* section, using Glue Studio, we can automatically generate a partitioning script by specifying partition keys for the **Data target – S3 bucket** node. In the following screenshot, `category` is specified as the partition key.

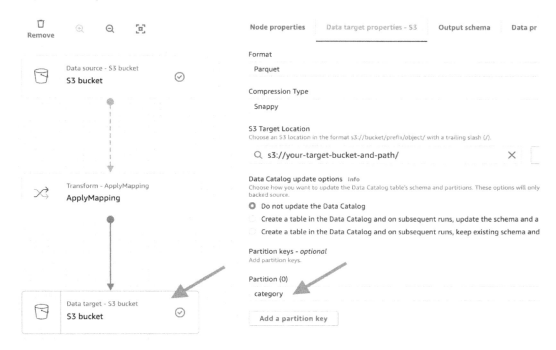

Figure 5.2 – Specifying partition key for Data target node

The following script is the partial code that is generated by Glue Studio based on this diagram. This script writes `snappy` compressed `Parquet` files with hive style partitioning as `category=<partition_value>`.

```
S3bucket_node3 = glueContext.write_dynamic_frame.from_options(
    frame=ApplyMapping_node2,
    connection_type="s3",
    format="glueparquet",
    connection_options={
        "path": "s3://your-target-bucket-and-path/",
        "partitionKeys": ["category"],
    },
```

```
    format_options={"compression": "snappy"},
    transformation_ctx="S3bucket_node3",
)
```

If you use Spark DataFrame for partitioning, you need to directly edit your Glue job script on Glue Studio. The following example shows the part of the job script that writes snappy compressed Parquet files with `category` based partitioning.

```
dataframe.write\
.partitionBy('category')\
.parquet(DST_S3_PATH). # The default compression type is
snappy.
```

The whole script is available at `https://github.com/PacktPublishing/Serverless-ETL-and-Analytics-with-AWS-Glue/blob/main/Chapter05/partitioning_by_dataframe.py`

> **Best practice to select partition keys**
>
> Please note that the number of partitions when you select partition keys for your data. The more number of partitions in a table increases, the higher the overhead of processing the partition metadata. Therefore, you should choose a low-cardinality column as a partition key. Also, note that avoid choosing a partition key that has many skewed values to lower the overhead of filtering values. Usually we use `year`, `month`, `day`, `category`, `region` and so on a partition key.

If you create a table in the Glue Data Catalog based on your data by the Glue Crawler, Athena DDLs and so on., you can define columns as partition keys in your table registered in Glue Data Catalog. Glue Data Catalog that we've seen in *Chapter2, Introduction to Important AWS Glue Features* supports partitioning columns.

The following output of AWS CLI `get-table` command shows a table metadata that is created based on the example dataset. You can see columns and the `category` partition key as follows.

```
$ aws glue get-table --database-name db_name --name product_
sales
{
    "Table": {
        "Name": "product_sales",
        "DatabaseName": "db_name",
        ...
        "StorageDescriptor": {
            "Columns": [
```

```
            {
                "Name": "product_name",
                "Type": "string"
            },
            {
                "Name": "price",
                "Type": "long"
            },
            ...
        },
        "PartitionKeys": [
            {
                "Name": "category",
                "Type": "string"
            }
        ],
...
```

To identify each partition column value for data retrieval by AWS Glue, Amazon Athena, Amazon EMR, and Amazon Redshift Spectrum you need to register the values of the partition key in your Glue Data Catalog table.

Registering partition values in AWS Glue Data Catalog

Primarily, there are four ways to reflect those partition column values in the Glue Data Catalog:

- **Glue DynamicFrame**: Adding partitions by Glue ETL jobs. An example of the script is shown at https://docs.aws.amazon.com/glue/latest/dg/update-from-job.html#update-from-job-partitions.

- **Spark DataFrame**: Running saveAsTable with partitionBy such as the following example:

```
# PySpark example
your_data_frame.write\
        .mode('overwrite')\
        .partitionBy('<partition_column>')\
        .option('path', 's3://your-bucket/path/')\
        .saveAsTable("db.table")
```

The preceding example registers a table that has a partition column such as `<partition_column>` in the Glue Data Catalog. It also writes the data to Amazon S3. The data is written into the s3 path, which is concatenated `s3://your-bucket/path/` with the pair of our specified partition column and its value, such as `s3://your-bucket/path/<partition_column>=<value>/`.

- Running the `ALTER TABLE ADD PARTITION` query by Amazon **Athena** or by Amazon **Redshift Spectrum**.

- Directly calling the **CreatePartition API** (`https://docs.aws.amazon.com/glue/latest/dg/aws-glue-api-catalog-partitions.html#aws-glue-api-catalog-partitions-CreatePartition`), which adds a partition to the Glue Data Catalog by specifying partition column names and the column value. If you add one or more partitions to the Glue Data Catalog, you can use the **BatchCreatePartition API** (`https://docs.aws.amazon.com/glue/latest/dg/aws-glue-api-catalog-partitions.html#aws-glue-api-catalog-partitions-BatchCreatePartition` or `https://docs.aws.amazon.com/glue/latest/webapi/API_CreateTable.html`).

Using the first two ways, you can write the data and add partition values to your Glue Data Catalog simultaneously. The other operations simply help in adding the partition values to the Glue Data Catalog. Therefore, you can operate these two operations after writing your data with the partitioning.

Partition Pruning AWS Glue

If you use the Glue DynamicFrame to read data from partitioned tables in the Glue Data Catalog, you can use data filtering queries that enable your Glue Spark job to avoid processing unnecessary partitions for your analysis. The DynamicFrame supports the following two types of data filtering queries:

- **Predicate pushdown**: This enables your Glue Spark job to filter partitions. This happens on the client (Spark job) side. This works as the following steps:

  I.   The Glue job firstly retrieves all partitions that are registered in the Glue Data Catalog, and it keeps them as a partitions list.

  II. The job filters the partitions in the list based on the specified predicate pushdown query

  III. The job reads the data located in the filtered partitions in *Step 2*.

- **Catalog-side predicate pushdown**: This is also a query to prune partitions as well as the predicate pushdown, however the pruning partitions happens on the sever (the Glue Data Catalog) side. This works as the following steps:

  I.   The Glue job requests the specified partitions registered in a table to the Glue Data Catalog.

  II.   The partitions list as a result of filtering on the server (Glue Data Catalog) side is returned returns the list to the job based on the request in *Step 1*

  III.   The job reads the data located in the specified partitions.

These predicate pushdowns contribute to making data retrieval faster compared to retrieving all data in your storage by the processing job.

You can operate *predicate pushdown* as mentioned by specifying the push_down_predicate option in DynamicFrame. You can also use this with **SparkSQL** by specifying partitions in the WHERE clause. In the following example, the DynamicFrame only reads the data in the partition whose category is grocery by setting category=='grocery' to push_down_predicate option.

```
# PySpark example of a pushdown predicate
glue_context.create_dynamic_frame.from_catalog(
    database="db_name",
    table_name="product_sales",
    push_down_predicate="category==grocery")
```

Also, you can operate *catalog-side predicate pushdown* by specifying catalogPartitionPredicate in a DynamicFrame. Please note that *partition indexes* in AWS Glue, which we'll see next needs to be enabled to use the catalog partition predicate. In the following example , the Glue DynamicFrame reads the data at the partition which is category == book.

```
# PySpark example of a catalog partition predicate
glue_context.create_dynamic_frame.from_catalog(
    database="db_name",
    table_name="product_sales",
    additional_
options={"catalogPartitionPredicate":"category=='grocery'"})
```

As discussed earlier, the catalog-side predicate pushdown partition prunes partitions on the Glue Data Catalog side instead of on processing job side. Catalog-side predicate pushdown can be much faster than using predicate pushdown on the job side if there are a lot of partitions such as over millions of partitions in your Amazon S3 bucket.

Running queries faster with partition indexes

Partition indexes (https://docs.aws.amazon.com/glue/latest/dg/partition-indexes.html) in AWS Glue is one of the functionalities in the Glue Data Catalog. This enables to reduce the query time to filter partitions in the Glue Data Catalog tables. Partition filtering works on the Glue Data Catalog side, instead of returning all partitions to a requester. Once you set the partition indexes to your table that has partitions, a requester (typically, a Glue job) only retrieves necessary partitions that you requested. If the partition index is not enabled, all partitions in the Glue Data Catalog table are returned to a requester and then the requester needs to choose partitions that you want to query. Using partition indexes can increase query performance and save costs such as requests to the Glue Data Catalog table.

Bucketing

Bucketing is a technique that is used to divide data into sub-data and to group rows based on one or more specified columns. Also, this can reduce your processed data by filtering any unnecessary data rows based on the bucketing information if you specify the bucketed columns in your queries. Bucketing can improve your query performance and then accelerate your analytic workloads, too.

You can also specify a bucketed column at table creation. When you set a column as the bucketed column, you should choose with high cardinality and that can be used often for filtering the data. The Glue Data Catalog supports bucketing. If you specify bucketing at table creation, then the bucketing columns are defined in the `StorageDescriptor` part of the Data Catalog. On the other hand, when Spark writes the data with bucketing, Spark adds the Spark format, which describes the bucketing information as parameters of the Data Catalog.

To write your data with bucketing on S3 with Glue ETL Spark jobs, you can mainly use the `bucketBy` method for a Spark DataFrame. MurmurHash (https://en.wikipedia.org/wiki/MurmurHash) is used in Spark and Glue by default. Please note that Glue DynamicFrameWriter doesn't support writing with bucketing in the writing process. For example, you can write the data using bucketing such as the product sales table that based on `data.json` by following examples of using a DataFrame. In this example, you need to pass the bucketed number and one or more columns to the `bucketBy` method:

```
# PySpark example of setting the bucketed number to 10 and
column to 'customer_id'
your_data_frame.write\
.bucketBy(10, 'customer_id')\
.parquet('s3://<your-bucket>/<path>/')
```

There are primarily two ways to reflect the bucket column values in the Glue Data Catalog for Glue ETL jobs to identify the columns as bucketed columns in the data retrieval phase:

- Spark DataFrame: Running `saveAsTable` with `bucketBy` as `dataframe.write. buckety(<number of buckets>,<bucketed columns>.saveAsTable("db. table")`. The Spark public document (`https://spark.apache.org/docs/3.1.1/ sql-data-sources-load-save-functions.html#bucketing-sorting- and-partitioning`) also shows how to register bucketed columns on your table by using `bucketBy` with `saveAsTable`.

- Running `CREATE TABLE` with `CLUSTERED BY (<bucketed columns>) INTO <number of buckets> BUCKETS`.

By using `saveAsTable` in a Spark DataFrame, you can write data with bucketing and add the bucketing information to your Glue Data Catalog simultaneously. The other option requires creating a new table and adding the bucketing information to the Data Catalog at the time of the new table creation.

> **Note**
>
> If you are creating a table using bucketing with Athena DDL, you can see the Athena DDL syntax at `https://docs.aws.amazon.com/athena/latest/ug/create- table.html`. In addition to the DDL, Athena CTAS can also be operated to define and register the bucketing information. An example of the CTAS query, including a definition of bucketing, is shown at `https://docs.aws.amazon.com/athena/latest/ug/ ctas-examples.html#ctas-example-bucketed`.

We've seen how we store data optimally, focusing on topics such as file formats, compression types, file splitability, and partitioning/bucketing. Next, we'll see the second topic, *Managing the number of files and each file size*, which you need to consider for optimizing your analytic queries.

Optimizing the number of files and each file size

The number of files and each file size are also related to the performance of your analytic workloads. In particular, the number of files and file sizes are related to the performance of the data retrieval phase by using an analytic engine in your analytic workloads. To understand the relationship between the number of files and the file size and the performance of the data retrieval process by an analytic engine, we'll look at how the engine generally retrieves data and returns the result as follows.

The basic process of data retrieval and returning a result is firstly getting a list of files, reading each file, processing the contents of the files based on your queries, and then returning the result. In particular, when processing data in Amazon S3, the analytic engine lists objects in your specified S3 bucket, gets objects, reads the contents, then processes and returns the result. When you use an AWS Glue ETL Spark job to process your data in the S3, in the data retrieval process, the Spark driver in the Glue job lists objects in the S3 bucket, then Spark Executors on the Glue job get objects based on the result listed by Spark Driver.

Therefore, the greater the number of files in your storage, the longer listing takes. In addition to this, if your data source is based on a lot of small files, it also takes longer to process data across multiple files because it needs more file I/O compared to the file I/O for a smaller number of files. Therefore, managing the number of files and file sizes is important for your data retrieval process by the analytic engine.

What is compaction?

We store various types of logs such as web access logs, application logs, and IoT device logs in storage such as Amazon S3. These logs are delivered by applications and devices continuously and periodically (in a relatively short period, from seconds to minutes). Furthermore, these logs often consist of a small file in the size of kilobytes or a few megabytes. Therefore, as the logs are delivered into your storage, the number of small files in your storage increases. Usually, this can cause a situation where *there are a lot of small files in your storage*, such as there being 100 million files and each file size being 1 KB.

If you directly run your analytic workloads for data that consists of a lot of small files, it's expected that the query time would increase because listing files in the data retrieval phase by an analytic engine takes a lot of time. Therefore, when running a processing job, you need to transform a lot of small-file data into data with the appropriate number of files, as well as the size of each file. This action to merge small files into larger ones and arrange the data is called **compaction**. Compaction is a necessary process to relax the a lot of *small files problem*, which increases query time and affects your analytic workloads. The following table shows the performance comparison of record count by a Spark DataFrame between non-compacted data and compacted data:

Data	Number of files and	Total data size	dataframe.count() speed (in seconds) (normalized by the speed of compacted data)
Non-compacted data	650,042 files	1.2 GiB	66.4
Compacted data	1 file	1.2 GiB	1.0

Table 5.3 – Comparison of the speed of record count by a Spark DataFrame between non-compacted data and compacted data (this speed is measured by seconds)

As you can see in the preceding table, counting records of compacted data is about 66 times faster than that of non-compacted data. Based on the result, we can see that compaction greatly contributes to increasing query performance if the compacted and non-compacted data have the same size.

In the following sections, we'll see how you can run compaction on your data with AWS Glue. AWS Glue provides flexible solutions to run compaction and basic compaction steps using Spark. In addition to Glue, you can also use the AWS Lake Formation automatic compaction functionality, which automatically runs compaction on your specified data. Additionally, we'll learn about Lake Formation's automatic compaction.

Compaction with AWS Glue ETL Spark jobs

You can process your data, merge the files, and store the data in columnar format using Glue ETL Spark jobs to optimize your analytic workloads. To build an automatic compaction, you essentially need to consider the following two key things in the compaction process:

- How you determine the number of files after the compaction process?
- How you control each file size through the compaction process?

You can control the number of output files in Glue ETL Spark job. Additionally, you can manage each file size by controlling the number of files when Glue job writes the data in your storage such as Amazon S3, and by specifying the file format and compression.

Essentially, Spark determines the number of output files based on the number of Spark partitions, which determines the amount of concurrency of processing data. The number of partitions is determined by input splits, such as data splitted size in EMRFS is defined as `fs.s3.block.size`, HDFS block size, and more. Additionally, the number is determined by the operations on your data in Spark such as `spark.sql.shuffle.partitions/spark.default.parallelism`, which defines the number of partitions after shuffling operations.

In a Glue job, by setting the number of partitions just before writing data with Spark, your Glue job writes the data with the same number of files as the number of partitions specified. You can control the number of partitions using the `repartition(<number>)` or `coalesce(<number>)` methods for a Glue DynamicFrame or Spark DataFrame. Please note that there is currently no option to specify the output file size in Spark when writing data. Therefore, to control the number of files and each file size by Spark, you need to control the number of partitions in your Spark application (Glue job).

The following steps show an example of compaction process by a Glue job:

1. Check the total size of input files and the number of files.
2. If possible, process a small part of the data with Spark and check the compression ratio of the output file size to the input file size (columnar formats such as Parquet and ORC are good as output file formats for analytic workloads).

3. Based on the compression ratio and each file size, compute and set the number of output partitions. It's good to start by setting 64 or 128 MB to efficiently process data with a Glue job.

4. Update the number of partitions by `repartition()` or `coalesce()_` method based on your input file size.

The compaction sample script is provided by AWS in the AWS provided GitHub's repository (`https://github.com/awslabs/aws-glue-blueprint-libs/blob/master/samples/compaction/compaction.py`). The compaction process in this script roughly works as follows.

1. Spark partition number and size is calculated by listing objects in a specified S3 folder in `get_partition_num_and_size` method.

2. If partition size control option (`enable_size_control`) is set to true, based on the calculated partition number and size, optimal file number per partition (`optimal_file_num`) is calculated.

3. The partition number is updated by `coalesce()` method with the calculated optimal file number. Then write the number of files.

Automatic Compaction with AWS Lake Formation acceleration

The Lake Formation acceleration feature automatically runs compaction on your data. This compaction is a background process and doesn't affect your analytic workloads. You don't have to implement a compaction Glue ETL job that reads your data and merges and compresses the data into a new one. To enable this feature, you need to create a table whose table type has **GOVERNED**. You can create a **GOVERNED** status table by checking the **Enable governed data access and management** box from **Create Table** in **Tables** in the Lake Formation console navigation pane, as shown in the following screenshot. After checking it, **Automatic compaction** will automatically be turned on. Once the **GOVERNED** status for a table has been enabled, Lake Formation starts monitoring your data and runs compaction jobs internally without interfering with concurrent queries:

Data management and security

 Governance choice cannot be changed after table creation
Governed tables support ACID (atomic, consistent, isolated, durable) transactions to guarantee data integrity with concurrent changes across multiple tables. Governed tables also come with automatic data layout optimizations by compacting small and delta files in order to maximize query performance.

☑ Enable governed data access and management
Check this option to enable transactional workloads and automatic data compaction.

☐ Automatic compaction
Enable this option to turn on automatic compaction on a governed table.

Figure 5.3 – Enabling governed status for a table

At the time of writing, this compaction feature is supported only for partitioned tables in the Parquet format. Next, we'll look at how to optimize our data layout with Amazon S3 functionalities.

Optimizing your storage with Amazon S3

So far, we've seen how we should store data optimally and how we can manage data to optimize data retrieval and accelerate the analytic workloads. The techniques primarily work on the data itself, such as storing data with columnar formats, data compaction, and more. Not only does it handle data itself optimally, but it's also important to think about optimization on the storage side.

Our data, such as logs of web access, device data, and so on, is continuously reported, and that data size grows over time. As the storage usage increases, the cost increases, too. To reduce the cost of storage usage, usually, we archive data that is not frequently or ever accessed. Generally, we can divide data into the following tiers based on the frequency of access to it:

- **Hot**: This is data that you usually access.

- **Warm**: This is data that you have relatively less access to or require less than hot data.

- **Cold**: This is data that you infrequently access or almost do not require.

Based on the three preceding tiers, usually, we select machines and configure replication policies.

Amazon S3 provides more flexible storage options that you can select. By selecting suitable options for your data and archiving your data effectively, you can reduce not only the storage cost but also the data retrieval time. In this section, we'll look at the S3 storage plans, the data life cycle that S3 also provides, and the way to archive or delete your unnecessary or infrequently accessed data with AWS Glue.

Selecting suitable S3 storage classes for your data

You can see the storage classes that S3 provides and the main usage of each storage class in the table under the *Comparing the Amazon S3 storage classes* section of the AWS documentation (`https://docs.aws.amazon.com/AmazonS3/latest/userguide/storage-class-intro.html`). Based on your data usage and access patterns, you should select a suitable class for your data. If you process the data with AWS Glue, Glue has options to exclude specific class objects and also has methods to change a storage class of objects. We'll see the options and methods in the *Excluding S3 storage classes, archiving, and deleting objects with AWS Glue* section.

Using S3 Lifecycle for managing object lifecycles

S3 Lifecycle runs automatic actions on your objects to manage objects in your storage based on your lifecycle configurations. You can set the lifecycle using the **Management** tab in your bucket view.

Firstly, you need to set the scope of automatic actions, such as **Limit the scope of this rule using one or more filters** (filter-based action) or **Apply to all objects in the bucket** (applying to all objects action), from **Choose a rule scope** in the following screenshot. If you select filter-based actions, you can set the filtering condition, such as **Prefix** or **Object tags**, as follows:

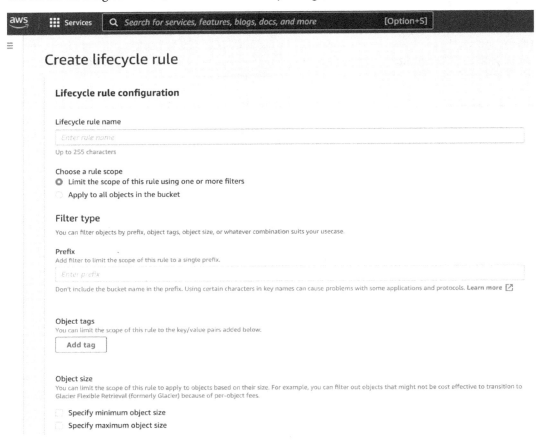

Figure 5.4 – The condition of automatic lifecycle actions

Then, you define the actual lifecycle actions on your objects with which the lifecycle configuration is applied. There are two types of provided actions:

- **Transition actions**: These are defined when objects move to another storage class. You can set the number of days after which to move an object to other storage classes such as *STANDARD-IA* class after an object is put on Amazon S3. If you have old data that you never use or infrequently access, such as data that has passed 30 days since the data creation, you should consider setting this action. By setting this action that moves old objects into archival storage classes, such as *STANDARD-IA*, you can decrease the storage cost of Amazon S3.

- **Expiration actions**: These are defined when objects expire or are deleted. You can set the number of days after which to expire or delete an object after the object is put on Amazon S3. If you have old data that was created some years ago, and you don't need to access the data, you can remove that data by setting this action. By removing unnecessary data, you can decrease not only the storage usage but also the cost of storage usage.

You can choose one or more rules, such as changing a current storage class or removing objects from the list, on the page shown in the following screenshot. The first two actions are *transition actions*, while the others are *expiration actions*:

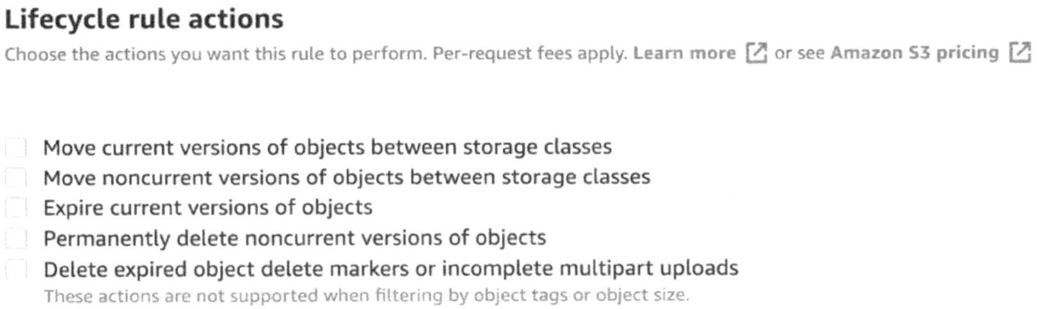

Lifecycle rule actions

Choose the actions you want this rule to perform. Per-request fees apply. **Learn more** [↗] or see **Amazon S3 pricing** [↗]

☐ Move current versions of objects between storage classes
☐ Move noncurrent versions of objects between storage classes
☐ Expire current versions of objects
☐ Permanently delete noncurrent versions of objects
☐ Delete expired object delete markers or incomplete multipart uploads
These actions are not supported when filtering by object tags or object size.

Figure 5.5 – The list of lifecycle actions

Please note that life cycle configurations are applied to not only new objects but all existing objects once you set the configuration. For more details about the S3 Lifecycle, please refer to https://docs.aws.amazon.com/AmazonS3/latest/userguide/object-lifecycle-mgmt.html.

By setting S3 Lifecycle rules, we can manage the data lifecycle. In particular, there are two actions that you can configure for your Amazon S3 bucket. These actions are Transition, which changes the data storage class, and Expiration, which expires or deletes the data. These actions are triggered days after the object's creation was set. Therefore, the S3 Lifecycle automatically archives your data and runs garbage collection without implementing custom code.

Next, we'll look at the functionalities of Glue for skipping data with a specific storage class, transitioning a storage class of your data, and deleting your data using Glue ETL jobs.

Excluding S3 storage classes, archiving, and deleting objects with AWS Glue

AWS Glue provides functionalities that are combined with S3 storage classes, and it can delete unnecessary objects. In particular, we'll see the following functionalities that Glue provides regarding archiving and deleting data:

- **Excluding S3 storage classes**: AWS Glue ETL jobs can process data across multiple storage classes excluding specific storage classes.

- **Transition of a storage class**: Transition a storage class of files in the specified S3 path or that is pointed to by the database and table in the Glue Data Catalog.

- **Purge objects**: Delete files in the specified S3 path or that are pointed to by the database and table in the Glue Data Catalog.

Now, let's take a look at them in detail.

Excluding S3 storage classes with the excludeStorageClasses option

You can filter the S3 storage classes in your AWS Glue ETL jobs to avoid failing to read data in specific classes such as `GLACIER` and `DEEP_ARCHIVE`. In particular, you can filter them by passing the `excludeStorageClasses` option to a DynamicFrame when creating it. For more details, please refer to `https://docs.aws.amazon.com/glue/latest/dg/aws-glue-programming-etl-storage-classes.html#aws-glue-programming-etl-storage-classes-dynamic-frame`.

Transitioning a storage class with the transition_s3_path or transition_table method

You can transition your file storage class to another class. When you want to archive specific partitions after running compaction on the data in the partitions, you can use this method and archive files in the partitions with `partitionPredicate`.

Here's a simple script demonstrating how to run the `transition_table` method to transition objects in the specific partition (`month=5`). After running the script, all objects in the `month=5` partition are transitioned to the Glacier storage class:

```
# PySpark script to transition objects in the month=5 partition
to GLACIER immediately.
glue_context.transition_table(
    database='db_name',
    table='table',
    transition_to='GLACIER'
    options={
```

```
    'retentionPeriod': 0,
    'partitionPredicate': '(month==5)'})
```

You can filter objects by not only partition predicates but also retention periods. For more details about transition operations in Glue, please refer to https://docs.aws.amazon.com/glue/latest/dg/aws-glue-api-crawler-pyspark-extensions-glue-context.html#aws-glue-api-crawler-pyspark-extensions-glue-context-transition_table.

Deleting objects with the purge_s3_path or purge_table method

You can delete your files from Glue ETL jobs with the purge_s3_path or purge_table method. When you want to delete objects in a specific partition after running compaction on the data in the partition, you can use this method and delete the files in the partition with partitionPredicate. Additionally, you can remove partition values from the Glue Data Catalog.

Here's a simple script demonstrating how to run the purge_table method to delete objects from the specific partition (month=5) and also delete the partition value from the Glue Data Catalog. After running the script, all objects in the month=5 partition are deleted and the partition value registered in Data Catalog is also deleted:

```
# PySpark script to delete objects in the month=5 partition
immediately.
glue_context.purge_table(
    database='db_name',
    table_name='purge_table',
    options={
        'partitionPredicate': '(month==5)',
        'retentionPeriod': 0})
```

You can filter objects by not only partition predicates but also retention periods. For more details about purge operations in Glue, please refer to https://docs.aws.amazon.com/glue/latest/dg/aws-glue-api-crawler-pyspark-extensions-glue-context.html#aws-glue-api-crawler-pyspark-extensions-glue-context-purge_table.

Summary

In this chapter, we learned how to design the data layout to accelerate our analytic workloads. In particular, we learned about it by focusing on three parts, including how we store our data optimally, how we manage the number of files and each file size, and how we optimize our storage by working with Amazon S3.

In the first part, we learned techniques to store our data optimally. These techniques include choosing file formats and compression types, understanding file splitability, and partitioning/bucketing. Then, we learned about data compaction to manage the number of files and each file size and to enhance analytic query performance. In the last part, we learned how to optimize our storage with Amazon S3 and Glue DynamicFrames. You can effectively use your storage by archiving, expiring, and deleting your data with Amazon S3 Lifecycle configurations and the Glue DynamicFrame methods.

Managing the data in your data lake with techniques introduced in this chapter will solve a lot of problems such as slow queries, analytic costs, storage costs, and more. In *Chapter 6, Data Management*, we'll see how we can manage data to match various use cases by diving into what kind of analysis we can do and who conducts the analysis by running queries.

Further reading

To learn more about what we've touched on in this chapter, please refer to the following resources:

- Apache Parquet: `https://parquet.apache.org/docs/`
- Apache ORC: `https://orc.apache.org/specification/ORCv0/`
- Apache Avro: `https://avro.apache.org`
- TPC-DS and its specification: `http://www.tpc.org/tpcds/` and `http://tpc.org/tpc_documents_current_versions/pdf/tpc-ds_v3.2.0.pdf`
- *Improve query performance using AWS Glue partition indexes*: `https://aws.amazon.com/blogs/big-data/improve-query-performance-using-aws-glue-partition-indexes/`
    - Video recording on YouTube: `https://youtu.be/jyfJ1X_RaCs`
- *Effective data lakes using AWS Lake Formation, Part 1: Getting started with governed tables*: `https://aws.amazon.com/blogs/big-data/part-1-effective-data-lakes-using-aws-lake-formation-part-1-getting-started-with-governed-tables/`
- Transitioning objects using Amazon S3 Lifecycle: `https://docs.aws.amazon.com/AmazonS3/latest/userguide/lifecycle-transition-general-considerations.html`
- Expiring objects using Amazon S3 Lifecycle: `https://docs.aws.amazon.com/AmazonS3/latest/userguide/lifecycle-expire-general-considerations.html`

6
Data Management

In the previous chapter, you learned how to optimize your data layout to accelerate performance in query engines and manage the data optimally to reduce costs. This is a really important topic, but it is just one aspect of a data lake. As the volume of data increases, a data lake is used by different stakeholders – not only data engineers and software engineers but also data analysts, data scientists, and sales and marketing representatives. Sometimes, the original data is not easy to use for these stakeholders because the raw data may not be structured well. To make business decisions based on data quickly and effectively, it is important to manage, clean up, and enrich the data so that these stakeholders can understand the data correctly, find insights from the data without any confusion, correlate them, and drive their business based on data.

In this chapter, you will learn how to manage, clean up, and enrich the data in typical data requirements, and how to achieve this using AWS Glue. AWS Glue provides various functionalities that allow you to implement ETL logic easily. In addition, Apache Spark has lots of capabilities for different data operations. With AWS Glue, you can take advantage of both, which will help you make your data lake effective in real-world use cases.

In this chapter, we will cover the following topics:

- Normalizing data
- Deduplicating records
- Denormalizing tables
- Securing data content
- Managing data quality

Technical requirements

For this chapter, you will need the following resources:

- An AWS account
- An AWS IAM role
- An Amazon S3 bucket

All the sample code needs to be executed in a Glue runtime (for example, the Glue job system, Glue Interactive Sessions, a Glue Studio notebook, a Glue Docker container, and so on). If you do not have any preferences, we recommend using a Glue Studio notebook so that you can easily start writing code. To use a Glue Studio notebook, follow these steps:

1. Open the AWS Glue console.
2. Click **AWS Glue Studio**.
3. Click **Jobs**.
4. Under **Create job**, click **Jupyter Notebook**, then **Create**.
5. For **Job name**, enter your preferred job name.
6. For **IAM Role**, choose an IAM role where you have enough permission.
7. Click **Start notebook job**.
8. Wait for the notebook to be started.
9. Write the necessary code and run the cells on the notebook.

Let's begin!

Normalizing data

Data normalization is a technique for cleaning data. There are different techniques for normalizing data that make it easy to understand and analyze. This section covers the following techniques and use cases:

- Casting data types and map column names
- Inferring schemas
- Computing schemas on the fly
- Enforcing schemas
- Flattening nested schemas
- Normalizing scale

- Handling missing values and outliers
- Normalizing date and time values
- Handling error records

Let's dive in!

Casting data types and map column names

In the context of data lakes, there can be a lot of different data sources. This may cause inconsistency in data types or column names. For example, when you want to join multiple tables where there is inconsistency, it can cause query errors or invalid calculations. To avoid such issues and make further analytics easier, it is a good approach to cast the data types and apply mapping to the data during the **extract, transform, load** (ETL) phase.

Let's create a simple DataFrame as an example:

```
from pyspark.sql import Row
product = [
    {'product_id': '00001', 'product_name': 'Heater', 'product_
price': '250'},
    {'product_id': '00002', 'product_name': 'Thermostat',
'product_price': '400'}
]

df_products = spark.createDataFrame(Row(**x) for x in product)
df_products.printSchema()
df_products.show()
```

The preceding code returns the following output. You will notice that there are three columns and that all of them are of the `string` type:

```
root
 |-- product_id: string (nullable = true)
 |-- product_name: string (nullable = true)
 |-- product_price: string (nullable = true)


+----------+------------+-------------+
|product_id|product_name|product_price|
+----------+------------+-------------+
|     00001|      Heater|          250|
```

```
|       00002|    Thermostat|             400|
+----------+------------+------------+
```

In natural analysis, you may want to calculate the average price for all products. To support such analysis use cases, the columns, such as `product_price`, should be converted from `string` into `integer`.

Apache Spark supports type casting in Spark DataFrames. You can cast the type as an integer and rename the column's name from `product_price` to `price` by running the following code:

```
from pyspark.sql.functions import col
df_mapped_dataframe = df_products \
    .withColumn("product_price", col("product_price").
cast('integer')) \
    .withColumnRenamed("product_price", "price")


df_mapped_dataframe.printSchema()
df_mapped_dataframe.show()
```

The preceding code returns the following output. You will notice that the column's name has been renamed to `price` and that the data type has been converted from `string` into `integer`, as expected:

```
root
 |-- product_id: string (nullable = true)
 |-- product_name: string (nullable = true)
 |-- price: integer (nullable = true)


+----------+------------+-----+
|product_id|product_name|price|
+----------+------------+-----+
|     00001|      Heater|  250|
|     00002|  Thermostat|  400|
+----------+------------+-----+
```

You can achieve the same thing with SQL syntax as well. The following code registers the `df_products` DataFrame as a Hive table and runs a `SELECT` query against the table:

```
df_products.createOrReplaceTempView("products")
df_mapped_sql = spark.sql("SELECT product_id, product_name,
INT(product_price) as price from products")
```

```
df_mapped_sql.printSchema()
df_mapped_sql.show()
```

The preceding code returns the following output. You will notice that you get the same result that you did with the DataFrame:

```
root
 |-- product_id: string (nullable = true)
 |-- product_name: string (nullable = true)
 |-- price: integer (nullable = true)


+----------+------------+-----+
|product_id|product_name|price|
+----------+------------+-----+
|     00001|      Heater|  250|
|     00002|  Thermostat|  400|
+----------+------------+-----+
```

In the preceding tutorial, you used a Spark DataFrame to cast column types and rename columns.

On the other hand, an AWS Glue DynamicFrame provides the ApplyMapping transform so that you can cast and apply the mapping of column names and data types. The following example shows how to use the ApplyMapping transform:

```
from pyspark.context import SparkContext
from awsglue.context import GlueContext
from awsglue import DynamicFrame

glueContext = GlueContext(SparkContext.getOrCreate())
dyf = DynamicFrame.fromDF(df_products, glueContext, "from_df")

dyf = dyf.apply_mapping(
    [
        ('product_id', 'string', 'product_id', 'string'),
        ('product_name', 'string', 'product_name', 'string'),
        ('product_price', 'string', 'price', 'integer')
    ]
)
df_mapped_dyf = dyf.toDF()
```

```
df_mapped_dyf.printSchema()
df_mapped_dyf.show()
```

The preceding code returns the following output. As you can see, you get the same result that you did with the DataFrame:

```
root
 |-- product_id: string (nullable = true)
 |-- product_name: string (nullable = true)
 |-- price: integer (nullable = true)

+----------+------------+-----+
|product_id|product_name|price|
+----------+------------+-----+
|     00001|      Heater|  250|
|     00002|  Thermostat|  400|
+----------+------------+-----+
```

As you have learned, you can use either a Spark DataFrame or Glue DynamicFrame for data type casting and column mapping.

Inferring schemas

Apache Spark can infer schemas from the content of data. With schema inference, you can create a DataFrame without passing the static schema structure.

When you read a CSV file without schema inference, you can set the `inferSchema` option to `False`. It is disabled by default. You can use the following code to create a DataFrame by reading from one sample CSV file located on Amazon S3:

```
df_infer_schema_false = spark.read.format("csv") \
    .option("header", True) \
    .option("inferSchema", False) \
    .load("s3://covid19-lake/static-datasets/csv/
CountyPopulation/County_Population.csv")
df_infer_schema_false.printSchema()
```

The preceding code returns the following output. You will notice that all of the columns are recognized as being of the string type:

```
root
 |-- Id: string (nullable = true)
 |-- Id2: string (nullable = true)
 |-- County: string (nullable = true)
 |-- State: string (nullable = true)
 |-- Population Estimate 2018: string (nullable = true)
```

When you set the inferSchema option to True, you must run following code:

```
df_infer_schema_true = spark.read.format("csv") \
    .option("header", True) \
    .option("inferSchema", True) \
    .load("s3://covid19-lake/static-datasets/csv/
CountyPopulation/County_Population.csv")
df_infer_schema_true.printSchema()
```

The preceding code returns the following output. You will notice that the Id2 and Population Estimate 2018 columns are registered as the integer type instead of the string type:

```
root
 |-- Id: string (nullable = true)
 |-- Id2: integer (nullable = true)
 |-- County: string (nullable = true)
 |-- State: string (nullable = true)
 |-- Population Estimate 2018: integer (nullable = true)
```

In this section, you learned that the inferSchema option manages the schema inference behavior to read CSV files.

It is a good idea to infer schemas from data when you do not want to define static schemas in advance and you want to define a schema from unpredictable data.

Computing schemas on the fly

A Spark DataFrame is a data representation in Apache Spark. It is powerful and widely used in a huge number of Spark clusters in various kinds of real-world use cases. A DataFrame is conceptually equivalent to a table, and it is optimized for relational database-like table operations such as aggregations and joins.

However, when you use a Spark DataFrame for ETL operations, you may face some typical issues. First, a DataFrame requires a schema to be provided before data is loaded. This can be a problem when you do not know or cannot predict the schema of the data in advance. Second, a DataFrame can have one schema per frame. This can be a problem when the same field in a frame has different types of values in multiple records. Even when you want to determine the type afterward, it is not possible. These issues often occur in messy data.

AWS Glue has a unique data representation called a DynamicFrame, which is similar to a Spark DataFrame. You can use it to convert a Spark DataFrame into a DynamicFrame and vice versa, but there are important differences between the two operations. First, in a Glue DynamicFrame, each record is self-describing. The Glue DynamicFrame computes a schema on-the-fly, so no schema is required initially. Second, a Glue DynamicFrame can have one schema per record, not per frame. The logical record in the DynamicFrame is called a DynamicRecord. When the same field in a DynamicFrame is of a different type in multiple DynamicRecords, the DynamicFrame allows you to determine the preferred types after loading the data.

Before trying DynamicFrame's on-the-fly schema feature, you need to upload a sample file to your S3 bucket and create a table on the Glue Data Catalog. Follow these steps:

1. Create a sample **JSON Lines (JSONL)** file:

    ```
    {"id":"aaa","key":12}
    {"id":"bbb","key":34}
    {"id":"ccc","key":56}
    {"id":"ddd","key":78}
    {"id":"eee","key":"90"}
    ```

2. Upload the sample file to your S3 bucket (replace the path with your S3 path):

    ```
    $ aws s3 cp sample.json s3://path_to_sample_data/
    ```

3. Create a Glue database:

    ```
    $ aws glue create-database --database-input Name=choice
    ```

4. Create a Glue crawler on s3://path_to_sample_data/:

    ```
    $ aws glue create-crawler --name choice --database
    choice --role GlueServiceRole --targets
    '{"S3Targets":[{"Path":"s3:// path_to_sample_data/"}]}'
    ```

5. Run the crawler (replace the IAM role with yours):

```
$ aws glue start-crawler --name choice
```

6. After running the crawler, you will see the `sample` table in the catalog.

Here's the schema of the `sample` table that was returned by the `get-table` AWS CLI command:

```
$ aws glue get-table --database-name choice --name sample
--query Table.StorageDescriptor.Columns --output table
--------------------
|      GetTable      |
+-------+----------+
| Name  |  Type    |
+-------+----------+
|  id   |  string  |
|  key  |  string  |
+-------+----------+
```

Now, we're all set to create a DynamicFrame. You can create a DynamicFrame from the table definition on the Glue Data Catalog by running the following code:

```
from pyspark.context import SparkContext
from awsglue.context import GlueContext

glueContext = GlueContext(SparkContext.getOrCreate())
dyf_sample = glueContext.create_dynamic_frame.from_catalog(
        database = "choice",
        table_name = "sample")
dyf_sample.printSchema()
```

The preceding code returns the following output. You will notice that the `key` column is registered as a `choice` type. This means that `key` could be either of the `int` or `string` type. This happened because the Spark DataFrame schema recognizes `key` as an `int` type, but the Glue Data Catalog recognizes `key` as a `string` type:

```
root
|-- id: string
|-- key: choice
|    |-- int
|    |-- string
```

There are five records in the sample JSONL file. The values in the key field in the first four records are all integers, while at the end of the file, there is one record with a `string` value in that column.

AWS Glue DynamicFrames allow you to determine the schema after loading the data by introducing the concept of a `choice` type. To query the `key` column or to save the frame, you need to resolve the `choice` type first using the `resolveChoice` transform method. For example, you can run the `resolveChoice` transform with the `cast:int` option to convert those `string` values into `int` values:

```
dyf_sample_resolved = dyf_sample.resolveChoice(specs =
[('key','cast:int')])
dyf_sample_resolved.printSchema()
```

The output of `printSchema` is as follows:

```
root
|-- id: string
|-- key: int
```

You will notice that the `key` column is now recognized as `int` instead of `choice` or `string`.

As you have learned, DynamicFrames have unique on-the-fly schema capabilities and the `choice` type allows you to determine the schema after data load. This would be useful for ETL workloads where your data can include different data types.

Enforcing schemas

In an Apache Spark DataFrame, you need to set a static schema per frame. Similarly, in DynamicFrames you can enforce a static schema using the `with_frame_schema` method.

Let's create a new DynamicFrame using the example data located on Amazon S3:

```
from pyspark.context import SparkContext
from awsglue.context import GlueContext

glueContext = GlueContext(SparkContext.getOrCreate())
dyf_without_schema = glueContext.create_dynamic_frame_from_
options(
    connection_type = "s3",
    connection_options = {
        "paths": ["s3://awsglue-datasets/examples/us-
legislators/all/events.json"]
```

```
    },
    format = "json"
)
dyf_without_schema.printSchema()
```

The schema is automatically recognized, as shown in the following code. You will notice that the start_date, end_date, and identifier columns are recognized as strings:

```
Root
|-- classification: string
|-- name: string
|-- end_date: string
|-- identifiers: array
|    |-- element: struct
|    |    |-- scheme: string
|    |    |-- identifier: string
|-- id: string
|-- start_date: string
|-- organization_id: string
```

Now, let's pass a static schema to the with_frame_schema method using the same data. Be careful not to pass the schema after schema computation. Do not execute the printSchema method before the with_frame_schema method since the printSchema method triggers schema computation and with_frame_schema is only available before schema computation:

```
from awsglue.gluetypes import Field, ArrayType, StructType,
StringType, IntegerType


dyf_without_schema_tmp = glueContext.create_dynamic_frame_from_
options(
    connection_type = "s3",
    connection_options = {
        "paths": ["s3://awsglue-datasets/examples/us-
legislators/all/events.json"]
    },
    format = "json"
)


schema = StructType([
```

```
        Field("id", StringType()),
        Field("name", StringType()),
        Field("classification", StringType()),
        Field("identifiers", ArrayType(StructType([
                Field("schema", StringType()),
                Field("identifier", IntegerType())
            ])),
        ),
        Field("start_date", IntegerType()),
        Field("end_date", IntegerType()),
        Field("organization_id", StringType()),
    ])

    dyf_with_schema = dyf_without_schema_tmp.with_frame_
    schema(schema)
    dyf_with_schema.printSchema()
```

The output of printSchema is now as follows:

```
root
|-- id: string
|-- name: long
|-- classification: string
|-- identifiers: array
|    |-- element: struct
|    |    |-- schema: string
|    |    |-- identifier: int
|-- start_date: int
|-- end_date: int
|-- organization_id: string
```

You will notice that the start_date, end_date, and identifier columns are now recognized as integers instead of strings. Schema enforcement for a DynamicFrame is useful when you want to use a DynamicFrame but you do not want to rely on on-the-fly schemas or schema inference.

Flattening nested schemas

When you process unstructured/semi-structured data, you may see a schema that includes a deep nested struct or an array generated from applications. Here's an example of a nested schema:

```
{
    "count": 2,
    "entries": [
        {
            "id": 1,
            "values": {
                "k1": "aaa",
                "k2": "bbb"
            }
        },
        {
            "id": 2,
            "values": {
                "k1": "ccc",
                "k2": "ddd"
            }
        }
    ]
}
```

Typically, for most query engines, a nested schema introduces additional complexity for analytics. Also, for humans, it is not easy to read. To overcome that, you can flatten the schema. AWS Glue's `Relationalize` transform helps you convert a deep nested schema into a flat schema:

```
from pyspark.context import SparkContext
from awsglue.context import GlueContext

glueContext = GlueContext(SparkContext.getOrCreate())

dyf = glueContext.create_dynamic_frame_from_options(
    connection_type = "s3",
    connection_options = {"paths": ["s3://path_to_nested_
json/"]},
    format = "json"
```

```
)
dyf.printSchema()
dyf.toDF().show()
```

The output of printSchema is now as follows:

```
root
|-- count: int
|-- entries: array
|     |-- element: struct
|     |     |-- id: int
|     |     |-- values: struct
|     |     |     |-- k1: string
|     |     |     |-- k2: string
```

The output of show is now as follows:

```
+-----+--------------------+
|count|             entries|
+-----+--------------------+
|    2|[{1, {aaa, bbb}},...|
+-----+--------------------+
```

Then, you can perform a Relationalize transform on this nested schema:

```
from awsglue.transforms import Relationalize
dfc_root_table_name = "root"
dfc = Relationalize.apply(
    frame = dyf,
    staging_path = "s3://your-tmp-s3-path/",
    name = dfc_root_table_name
)
dfc.keys()
```

The output of keys is now as follows:

```
dict_keys(['root', 'root_entries'])
```

The Relationalize transform returns a DynamicFrameCollection object. Now, you have two DynamicFrames inside this collection. Let's extract both:

```
dyf_flattened_root = dfc.select(dfc_root_table_name)
dyf_flattened_root.printSchema()
dyf_flattened_root.toDF().show()
```

The output is now as follows:

```
root
|-- count: int
|-- entries: long


+-----+-------+
|count|entries|
+-----+-------+
|    2|      1|
+-----+-------+
```

Then, extract the second DynamicFrame inside the collection:

```
dyf_flattened_entries = dfc.select('root_entries')
dyf_flattened_entries.printSchema()
dyf_flattened_entries.toDF().show()
```

The output is now as follows:

```
root
|-- id: long
|-- index: int
|-- entries.val.id: int
|-- entries.val.values.k1: string
|-- entries.val.values.k2: string

+---+-----+--------------+---------------------+---------------------+
| id|index|entries.val.id|entries.val.values.k1|entries.val.values.k2|
+---+-----+--------------+---------------------+---------------------+
|  1|    0|             1|                  aaa|                  bbb|
|  1|    1|             2|                  ccc|                  ddd|
+---+-----+--------------+---------------------+---------------------+
```

Figure 6.1 – Relationalized DynamicFrame

If you want to rejoin these two DynamicFrames, run the following code:

```
df_flattened_root = dyf_flattened_root.toDF()
df_flattened_entries = dyf_flattened_entries.toDF()

df_joined = df_flattened_root.join(df_flattened_entries)
df_joined.printSchema()
df_joined.show()
```

The output will be as follows:

```
root
 |-- count: integer (nullable = true)
 |-- entries: long (nullable = true)
 |-- id: long (nullable = true)
 |-- index: integer (nullable = true)
 |-- entries.val.id: integer (nullable = true)
 |-- entries.val.values.k1: string (nullable = true)
 |-- entries.val.values.k2: string (nullable = true)

+-----+-------+---+-----+--------------+---------------------+---------------------+
|count|entries| id|index|entries.val.id|entries.val.values.k1|entries.val.values.k2|
+-----+-------+---+-----+--------------+---------------------+---------------------+
|    2|      1|  1|    0|             1|                  aaa|                  bbb|
|    2|      1|  1|    1|             2|                  ccc|                  ddd|
+-----+-------+---+-----+--------------+---------------------+---------------------+
```

Figure 6.2 – Rejoined DynamicFrame

In this section, you learned that Relationalize returns a collection of DynamicFrames from deep nested data. It is useful for flattening nested data.

Normalizing scale

In the context of mathematics, **machine learning** (**ML**), or statistics, normalization is commonly used to prepare data on the same scale. Imagine that you have an Amazon review dataset and that each review has a star rating for an item. The value of the rating is 1 to 5 in the original data. On the other hand, most ML algorithms expect a value between 0 to 1. If you prefer to rescale data, you can use any typical normalization method, such as min-max normalization, mean normalization, and Z-score normalization.

AWS Glue DataBrew supports mechanisms such as mean normalization and Z-scale normalization. You can easily scale and normalize the values with a GUI.

Handling missing values and outliers

Real-world data typically includes missing values or outliers, and they sometimes cause invalid trends in analysis or unexpected results in ML.

With AWS Glue jobs, you can use the `FillMissingValues` transform to handle missing values in the dataset. The `FillMissingValues` transform has been built on top of an ML algorithm. It detects null values and empty strings as missing values in a specific column and adds a new column with values that are automatically predicted by the ML algorithm, such as linear regression and random forest.

With AWS Glue DataBrew, you can fill missing values with predefined sets such as average, median, custom value, empty string, last valid value, and others. You can also detect outliers and replace them with the rescaled values.

Normalizing date and time values

Real-world data uses different notations of date and time. In the US, It is common to use the `MM/dd/yyyy` format (for example, 12/25/2021), whereas in Europe, it is common to use the `dd/MM/yyyy` format (for example, 25/12/2021). Since they can be confused with each other, it is important to convert international use cases into a unified format.

Unix time (also known as epoch time or POSIX time) is used in various systems. It is the number of seconds that have elapsed since the Unix epoch, excluding leap seconds. The Unix time of 00:00, December 25, 2021, in UTC is `1640390400`. Since it is hard for a human to read, typically, it is converted into a human-readable timestamp format in queries or dashboards.

ISO 8601 is an international standard that covers the worldwide exchange and communication of date- and time-related data. For example, the ISO 8601 format for the date and time of 00:00, December 25, 2021, in UTC is `2021-12-25T00:00:00+00:00`.

In the case of international use cases, it is important to choose a timezone to show the data. Usually, an application needs to adjust the end user's timezone. If you expect all the end users to be in a specific timezone, it may be also okay to store the timestamp within that specific timezone.

With AWS Glue, you can use any of Spark's or Glue DynamicFrame's methods to convert a specific date and time format into a timestamp type. Spark has various date and time functions, including `unix_timestamp`, `date_format`, `to_unix_timestamp`, `from_unixtime`, `to_date`, `to_timestamp`, `from_utc_timestamp`, and `to_utc_timestamp`.

The following is an example DataFrame that includes a timestamp record:

```
df_time_string = spark.sql("SELECT '2021-12-25 00:00:00' as
timestamp_col")
df_time_string.printSchema()
df_time_string.show()
```

The schema and the following data are returned by the preceding code:

```
root
 |-- timestamp_col: string (nullable = false)


+-------------------+
|      timestamp_col|
+-------------------+
|2021-12-25 00:00:00|
+-------------------+
```

Now, let's convert the data type from a string type into a timestamp type using the DataFrame's to_timestamp method:

```
from pyspark.sql.functions import to_timestamp, col

 df_time_timestamp = df_time_string.withColumn(
     "timestamp_col",
     to_timestamp(col("timestamp_col"), 'yyyy-MM-dd HH:mm:ss')
)
df_time_timestamp.printSchema()
df_time_timestamp.show()
```

The printSchema output is shown in the following code block. You will notice that the timestamp_col column is now recognized as timestamp instead of string:

```
root
 |-- timestamp_col: timestamp (nullable = true)


+-------------------+
|      timestamp_col|
+-------------------+
```

```
|2021-12-25 00:00:00|
+-------------------+
```

AWS Glue DynamicFrame also has the `ApplyMapping` transformation for casting values, including timestamps. The following code initiates Glue-related classes and converts the sample DataFrame into a DynamicFrame:

```
from pyspark.context import SparkContext
from awsglue.context import GlueContext
from awsglue import DynamicFrame

glueContext = GlueContext(SparkContext.getOrCreate())
dyf = DynamicFrame.fromDF(df_time_string, glueContext, "from_
df")
```

The following code finds columns whose names contain `timestamp_col` dynamically and converts the `string` value in the column into the `timestamp` type:

```
mapping = []
for field in dyf.schema():
    if field.name == 'timestamp_col':
        mapping.append((
            field.name, field.dataType.typeName(),
            field.name, 'timestamp'
        ))
    else:
        mapping.append((
            field.name, field.dataType.typeName(),
            field.name, field.dataType.typeName()
        ))

dyf = dyf.apply_mapping(mapping)

df_time_timestamp_dyf = dyf.toDF()
df_time_timestamp_dyf.printSchema()
df_time_timestamp_dyf.show()
```

You will see the same result that you saw previously:

```
root
 |-- timestamp_col: timestamp (nullable = true)


+-------------------+
|      timestamp_col|
+-------------------+
|2021-12-25 00:00:00|
+-------------------+
```

Another typical date and time handling operation is to extract some values, such as the year, month, and day from the timestamp column dynamically. This is commonly done when you want to partition data into data lake storage based on the timestamp.

The following code extracts the year, month, and day values from the timestamp_col column:

```
from pyspark.sql.functions import year, month, dayofmonth

df_time_timestamp_ymd = df_time_timestamp \
    .withColumn('year', year("timestamp_col"))\
    .withColumn('month', month("timestamp_col"))\
    .withColumn('day', dayofmonth("timestamp_col"))
df_time_timestamp_ymd.printSchema()
df_time_timestamp_ymd.show()
```

The preceding code returns the following output. You will notice that the DataFrame has three additional columns – year, month, and day – and that those columns contain the values that were extracted from the timestamp_col column:

```
root
 |-- timestamp_col: timestamp (nullable = true)
 |-- year: integer (nullable = true)
 |-- month: integer (nullable = true)
 |-- day: integer (nullable = true)


+-------------------+----+-----+---+
|      timestamp_col|year|month|day|
+-------------------+----+-----+---+
```

```
|2021-12-25 00:00:00|2021|   12| 25|
+-------------------+----+-----+---+
```

With a Glue DataFrame, you can achieve the same by running the following code using the map function:

```
def add_timestamp_column(record):
    dt = record["timestamp_col"]
    record["year"] = dt.year
    record["month"] = dt.month
    record["day"] = dt.day
    return record


dyf = dyf.map(add_timestamp_column)
df_time_timestamp_dyf_ymd = dyf.toDF()
df_time_timestamp_dyf_ymd.printSchema()
df_time_timestamp_dyf_ymd.show()
```

The preceding code returns the following output:

```
root
 |-- timestamp_col: timestamp (nullable = true)
 |-- year: integer (nullable = true)
 |-- month: integer (nullable = true)
 |-- day: integer (nullable = true)


+-------------------+----+-----+---+
|      timestamp_col|year|month|day|
+-------------------+----+-----+---+
|2021-12-25 00:00:00|2021|   12| 25|
+-------------------+----+-----+---+
```

In this section, you learned that you can easily normalize the date/time format in both a Spark DataFrame and a Glue DynamicFrame. You can also extract year/month/value values from a timestamp. This is useful for time series data, as well as data layouts that use time-based partitioning.

Handling error records

If the data is corrupted, Apache Spark or AWS Glue may not be able to read the records successfully. This can cause missing values and invalid results.

If you want to manage such situations, the Glue DynamicFrame class can detect error records. The following code detects the error records and aborts the job when the error rate exceeds the threshold:

```
import sys
from pyspark.context import SparkContext
from awsglue.context import GlueContext
from awsglue.job import Job
from awsglue.utils import getResolvedOptions


ERROR_RATE_THRESHOLD = 0.2
glue_context = GlueContext(SparkContext.getOrCreate())


dyf = glue_context.create_dynamic_frame.from_options(
    connection_type = "s3",
    connection_options = {'paths': ['s3://your_input_data_
path/']},
    format = "csv",
    format_options={'withHeader': False}
)
dataCount = dyf.count()
errorCount = dyf.errorsCount()
errorRate = errorCount/(dataCount+errorCount)
print(f"error rate: {errorRate}")
if errorRate > ERROR_RATE_THRESHOLD:
    raise Exception(f"error rate {errorRate} exceeded
threshold: {ERROR_RATE_THRESHOLD}")


errorDyf = dyf.errorsAsDynamicFrame()
glue_context.write_dynamic_frame_from_options(
    frame=errorDyf,
    connection_type='s3',
    connection_options={'path': 's3://your_error_frame_path/'},
```

```
    format='json'
)
```

In this section, you learned that Glue provides a set of capabilities that can help you handle typical error records. Based on your requirements, you can trigger an exception when the error rate exceeds the predefined threshold.

Deduplicating records

When you start analyzing the business data, you may find that it's incorrect and that there are multiple different notations of the same record.

The following example table contains duplicates:

customer_name	email	phone
Barbara Gordon	gordon@example.com	117.835.2584
Gordon, Barbara	gordon@example.com	117-835-2584
Rebecca Thompson	thompson@example.net	001-469-964-3897x9041
Rachel Gilbert	gilbert@example.com	001-510-198-4613x23986
Gilbert, R.	gilbert@example.com	
Tanya Fowler	tanya@example.net	(067)150-0263

Figure 6.3 – Customer table with duplicates

As you may have noticed, there are only four unique records in the preceding table. Two records have two different notations, which causes duplication. If you analyze the data with these kinds of duplicated records, the result may include unexpected bias, so you will get an incorrect result.

With AWS Glue, you can use the FindMatches transform to find duplicated records. FindMatches is one of the ETL transforms provided in the Glue ETL library. With the FindMatches transform, you can match records and identify and remove duplicate records based on the ML model.

Let's look at the end-to-end matching process:

1. Register a table definition for your data in AWS Glue Data Catalog. You can use a Glue crawler, DDL, or the Glue catalog API to catalog your data.

2. Create new Glue ML transforms using FindMatches. You need to choose the table created in *step 1*, give primary keys, and tune the balance between **Recall** and **Precision** and **Lower cost** and **Accuracy**.

3. Train the `FindMatches` model by providing a labeling file that represents a perfect mapping of the records. You can estimate the quality of the model by reviewing the match quality metrics and uploading better labeling files if you want to improve the quality.

4. Create and run an AWS Glue ETL job that uses your `FindMatches` transform.

You can find detailed steps in *Integrate and deduplicate datasets using AWS Lake Formation* (`https://aws.amazon.com/blogs/big-data/integrate-and-deduplicate-datasets-using-aws-lake-formation-findmatches/`).

Once you have completed the preceding steps, you will see the results shown in the following table:

customer_name	email	phone	match_id
Barbara Gordon	gordon@example.com	117.835.2584	1
Gordon, Barbara	gordon@example.com	117-835-2584	1
Rebecca Thompson	thompson@example.net	001-469-964-3897x9041	2
Rachel Gilbert	gilbert@example.com	001-510-198-4613x23986	3
Gilbert, R.	gilbert@example.com		3
Tanya Fowler	tanya@example.net	(067)150-0263	4

Figure 6.4 – Deduplicated customer table

After matching the datasets, you will see that the result table represents the source table's structure and data, as well as one more column: `match_id`. Each of the matched records displays the same `match_id` value. By utilizing these `match_id` values, you can filter with only distinct values to get unique records.

In this section, you learned that you can easily take advantage of the ML model in the `FindMatches` transform and deduplicate the same records efficiently.

Denormalizing tables

In this section, we will look at an example use case. There is a fictional e-commerce company that sells products and has a website that allows people to buy these products. There are three tables stored in the web system – two dimension tables, `product` and `customer`, and one fact table, `sales`. The `product` table stores the product's name, category, and price. The `customer` table stores individual customer names, email addresses, and phone numbers. These email addresses and phone numbers are sensitive pieces of information that need to be handled carefully. When a customer buys a product, that activity is recorded in the `sales` table. One new record is inserted into the `sales` table every time a customer buys a product.

The following is the `product` dimension table:

product_id	product_name	category	price
11	Introduction to Cloud	Ebooks	15
12	Best practices on data lakes	Ebooks	25
21	Data Quest	Video games	30
22	Final Shooting	Video games	20

Figure 6.5 – Product table

The following code can be used to populate the preceding sample data in a Spark DataFrame:

```
df_product = spark.createDataFrame([
    (11, "Introduction to Cloud", "Ebooks", 15),
    (12, "Best practices on data lakes", "Ebooks", 25),
    (21, "Data Quest", "Video games", 30),
    (22, "Final Shooting", "Video games", 20)
], ['product_id', 'product_name', 'category', 'price'])
df_product.show()
df_product.createOrReplaceTempView("product")
```

The preceding code returns the following output:

```
+----------+--------------------+-----------+-----+
|product_id|        product_name|   category|price|
+----------+--------------------+-----------+-----+
|        11|Introduction to C...|     Ebooks|   15|
|        12|Best practices on...|     Ebooks|   25|
|        21|          Data Quest|Video games|   30|
|        22|      Final Shooting|Video games|   20|
+----------+--------------------+-----------+-----+
```

Figure 6.6 – DataFrame for the Product table

The following is the `customer` dimension table:

uid	customer_name	email	phone
A103	Barbara Gordon	gordon@example.com	117.835.2584
A042	Rebecca Thompson	thompson@example.net	001-469-964-3897x9041
A805	Rachel Gilbert	gilbert@example.com	001-510-198-4613x23986
A404	Tanya Fowler	tanya@example.net	(067)150-0263

Figure 6.7 – Customer table

The following code can be used to populate the preceding sample data in a Spark DataFrame:

```
df_customer = spark.createDataFrame([
    ("A103", "Barbara Gordon", "gordon@example.com",
"117.835.2584"),
    ("A042", "Rebecca Thompson", "thompson@example.net", "001-
469-964-3897x9041"),
    ("A805", "Rachel Gilbert", "gilbert@example.com", "001-510-
198-4613x23986"),
    ("A404", "Tanya Fowler", "tanya@example.net", "(067)150-
0263")
], ['uid', 'customer_name', 'email', 'phone'])
df_customer.show(truncate=False)
df_customer.createOrReplaceTempView("customer")
```

The preceding code returns the following output:

```
+----+----------------+--------------------+----------------------+
|uid |customer_name   |email               |phone                 |
+----+----------------+--------------------+----------------------+
|A103|Barbara Gordon  |gordon@example.com  |117.835.2584          |
|A042|Rebecca Thompson|thompson@example.net|001-469-964-3897x9041 |
|A805|Rachel Gilbert  |gilbert@example.com |001-510-198-4613x23986|
|A404|Tanya Fowler    |tanya@example.net   |(067)150-0263         |
+----+----------------+--------------------+----------------------+
```

Figure 6.8 – DataFrame for the Customer table

The following is the `sales` fact table (the `purchased_by` field is the foreign key for the `uid` field in the `customer` table):

product_id	product_by	purchased_at
21	A042	2022-03-30T01:30:00Z
22	A805	2022-04-01T02:00:00Z
11	A103	2022-04-21T11:40:00Z
12	A404	2022-04-28T08:20:00Z

Figure 6.9 – Sales table

The following code can be used to populate the preceding sample data in a Spark DataFrame:

```
df_sales = spark.createDataFrame([
    (21, "A042", "2022-03-30T01:30:00Z"),
```

```
    (22, "A805", "2022-04-01T02:00:00Z"),
    (11, "A103", "2022-04-21T11:40:00Z"),
    (12, "A404", "2022-04-28T08:20:00Z")
], ['product_id', 'purchased_by', 'purchased_at'])
df_sales.show(truncate=False)
df_sales.createOrReplaceTempView("sales")
```

The preceding code returns the following output:

```
+---------+------------+--------------------+
|ticket_id|purchased_by|purchased_at        |
+---------+------------+--------------------+
|216      |A042        |2022-03-30T01:30:00Z |
|217      |A805        |2022-04-01T02:00:00Z |
|218      |A103        |2022-04-21T11:40:00Z |
+---------+------------+--------------------+
```

Figure 6.10 – DataFrame for the Sales table

These tables are well-designed and normalized in the context of relational databases. However, this means that the data analyst always needs to join the tables for analysis.

For example, if you want to find the names of customers who bought products in April, you need to join the customer and sales tables, then filter with the purchased_at column:

```
Spark.sql("SELECT customer_name, purchased_at FROM sales JOIN
customer ON sales.purchased_by=customer.uid WHERE purchased_at
LIKE '2022-04%'").show()
```

The preceding code returns the following output:

```
+---------------+--------------------+
| customer_name|        purchased_at|
+---------------+--------------------+
|Rachel Gilbert|2022-04-01T02:00:00Z|
|Barbara Gordon|2022-04-21T11:40:00Z|
|  Tanya Fowler|2022-04-28T08:20:00Z|
+---------------+--------------------+
```

Figure 6.11 – Customers who bought products in April

It is not critical when the tables are small, but if the tables are large, you will spend an unnecessarily long time joining tables. In addition, joins can cause huge memory consumption in well-known analytic engines, including Apache Spark, and it sometimes causes **out-of-memory** (**OOM**) errors.

Denormalization is one of the typical optimization techniques that has fewer joins and simpler queries. Once you denormalize the table by joining the source tables in advance, you won't need to join the tables in the analysis phase, and your query syntax will be simpler. The disadvantage of denormalization is that you will need to have some redundancy in the data, and you will have to think about how to keep the denormalized table up to date.

Let's denormalize the four preceding tables into a destination table:

```
df_product_sales = df_product.join(
    df_sales,
    df_product.product_id == df_sales.product_id
)
df_destination = df_product_sales.join(
    df_customer,
    df_product_sales.purchased_by == df_customer.uid
)
df_destination.createOrReplaceTempView("destination")
df_destination.select('product_
name','category','price','customer_
name','email','phone','purchased_at').show()
```

The following table shows some of the columns in the destination table:

```
+-------------------+------------+-----+----------------+-------------------+------------------+--------------------+
|       product_name|    category|price|   customer_name|              email|             phone|         purchased_at|
+-------------------+------------+-----+----------------+-------------------+------------------+--------------------+
|         Data Quest|Video games |   30|Rebecca Thompson|thompson@example.net|001-469-964-3897x...|2022-03-30T01:30:00Z|
|     Final Shooting|Video games |   20|   Rachel Gilbert|gilbert@example.com|001-510-198-4613x...|2022-04-01T02:00:00Z|
|Introduction to C...|      Ebooks|   15|  Barbara Gordon|  gordon@example.com|       117.835.2584|2022-04-21T11:40:00Z|
|Best practices on...|      Ebooks|   25|    Tanya Fowler|   tanya@example.net|    (067)150-0263|2022-04-28T08:20:00Z|
+-------------------+------------+-----+----------------+-------------------+------------------+--------------------+
```

Figure 6.12 – DataFrame for the Destination table

Once you have created the destination table, you can easily find the name of people who purchased products in April without joining all the relevant tables every time:

```
spark.sql("SELECT product_name,category,price,customer_
name,email,phone,purchased_at FROM destination WHERE purchased_
at LIKE '2022-04%'").show()
```

Here is the output of the preceding code:

```
+-------------------+------------+-----+----------------+-------------------+------------------+--------------------+
|       product_name|    category|price|   customer_name|              email|             phone|         purchased_at|
+-------------------+------------+-----+----------------+-------------------+------------------+--------------------+
|     Final Shooting|Video games |   20|Rachel Gilbert|gilbert@example.com|001-510-198-4613x...|2022-04-01T02:00:00Z|
|Introduction to C...|      Ebooks|   15|Barbara Gordon|  gordon@example.com|       117.835.2584|2022-04-21T11:40:00Z|
|Best practices on...|      Ebooks|   25|    Tanya Fowler|   tanya@example.net|    (067)150-0263|2022-04-28T08:20:00Z|
+-------------------+------------+-----+----------------+-------------------+------------------+--------------------+
```

Figure 6.13 – The Destination table with customers who bought products in April 2022

In this section, you learned that tables can be denormalized by joining them. This is useful for optimizing performance in analytics workloads as it avoids having multiple joins in queries. However, if you denormalize the tables and store them on data lakes, they will be a little bit harder to maintain because you need to repeat the denormalization process whenever the source tables are changed. You will need to decide on a direction based on your workload.

Securing data content

In the context of a data lake, security is a "job zero" priority. In *Chapter 8*, *Data Security*, we will dive deep into security. In this section, we cover basic ETL operations that secure data. The following common techniques can be used to hide confidential values from data:

- Masking values
- Hashing values

In this section, you will learn how to mask/hash values that are included in your data.

Masking values

In business data lakes, the data can contain sensitive data, such as people's names, phone numbers, credit card numbers, and so on. Data security is an important aspect of data lakes. There are different approaches to handling such data securely. It is a good idea to just drop the sensitive data when you collect the data from data sources when you won't use the sensitive data in analytics. It is also common to manage access permissions on certain columns or records of the data. Another approach is to mask the data entirely or partially when you want to keep it confidential but also keep the same format – for example, the number of digits or characters.

With AWS Glue, you can mask a specific column using Spark DataFrame's `withColumn` method by replacing the text based on a regular expression:

```
from pyspark.sql.functions import regexp_replace
df_masked = df_destination.withColumn("phone", regexp_
replace("phone", r'(\d)', '*'))
df_masked.select('product_name','category','price','customer_
name','email','phone','purchased_at').show()
```

Once you have masked the data, you will see the following output. You will notice that only the numbers have been replaced in the `phone` column:

```
+-------------------+-------------+-----+-----------------+------------------------+--------------------+-------------------+
|       product_name|     category|price|    customer_name|                   email|               phone|       purchased_at|
+-------------------+-------------+-----+-----------------+------------------------+--------------------+-------------------+
|         Data Quest| Video games |   30|Rebecca Thompson |thompson@example.net|***-***-***-****x...|2022-03-30T01:30:00Z|
|     Final Shooting| Video games |   20|  Rachel Gilbert | gilbert@example.com|***-***-***-****x...|2022-04-01T02:00:00Z|
|Introduction to C...|      Ebooks|   15|  Barbara Gordon |   gordon@example.com|     ***.***.****|2022-04-21T11:40:00Z|
|Best practices on...|      Ebooks|   25|    Tanya Fowler |    tanya@example.net|   (***)***-****|2022-04-28T08:20:00Z|
+-------------------+-------------+-----+-----------------+------------------------+--------------------+-------------------+
```

Figure 6.14 – The Destination table contains masked phone numbers

In terms of **personally identifiable information (PII)** data, AWS Glue has a native capability for detecting the PII data dynamically based on the data. At the time of writing, it can detect the following 16 entities:

- ITIN (US)
- Email
- Passport Number (US)
- US Phone
- Credit Card
- Bank Account (US, Canada)
- US Driving License
- IP Address
- MAC Address
- DEA Number (US)
- HCPCS Code (US)
- National Provider Identifier (US)
- National Drug Code (US)
- Health Insurance Claim Number (US)
- Medicare Beneficiary Identifier (US)
- CPT Code (US)

If you want to detect the PII data and mask it based on the detected result, you can use the following code:

```
entities_filter = [] # Empty list means we detect all entities.
sample_fraction = 1.0 # 100%
```

```
threshold_fraction = 0.8 # At least 80% of rows for a given
column should contain the same entity in order for the column
to be classified as that entity.
transformation_ctx = ""
stage_threshold = 0
total_threshold = 0


recognizer = EntityRecognizer()
results = recognizer.classify_columns(frame=dyf, entities_
filter=entities_filter, sample_fraction=sample_fraction,
threshold_fraction=threshold_fraction, stageThreshold=stage_
threshold, totalThreshold=total_threshold)


for key in results:
    for recognized_value in results[key]:
        # Mask CREDIT_CARD, PHONE_NUMBER and IP_ADDRESS columns
        if recognized_value in ["CREDIT_CARD", "PHONE_NUMBER",
"IP_ADDRESS"]:
            df = df.withColumn(key, regexp_replace(key,
r'(\d)', '*'))
```

In this section, you learned that, with AWS Glue, you can easily mask your data. Glue's PII detection helps you dynamically choose the confidential columns and mask them.

Hashing values

Another way to keep data secure but still make some analytic queries available is hashing. **Hashing** is the process of passing data to a hash function and converting it into the result. Hashed data is always the same length, regardless of the amount of original data. MD5 is one of the common hash mechanisms for returning a 128-bit checksum as a hex string of the value. SHA2 returns a checksum from the SHA-2 family (for example, SHA-224, SHA-256, SHA-384, or SHA-512) as a hex string of the value.

These hashing algorithms are one-way, which means they can't be reversed. One possible way to retrieve the original value from a hashed result is to brute-force it. A brute-force attack is commonly performed by generating all the possible values, making a hash of them, and then comparing the generated hashes with the original hash result.

Let's compute a hash for one column in the table. Apache Spark supports hashing algorithms such as MD5, SHA, SHA1, SHA2, CRC32, and xxHash. Here, we will use SHA2 to hash the email column:

```
from pyspark.sql.functions import sha2
df_hashed = df_masked.withColumn("email", sha2("email", 256))
```

```
df_hashed.select('product_name','category','price','customer_
name','email','phone','purchased_at').show()
```

Once you have hashed the data, you will see the following output. You will notice that only the email addresses have been hashed in the `email` column:

```
+------------------+-----------+-----+----------------+--------------------+--------------------+--------------------+
|      product_name|   category|price|   customer_name|               email|               phone|        purchased_at|
+------------------+-----------+-----+----------------+--------------------+--------------------+--------------------+
|        Data Quest|Video games|   30|Rebecca Thompson|97cf4c3dfff3a1245...|***-***-***-****x...|2022-03-30T01:30:00Z|
|    Final Shooting|Video games|   20|  Rachel Gilbert|2857f8c8a7b8c1b7f...|***-***-***-****x...|2022-04-01T02:00:00Z|
|Introduction to C...|     Ebooks|   15|  Barbara Gordon|b8ba2a41ce2d45a99...|        ***.***.****|2022-04-21T11:40:00Z|
|Best practices on...|     Ebooks|   25|    Tanya Fowler|b822364443d400f56...|       (***)***-****|2022-04-28T08:20:00Z|
+------------------+-----------+-----+----------------+--------------------+--------------------+--------------------+
```

Figure 6.15 – The Destination table with hashed email addresses

If you want to integrate PII detection with hashing, you can use the following code:

```
entities_filter = [] # Empty list means we detect all entities.
sample_fraction = 1.0 # 100%
threshold_fraction = 0.8 # At least 80% of rows for a given
column should contain the same entity in order for the column
to be classified as that entity.
transformation_ctx = ""
stage_threshold = 0
total_threshold = 0

recognizer = EntityRecognizer()
results = recognizer.classify_columns(frame=dyf, entities_
filter=entities_filter, sample_fraction=sample_fraction,
threshold_fraction=threshold_fraction, stageThreshold=stage_
threshold, totalThreshold=total_threshold)

for key in results:
    for recognized_value in results[key]:
        # Hash DRIVING_LICENSE, PASSPORT_NUMBER, and USA_ITIN
columns using SHA-2
        if recognized_value in ["DRIVING_LICENSE", "PASSPORT_
NUMBER", "USA_ITIN"]:
            df = df.withColumn(key, sha2(key, 256))
```

In this section, you learned that, similar to masking, you can easily hash your data with AWS Glue.

Managing data quality

When you build a modern data architecture from different data sources, the incoming data may contain incorrect, missing, or malformed data. This can make data applications fail. It can also result in incorrect business decisions due to incorrect data aggregations. However, it can be hard for you to evaluate the quality of the data if there is no automated mechanism. Today, it is important to manage data quality by applying predefined rules and verifying if the data meets those criteria or not.

Different frameworks can be used to monitor data quality. In this section, we will introduce two mechanisms: AWS Glue DataBrew data quality rules and DeeQu.

AWS Glue DataBrew data quality rules

Glue DataBrew data quality rules allow you to manage data quality to detect typical data issues easily. In this section, we will use a human resources dataset (`https://eforexcel.com/wp/downloads-16-sample-csv-files-data-sets-for-testing/`).

Follow these steps to manage data quality with Glue DataBrew:

1. Create a data quality ruleset against your dataset.
2. Create data quality rules. You can define multiple data quality rules here – for example, a rule to make sure that row count is correct and expected, there are no duplicate records, and so on.
3. Create and run a profile job with the ruleset.
4. Inspect the data quality rule's validation results.

If data quality issues are detected by the rules, you can run DataBrew jobs to clean up the data and rerun the data quality checks.

You can find detailed steps in *Enforce customized data quality rules in AWS Glue DataBrew* (`https://aws.amazon.com/blogs/big-data/enforce-customized-data-quality-rules-in-aws-glue-databrew/`).

DeeQu

DeeQu, an open source data quality library, addresses data quality monitoring requirements and can scale to large datasets. DeeQu is built on top of Apache Spark to define "unit test for data." With DeeQu, you can populate data quality metrics and define data quality rules easily.

DeeQu version 2.x runs with Spark 3.1, as well as with AWS Glue 3.0 jobs. Follow these steps before running any DeeQu code:

1. Download the DeeQu 2.x JAR file from the Maven repository (`https://mvnrepository.com/artifact/com.amazon.deequ/deequ/2.0.0-spark-3.1`).

2. Download the PyDeeQu 1.0.1 Wheel file from pypi.org (`https://pypi.org/project/pydeequ/`).

3. Upload the JAR file and the Wheel file to your S3 bucket.

4. Configure library dependencies. When you use the Glue job system, configure the `--extra_jars` and `--extra_py_files` parameters with the S3 paths of the JAR/Wheel files. When you use Glue Studio Notebook or Glue Interactive Sessions, configure `%extra_jars` and `%extra_py_files`, like so:

   ```
   %extra_jars s3://path_to_your_lib/deequ-2.0.0-spark-
   3.1.jar
   %extra_py_files s3://path_to_your_lib/pydeequ-1.0.1-py3-
   none-any.whl
   ```

5. First, let's initialize `SparkSession` and generate some sample data:

   ```
   import pydeequ
   from pyspark.sql import SparkSession
   spark = (SparkSession
       .builder
       .config("spark.jars.packages", pydeequ.deequ_maven_
   coord)
       .config("spark.jars.excludes", pydeequ.f2j_maven_
   coord)
       .getOrCreate())
   df = spark.createDataFrame([
       (1, "Product A", "awesome thing.", "high", 2),
       (2, "Product B", "available at http://producta.
   example.com", None, 0),
       (3, None, None, "medium", 6),
       (4, "Product D", "checkout https://productd.example.
   org", "low", 10),
       (5, "Product E", None, "high", 18)
   ], ['id', 'productName', 'description', 'priority',
   'numViews'])
   ```

6. Now, let's run the analyzer to measure the metrics in the sample data:

```
from pydeequ.analyzers import *
analysisResult = AnalysisRunner(spark) \
    .onData(df) \
    .addAnalyzer(Size()) \
    .addAnalyzer(Completeness("id")) \
    .addAnalyzer(Completeness("productName")) \
    .addAnalyzer(Maximum("numViews")) \
    .addAnalyzer(Mean("numViews")) \
    .addAnalyzer(Minimum("numViews")) \
    .run()
analysisResult_df = AnalyzerContext.
successMetricsAsDataFrame(spark, analysisResult)
analysisResult_df.show()
```

The preceding code returns the following output:

```
+-------+-----------+------------+-----+
| entity|   instance|        name|value|
+-------+-----------+------------+-----+
|Dataset|          *|        Size|  5.0|
| Column|         id|Completeness|  1.0|
| Column|productName|Completeness|  0.8|
| Column|   numViews|     Maximum| 18.0|
| Column|   numViews|        Mean|  7.2|
| Column|   numViews|     Minimum|  0.0|
+-------+-----------+------------+-----+
```

7. Now, let's apply a verification check to understand if the data meets the predefined quality rules:

```
from pydeequ.checks import *
from pydeequ.verification import *

check = Check(spark, CheckLevel.Warning, "Review Check")

checkResult = VerificationSuite(spark) \
    .onData(df) \
    .addCheck(
```

```
            # we expect 5 row
            check.hasSize(lambda x: x == 5) \
            # should never be NULL
            .isComplete("id") \
            # should not contain duplicates
            .isUnique("id") \
            # should never be NULL
            .isComplete("productName") \
            # should only contain the values "high",
"medium", and "low"
            .isContainedIn("priority", ["high", "medium",
"low"]) \
            # should not contain negative values
            .isNonNegative("numViews") \
            # at least half of the descriptions should
contain a url
            .containsURL("description", lambda x: x >= 0.5) \
            # half of the items should have less than 10
views
            .hasApproxQuantile("numViews", ".5", lambda x: x
<= 10)) \
        .run()

checkResult_df = VerificationResult.
checkResultsAsDataFrame(spark, checkResult)
checkResult_df.show()
```

The preceding code returns the following output:

```
+-----------+-----------+-----------+------------------
--+----------------+--------------------+
|      check|check_level|check_status|
constraint|constraint_status|  constraint_message|
+-----------+-----------+-----------+------------------
--+----------------+--------------------+
|Review Check|    Warning|
Warning|SizeConstraint(Si...|          Success|
|
|Review Check|    Warning|
Warning|CompletenessConst...|          Success|
```

```
|
|Review Check|       Warning|
Warning|UniquenessConstra...|                    Success|
|
|Review Check|       Warning|
Warning|CompletenessConst...|                    Failure|Value: 0.8
does n...|
|Review Check|       Warning|
Warning|ComplianceConstra...|                    Success|
|
|Review Check|       Warning|
Warning|ComplianceConstra...|                    Success|
|
|Review Check|       Warning|
Warning|containsURL(descr...|                    Failure|Value: 0.4
does n...|
|Review Check|       Warning|
Warning|ApproxQuantileCon...|                    Success|
|
+------------+-----------+------------+-------------------
--+------------------+--------------------+
```

When you want to see all the messages provided by the verification, you can run the following code:

```
checkResult_df.show(truncate=False)
```

The preceding code returns the following output:

Figure 6.16 – DeeQU data quality check result

In this section, you learned that Glue DataBrew and DeeQu help you analyze and validate data quality in your dataset.

Summary

In this chapter, you learned how to manage, clean up, and enrich your data using various functionalities available on AWS Glue and Apache Spark. In terms of normalizing data, you looked at several techniques, including schema enforcement, timestamp handling, and others. To deduplicate records, you experimented with using ML transforms with a sample dataset, while to denormalize tables, you joined multiple tables and enriched the data to optimize the analytic workload. When learning about masking and hashing values, you performed basic ETL to improve security. Moreover, you learned that Glue PII Detection helps you choose confidential columns dynamically. Finally, you learned how to manage data quality with Glue DataBrew data quality rules and DeeQu.

In the next chapter, you will learn about the best practices for managing metadata on data lakes.

7

Metadata Management

Just as with relational databases, AWS Glue relies on the concepts of databases and tables to organize and manage datasets. That said, these concepts are quite different in their execution. In a relational database, the data to be stored and its descriptors (such as the schema and comments, also known as metadata) are stored and managed together: there is no way to store data without describing it first, and there is no way to add metadata to already written data.

In big data environments, the storage and metadata layers are decoupled. There is no centralized storage system because of dataset size limitations, and data is typically dumped without a format onto distributed, large-scale systems such as Apache Hadoop or **Amazon S3**. This means we as users have to bring the metadata to the data wherever it is stored, cataloging the data and specifying its location, how to read it, and how to understand it (its schema).

In the case of Glue, this centralized cataloging entity is known as the Glue Data Catalog, a serverless metadata repository for all datasets in your data lake. In this chapter, we will cover all aspects of the Data Catalog, including the following topics:

- Populating metadata – creating, updating, and deleting entries within the Data Catalog
- Maintaining metadata – automation and management features for the Data Catalog
- Partition management – avoiding low query execution times by managing your partitions
- Versioning and rollback – dealing with version management and changes within the Data Catalog
- Lineage –understanding the flow of data within your data lake

Upon completion of this chapter, you will know how to operate, manage, and maintain a successful catalog, enabling you to process data stored as described in previous chapters.

Technical requirements

This chapter requires the following:

- The AWS **command-line interface (CLI)** (https://aws.amazon.com/cli/) installed in your environment

- A Python interpreter and the boto3 library (https://aws.amazon.com/sdk-for-python/) installed in your environment

Populating metadata

The first step of any Data Catalog is to populate it with databases and tables. AWS Glue provides both manual and automatic options for doing so, the latter being particularly useful to avoid the cumbersomeness of defining datasets from scratch. This section will explain how the Data Catalog works and will demonstrate how to interact with it in different ways.

Glue Data Catalog API

Just as in other AWS services, AWS Glue offers a fully fledged **application programming interface (API;** https://docs.aws.amazon.com/glue/latest/dg/aws-glue-api.html) to interact with it, which includes the Data Catalog. Thus, operations such as creating a database or a table can be done through said API or any of its containers, such as the AWS CLI or any of the **software development kits (SDKs)**.

For instance, let's start populating our catalog manually. The first step is to create a database, which we can do using the AWS CLI. The CLI *Command Reference* page (https://docs.aws.amazon.com/cli/latest/reference/glue/index.html) has a complete list of all available CLI commands for Glue, which follow the same notation as API calls. In this case, the CreateDatabase API call is mirrored with the create-database CLI command, so that's what we will use, as illustrated in the following code snippet:

```
aws glue create-database --database-input
"{\"Name\":\"sampledb\"}"
```

Next, let's create a table inside the database, for which we can use the create-table command. Please note that the create-table operation requires passing a TableInput object that determines all the properties of the table. This object can be defined using **JavaScript Object Notation (JSON)** notation. In this case, our table will have three columns (name and surname—of type string; and **identifier (ID)**—of type int) and will be stored in S3 as JSON files. The code is illustrated in the following snippet:

```
aws glue create-table \
        --database-name sampledb \
```

```
    --table-input   '{"Name":"sampletable", \
        "StorageDescriptor":{ "Columns":[ \
        {"Name":"name", "Type":"string"}, \
        {"Name":"surname", "Type":"string"}, \
        {"Name":"id", "Type":"int"}], \
        "Location":"s3://sample-path/", \
        "SerdeInfo":{"SerializationLibrary":"org.openx.
data.jsonserde.JsonSerDe"}}, \
        "Parameters":{"classification":"json"}}
```

Finally, let's use an AWS SDK to create a partition inside the table. AWS offers a wide variety of SDKs (https://aws.amazon.com/tools/), but the easiest one to use is probably the Python one, also known as boto3 (https://boto3.amazonaws.com/v1/documentation/api/latest/reference/services/glue.html). The official documentation (https://boto3.amazonaws.com/v1/documentation/api/latest/reference/services/glue.html) lists all available methods and how to use them, but they follow a very similar structure to that of the **REpresentational State Transfer (REST)** API. This time, we'll be using the create_partition method, which similarly to before requires passing a PartitionInput object that defines the properties of the partition. The code is illustrated in the following snippet:

```python
import boto3
glue_client = boto3.client('glue')


response = glue_client.create_partition(
    DatabaseName='sampledb',
    TableName='sampletable',
    PartitionInput={
        'Values': [
            '2019',
        ],
        'StorageDescriptor': {
            'Columns': [
                {
                    'Name': 'name',
                    'Type': 'string'
                },
                {
                    'Name': 'surname',
```

```
                            'Type': 'string'
                },
                {
                    'Name': 'id',
                    'Type': 'int'
                }
            ],
            'Location': 's3://sample-path/year=2019/',
            'SerdeInfo': {
                'SerializationLibrary': 'org.openx.data.
jsonserde.JsonSerDe'
            }
        },
        'Parameters': {
            'classification': 'json'
        }
    }
}
```

The three methods we used to create the database, table, and partition respectively are equivalent and exchangeable, and there are no differences as to how they are represented in the AWS backend. This is because, as stated before, both the AWS CLI and any of the SDKs use REST API calls internally to interface with the service endpoint.

In this section, we discussed different ways of interacting with the Glue API through the AWS API. Next, we will discuss interacting with the API through **Structured Query Language** (**SQL**) statements, which might be more natural for data engineers or database administrators.

DDL statements

The most natural way of interacting with the Data Catalog for the majority of users is to use **Data Definition Language** (**DDL**) statements since that is similar to a relational SQL database. AWS Glue, however, does not offer any SQL interface to interact with the catalog directly—the only way to interact with it is through API calls.

Because of this limitation, several external services and applications have been developed as a translation layer between the SQL language and the necessary API calls to run DDL statements on the Glue Data Catalog. In the following sections, we will cover services providing this capability within AWS; however, with the API being an open specification, literally any third-party SQL interpreter could interact with the Data Catalog.

Apache Hive

Apache Hive (`https://hive.apache.org/`) is an open source project that delivers a data warehouse designed for the Hadoop ecosystem, allowing users to explore and query large datasets using a variation of the **American National Standards Institute (ANSI)** SQL language, known as the **Hive Query Language (HiveQL)**. Hive relies on the Hive Metastore, a single-node metadata repository that holds all information about Hive tables—a concept very similar to that of the Glue Data Catalog.

Even though Hive is not directly related to Glue, AWS offers Hive-Glue compatibility through **Amazon Elastic MapReduce (Amazon EMR)** clusters (`https://aws.amazon.com/emr/`). This allows users to use the Glue Data Catalog in place of the Hive metastore, which effectively lets them run SQL queries on Glue tables through Hive. Users can launch EMR clusters with Hive-Glue compatibility (`https://docs.aws.amazon.com/emr/latest/ReleaseGuide/emr-hive-metastore-glue.html`) by adding the following configuration property at launch time:

```
[
   {
     "Classification": "hive-site",
     "Properties": {
       "hive.metastore.client.factory.class": "com.amazonaws.
glue.catalog.metastore.AWSGlueDataCatalogHiveClientFactory",
       "hive.metastore.schema.verification": "false"
     }
   }
]
```

This property effectively replaces the Hive metastore factory class with a custom one developed by AWS, which will interact with the Glue Data Catalog instead.

Once an EMR cluster has been launched with Glue Data Catalog integration, we can start Hive and begin running queries, as follows:

```
$> hive
```

Now let's repeat the operations we ran through the REST API, this time with Hive. We start by creating a new database, like so:

```
CREATE DATABASE sampledb;
```

We can then create our table within the database, as follows:

```
CREATE EXTERNAL TABLE sampledb.sampletable (
        name STRING,
        surname STRING,
        id INT
)
PARTITIONED BY (year STRING)
ROW FORMAT SERDE 'org.openx.data.jsonserde.JsonSerDe'
LOCATION 's3://sample-path/'
```

And finally, we can add a partition to the created table, like so:

```
ALTER TABLE sampledb.sampletable
ADD PARTITION (year='2019')
LOCATION 's3://sample-path/year=2019/'
```

Please note that in order for these operations to succeed, the **Identity and Access Management** (**IAM**) role attached to the EMR cluster's nodes must have the necessary permissions to do the equivalent Glue actions. For instance, when creating a database, the IAM role must have explicit permission to perform the `CreateDatabase` action (https://docs.aws.amazon.com/glue/latest/dg/aws-glue-api-catalog-databases.html#aws-glue-api-catalog-databases-CreateDatabase). This also extends to additional Glue Data Catalog features such as **Key Management Service** (**KMS**) encryption.

Even though this Hive-Glue compatibility works, there are certain limitations and considerations to take into account when using it, the most notable being these:

- Hive **atomicity, consistency, isolation, and durability** (**ACID**) transactions (which enable operations such as `DELETE` or `UPDATE`) are not supported.

- Hive cannot rename tables, as tables in Glue cannot be renamed.

- Even though users could theoretically create Hive-managed tables and they would appear in the Data Catalog, these tables are not accessible as their data would be stored in the Hive cluster's local **Hadoop Distributed File System** (**HDFS**) storage. Therefore, it is recommended to use the `EXTERNAL` keyword for all of your tables.

To check a complete list of limitations and considerations, please refer to the public AWS documentation (https://docs.aws.amazon.com/emr/latest/ReleaseGuide/emr-hive-metastore-glue.html).

Apache Spark™

Apache Spark (`https://spark.apache.org/`) is an open source framework for big data processing, enabling the processing of large datasets over a cluster of compute nodes. Spark is part of the Hadoop ecosystem and, as such, is capable of interacting with tables defined in a Hive metastore. Spark is also offered as part of Amazon EMR clusters, and just as with Hive, AWS has developed specific integrations to enable Spark to interact with tables defined in the Glue Data Catalog.

Spark offers two main ways of dealing with data: programmatically via code, or through its own implementation of the ANSI SQL language, Spark SQL. Spark SQL is also the name of Spark's SQL libraries and modules, which enable the use of SQL queries within Spark code and provide the `spark-sql` **read-eval-print loop** (**REPL**) environment where queries can be executed interactively through a command-line terminal.

As the AWS documentation (`https://docs.aws.amazon.com/emr/latest/ReleaseGuide/emr-spark-glue.html`) describes (and similarly to Hive), Glue-Spark integration can be enabled by passing the following configuration property to the EMR cluster at launch time:

```
[
  {
    "Classification": "spark-hive-site",
    "Properties": {
      "hive.metastore.client.factory.class": "com.amazonaws.
glue.catalog.metastore.AWSGlueDataCatalogHiveClientFactory"
    }
  }
]
```

This property effectively replaces the Hive metastore factory class used by Spark with a custom-developed one that interacts with the Glue Data Catalog instead.

Once the cluster has been launched, we can run SQL queries easily by starting a REPL, as follows:

```
$ spark-sql
```

We can then repeat the same steps we did in the *Glue Data Catalog API* section within the REPL, as illustrated here:

```
CREATE DATABASE sampledb;
```

Creating a table is done in a similar fashion, as we can see here:

```
CREATE EXTERNAL TABLE sampledb.sampletable (
        name STRING,
        surname STRING,
        id INT
)
PARTITIONED BY (year STRING)
ROW FORMAT SERDE 'org.openx.data.jsonserde.JsonSerDe'
LOCATION 's3://sample-path/'
```

And finally, adding a partition is again very similar, as illustrated here:

```
ALTER TABLE sampledb.sampletable
ADD PARTITION (year='2019')
LOCATION 's3://sample-path/year=2019/'
```

Just as with Hive, the IAM role attached to the EMR cluster nodes will need specific Glue permissions to perform operations that stem from the executed commands.

Spark has very few differences from Hive in its SQL implementation and interpretation. Because of this, many of the Hive-Glue integration limitations mentioned earlier also apply to Spark. For a full list of them, please check the public AWS documentation (https://docs.aws.amazon.com/emr/latest/ReleaseGuide/emr-spark-glue.html).

Amazon Athena

Athena (https://aws.amazon.com/athena/) is a serverless query service designed to enable SQL querying over datasets stored in S3. Since its purpose overlaps heavily with that of Glue's, Athena was designed from an early stage to work with the Glue Data Catalog, enabling SQL querying over Glue tables.

That said, there are certain limitations to the SQL queries you can run on the Data Catalog. Athena is based on Presto (https://prestodb.io/), an open source SQL engine developed by Facebook engineers, and thus it will be as powerful—in terms of querying—as Presto is. Presto uses PrestoSQL, an ANSI-compatible implementation of the SQL language that covers most, but not all, SQL language operations.

Let's try to use Athena to perform the same operations we performed with the API: creating a database, creating a table inside the database, and adding a partition to said table. If you followed the instructions for Apache Hive, you'll find that the queries are pretty much identical—since they are quite simple and both engines support the SQL language, there's not much variation.

Creating a database can be done with a simple statement, as follows:

```
CREATE DATABASE sampledb;
```

When creating a table, there are two things to pay particular attention to, as outlined here:

- The EXTERNAL keyword must be added to the CREATE TABLE statement. This is inherited from the Apache Hive concept of managed tables and external tables; however, in Athena (and Glue), all tables are considered external.

- Just as we provided the location, **serializer/deserializer (SerDe)**, and classification information to the table definition JSON in the previous examples, we need to tell Athena all this information in a similar fashion.

The CREATE TABLE statement then looks like this:

```
CREATE EXTERNAL TABLE sampledb.sampletable (
        name STRING,
        surname STRING,
        id INT
)
PARTITIONED BY (year string)
ROW FORMAT SERDE 'org.openx.data.jsonserde.JsonSerDe'
LOCATION 's3://sample-path/'
```

Finally, let's add a partition to the created table, just as before, like so:

```
ALTER TABLE sampledb.sampletable
ADD PARTITION (year='2019')
LOCATION 's3://sample-path/year=2019/'
```

Note how in this case, we didn't have to specify the schema or the serialization properties of the partition. This is because Athena (by Presto's design) expects all partitions of a table to have the same schema and properties as the table itself. Therefore, complex scenarios where the partition schema and properties evolve over time cannot use Athena or should update the table schema itself rather than individual partitions.

Another important thing to note is Athena includes an automatic partition detection mechanism built into the MSCK REPAIR TABLE statement. This will automatically identify all partitions in the location specified by the table's location property, provided that they follow the Hive-style partitioning format (https://docs.aws.amazon.com/athena/latest/ug/partitions. html). If not following this format, partitions will have to be added manually.

Glue crawlers

If you followed the previous sections, you will have seen that populating metadata within the Data Catalog is not a hard task. That said, it can quickly turn into an extremely repetitive task for large data lakes, sometimes becoming unfeasible for a single engineer to map all datasets to Glue tables manually. Because of this, AWS developed Glue crawlers.

A Glue crawler is an AWS entity that will scan the contents of a given data location, automatically infer a schema from it, and define it as a table in the Glue Data Catalog. Crawlers are recursive, which means they will work with complex nested structures such as table partitions and can be run periodically to add partitions or update the schema of a table with new fields populated by new incoming data.

Because of their usefulness and ease of use, crawlers are the recommended way of populating the Data Catalog, even for small and simple setups. Running a crawler will have a small cost and finish within a matter of minutes, which is more comfortable and less error-prone than—for instance—defining a `TableInput` object to be used with a CLI command.

Crawlers can automatically infer schema from data stored in the following silos:

- Amazon S3 buckets and prefixes
- Amazon DynamoDB tables
- Amazon Redshift clusters
- **Amazon Relational Database Service (Amazon RDS)** databases
- Non-RDS-hosted relational databases (MariaDB, SQL Server, MySQL, Oracle, and PostgreSQL)
- MongoDB and Amazon DocumentDB databases

Depending on which silo is being crawled, the crawler will behave and perform in different ways, which we'll discuss in the following sections.

Crawler behavior

The way a crawler determines the existence of a table or partition is critical to understanding how it works and avoiding possible issues. A malfunctioning crawler can break your data pipelines by updating a table definition with the wrong one, or it can pollute the Data Catalog by creating thousands of tables that should have been partitions of a large dataset.

The way a crawler works depends on what type of data store it is analyzing, with behaviors falling into one of the three following categories:

- **Java Database Connectivity (JDBC) data stores and document stores**: For relational datasets (Amazon Redshift, Amazon Aurora, MariaDB, SQL Server, MySQL, Oracle, and PostgreSQL) and document stores (MongoDB and DocumentDB), the crawler will simply list databases and tables, and then retrieve each table's schema by describing it. There isn't much complexity to this setup, as it is pretty much just copying information over to the Data Catalog.

- **DynamoDB tables**: By default, the crawler will scan all items in the specified table, at a rate specified by the user when creating a crawler. This rate is specified as a percentage of the total read capacity units of the table. Alternatively, if all records in the table can be assumed to have a similar schema, the user can configure the crawler to only analyze a sample of the table to avoid consuming unnecessary read capacity units.

- **S3 datasets**: Given an S3 location, the crawler will recursively analyze objects located within it and compare their schema. S3 crawler behavior is sufficiently complex that we've separated it into its own section, located right after this one.

Let's discuss crawler behavior for S3 datasets in detail.

S3 crawler behavior

As stated earlier, when crawling an S3 location, the crawler will read the contents of objects located within it to infer their schema. Schema inference is carried out by an entity known as a classifier.

A **classifier** is a piece of software that the crawler will execute to determine which format a file has been written in. For instance, the **comma-separated values (CSV)** classifier determines whether a particular file is written in CSV format or not. Determining the format is important for two reasons, as outlined here:

- The crawler needs to know how to read the file. Plain-text file formats such as CSV are easy to read, but more complex file formats such as Parquet require the use of specific libraries.

- The SerDe information will be written on the resulting table in the Data Catalog, which will let other services and applications read the data properly. The term *SerDe* refers to the Java classes to be used to serialize (write) and deserialize (read) information to and from the file.

A crawler has a set of built-in classifiers (one per each supported format), and users can also define custom classifiers for certain file formats that require more tuning. A list of built-in classifier formats can be found in the public documentation (`https://docs.aws.amazon.com/glue/latest/dg/add-classifier.html#classifier-built-in`).

Whenever a crawler needs to infer the schema for an S3 object, it will run the object through all the classifiers, starting with the custom ones and following with the built-in ones. Upon completing its analysis, each classifier returns a value between 0 and 1 that determines the certainty the classifier has that the file belongs to its format. The first classifier to report a certainty of 1 is used as the format. If no classifier returns a 1 value, the one with the higher certainty value is used. Finally, if all classifiers return a 0 value, the crawler will set the table format to UNKNOWN and the schema inference process will fail.

The selected classifier will then be used to determine the file's schema, based on the file format. The way this happens depends on how much schema information the file format inherently has, as explained in more detail here:

- For structured file formats (such as Avro, **Optimized Row Columnar (ORC)**, or Parquet), the schema will simply be extracted from the file's own metadata.

- Semi-structured formats (such as JSON, **Extensible Markup Language (XML)**, or Ion) will have their schema inferred by reading a sample of records and inferring the types of the fields.

- Log files (such as Apache logs, Linux kernel logs, or AWS CloudTrail logs) will all rely on the use of predefined Grok patterns (`https://www.elastic.co/guide/en/logstash/current/plugins-filters-grok.html`).

- Non-structured file formats (such as CSV or any of its variations) will split records by a defined separator character and try to match each resulting field to the best-matching data type. Field names will be taken from the file's header, if available.

Now, let's go back to the crawler's behavior. When crawling an S3 path, the crawler looks at the contents of the path as a recursive tree, where each node in the tree is a subfolder (or S3 prefix) within the target S3 path. The crawler will then navigate to the deepest node in the tree and analyze the schema of all S3 objects present there, using the classifier process explained before.

After all schemas are determined, the crawler will then start comparing all identified schemas and group them together by similarity. There are four key factors to take into consideration here, as outlined here:

- File compatibility
- Schema similarity
- File similarity
- File group quantity

The following sub-sections will discuss each one of these factors.

File compatibility

Files must be compatible. This means they must use the same compression format and belong to the same format. For instance, a path with JSON and CSV files will not result in a single unified table as there would be no way to read all files simultaneously with the same SerDe information.

Schema similarity

Two schemas are considered similar if they have more than 70% fields present in both of them. For instance, take these two records:

```
{"id":1,"first_name":"Henrik","last_
name":"Paddington","country":"Ireland","city":"Dublin"}
{"id":2,"first_name":"James","last_
name":"Smith","country":"Canada","language":"English"}
```

In this example, the two records have five fields and only one of the fields is different, so there would be a 20% difference or 80% similarity between them. The crawler would consider them similar and place them in a group together. Now, let's look at an opposite example, as follows:

```
{"id":1,"car_brand":"Toyota","model":"Yaris"}
{"id":2,"car_brand":"Audi","year":2009}
```

This time, there's a total of three fields, with the last one being different for both (66% similar). The crawler would consider them different schemas, therefore placing them into separate groups.

File similarity

The same 70% similarity is applied to the number of files belonging to each schema in the tree node. Let's exemplify this with two schemas, as follows:

```
Schema A
{"id":1,"first_name":"Henrik","last_
name":"Paddington","country":"Ireland","city":"Dublin"}


Schema B
{"id":2,"car_brand":"Audi","year":2009}
```

These two schemas only share one field out of six different ones, so they would be considered different by the crawler. Now, let's assume two different distributions, as follows:

1. The S3 path contains eight files with schema A and two with schema B. This would mean 80% of files belong to the same schema, which is larger than the 70% threshold. The crawler will consider the path to have schema A, and ignore files with schema B. The ignored files will be notified to the user through CloudWatch logging, with the following message:

   ```
   INFO: Some files do not match the schema detected.
   Remove or exclude the following files from the crawler
   (truncated to first 200 files):
   ```

```
sample-path/B1.json
sample-path/B2.json
```

The table creation process is also part of the logs and helps identify table creation API calls for audit purposes. You can see an example of this here:

```
INFO: Created table sampledata in database sampled
```

2. The S3 path contains seven files with schema A and three files with schema B. This time, the 70% threshold would be hit, meaning the contents of the path are not homogeneous enough for the crawler to assume a single schema. In this case, the crawler would create an individual table for each file within the S3 path. The threshold being hit is notified to the user via CloudWatch logging. The code is illustrated in the following snippet:

```
[main] INFO com.amazonaws.services.glue.statetree.
detector.streaming.S3StreamingPartitionDetector - Minimum
frequency threshold surpassed for aggregated file set:
sample-path/
 [main] INFO com.amazonaws.services.glue.statetree.
detector.streaming.S3StreamingPartitionDetector -
Clustered Schema Count: 7
```

This would also include details of conflicting schemas and their fields, alongside table creation messages for each table created.

File group quantity

There can only be a maximum of five groups at all times. A sixth group appearing will immediately stop the crawling process, with the crawler creating tables and/or partitions based on the information it had read up until that point.

The crawler will then apply this logic to each level in the recursive tree, going from its bottom to its top and comparing the resulting schemas at each level. At every level, only files or partitions (subfolders or prefixes) can be present. Having both at the same level will result in the schema detection process being interrupted, with the crawler writing results to the Data Catalog based on information obtained up until that point. Partitions must also follow Hive-style naming for them to be recognized as partitions of a table rather than individual tables within a common directory.

Now that we understand how schema detection behaves for different file formats, we'll discuss how crawlers are executed and the different stages they go through.

Crawler life cycle

In order to troubleshoot and manage crawlers, it is important to understand how they are executed. Crawlers have four different states, which cycle in the following order:

1. **Ready**: The crawler is waiting to be executed.

2. **Starting**: The crawler is waiting for **data processing units (DPUs)** to be allocated to run.

3. **Running**: The crawler is analyzing its target.

4. **Stopping**: The crawler is writing results to the Data Catalog.

Just as with many other AWS resources, the crawler generates AWS CloudWatch logs on every execution. Crawler logs include messages about the crawler's state changes, issues encountered during execution, and results written to the Data Catalog.

Every instance of a crawler's execution is known as a **crawl** and has an associated crawl ID that can be used to identify an execution uniquely. This crawl ID is not exposed by the AWS web **user interface (UI)** and can only be retrieved in the following two ways:

- By checking the crawler's execution logs. Every log message written by a crawler execution will be preceded by its crawl ID, as in the following code example:

```
[e6021d6f-8fc6-4ac7-96a2-07dee35ccf14] BENCHMARK: Running
Start Crawl for Crawler cases-ddb
```

In the preceding example, the e6021d6f-8fc6-4ac7-96a2-07dee35ccf14 string would be the crawl ID.

- By using the Glue REST API (in any of its forms). The GetCrawler API call returns a Crawler object containing a LastCrawl object, which contains the MessagePrefix property. This is the same as the crawl ID.

Retrieving the crawl ID can also be useful in instances when contacting AWS Premium Support is necessary, as this will help AWS engineers easily identify an execution.

Even though all crawlers execute the same four stages, they can be configured to modify the behavior of each stage. In the following section, we will discuss these configuration options.

Crawler configuration

Crawlers have several configuration options that are critical to their functioning. These options determine the way crawlers behave when updating, deleting, or comparing the schema of tables. In the following sub-sections, we will go over each category of configuration options.

Catalog update behavior

The following options modify what a crawler does when writing results to an already existing Data Catalog table. They are listed based on their name in the Glue console web UI:

- **Update the table definition in the Data Catalog**: This will update all properties of the table that have changed, including changes such as removing columns or changing data types. Because of the drastic changes it can apply, it is not recommended for production setups unless no changes to incoming data can be guaranteed.

- **Add new columns only**: This option will only add new columns if they are discovered in subsequent crawls. Recommended for setups where there is constant data ingestion with evolving fields, such as streaming data coming from a changing REST API endpoint, for instance.

- **Ignore the change and don't update the table**: As the name implies, this will ignore all schema and property changes. Only new partitions will be added to the table. Recommended for most setups, as once the schema has been verified to work, it can be kept stable.

When configuring these options through the API, they can be found in the `SchemaChangePolicy` object inside the `Crawler` object. The `UpdateBehavior` property can be configured to the following values:

- `UPDATE_IN_DATABASE` for the **Update the table definition in the Data Catalog** option

- `LOG` for the **Ignore the change and don't update the table** option

In order to achieve the **Add new columns only** setting, the `UpdateBehavior` option must be set to `UPDATE_IN_DATABASE` and the following section should be added to the crawler definition object:

```
"Configuration": "{\"Version\":1.0, \"CrawlerOutput\":
{\"Tables\":{\"AddOrUpdateBehavior\":\"MergeNewColumns\"}}}
```

Catalog deletion behavior

In regard to what happens when the crawler doesn't find an object already defined in the catalog, there are again three possible options, as follows:

- **Delete tables and partitions from the Data Catalog**: As the name implies, anything that's not found will be deleted. Not recommended for production setups as it can result in accidental deletion.

- **Ignore the change and don't update the table in the Data Catalog**: Nothing will happen.

- **Mark the table as deprecated in the Data Catalog**: The table will be deprecated instead of deleted. A deprecated table is marked by a custom property added to the table parameters; however, it has no practical effect other than notifying users, and deprecated tables can still be queried and accessed normally. A deprecated table will have the following property in its definition:

```
"Parameters": {
            "DEPRECATED_BY_CRAWLER": "1642411200907"
}
```

Here, the value of the DEPRECATED_BY_CRAWLER property is the timestamp of the deprecation.

Table schema inheritance

When dealing with a partitioned table's schema, we would typically assume the partitions of a table will have the same schema as the table itself. This assumption has, however, been challenged with the appearance of technologies such as REST APIs and streaming, where a schema can be evolving over time. A common use case would be REST API logs, where a new method or property might be added one day. If a table is partitioned by day, for example, usage logs for that API will suddenly have a new column starting on that partition, which means the partition-level schemas and the table-level one are not the same.

Some frameworks and query engines assume schema equality between a table and its partitions, and some provide the flexibility of having different schemas. As described in previous sections, Amazon Athena is an example of such a service, whereby trying to query a table with different table-level and partition-level schemas will result in an error. In order to tackle this issue automatically, crawlers can update the schema of the table's partitions every time they run with the schema of the table itself.

This can be configured by enabling the **Update all new and existing partitions with metadata from the table** option in the console, or by defining the option in the CrawlerOutput section of the Crawler object, like so:

```
"CrawlerOutput": {
      "Partitions": {"AddOrUpdateBehavior": "InheritFromTable"
}
  }
```

Crawler behavior modification

As described earlier, schema similarity is one of the factors that a crawler considers in order to differentiate tables from partitions automatically. This, however, can result in situations where the crawler assumes two different schemas should be different tables, even though the user might want to have them as partitions. This is only possible as long as the schemas are compatible—that is, they don't overlap or cause conflicts between each other.

Take these two schemas, for instance:

```
Schema A
{"id":1,"first_name":"Henrik","last_name":"Paddington"}


Schema B
{"car_brand":"Audi","year":2009}
```

Even though the schemas are not similar, they could still be unified under a single table if we combine them, as follows:

```
Unified schema
{"id":1,"first_name":"Henrik","last_name":"Paddington","car_
brand":"Audi","year":2009}
```

Querying the table would simply return NULL values on records that don't have the column, and this would allow us to query everything under a single table rather than having to query two separate tables and join the results.

In order to tackle these situations, crawlers can be configured to ignore the similarity threshold and combine schemas whenever possible. This can be achieved by enabling the **Create a single schema for each S3 path** option in the console, or by adding the following property to the crawler definition object:

```
{
    "Version": 1.0,
    "Grouping": {
        "TableGroupingPolicy": "CombineCompatibleSchemas"}
}
```

Keep in mind not all schemas are compatible. Take the following example:

```
Schema A
{"id":1,"first_name":"Henrik","last_name":"Paddington"}


Schema B
{"id":true, "car_brand":"Audi","year":2009}
```

In this case, both Schema A and Schema B have an id field; however, in Schema A, it would be of an integer type, whereas in Schema B it would be of a Boolean type. This would cause a direct conflict when defining a table, so the crawler would be unable to combine them.

On the other hand, the opposite can also happen. If two schemas are similar enough, they will be combined into a single table even if the user would expect them to be different tables being crawled at the same time. In order to avoid this, crawlers can be configured to take a given level of the recursive tree as the table level 1, at which schema merging will stop and tables will be output.

This can be configured through the **Table Level** option in the AWS console, or by adding the following property to the crawler definition object:

```
{

    "Version": 1.0,
    "Grouping" = {
            TableLevelConfiguration = 2
    }
}
```

Before schemas can be compared and analyzed in the ways described earlier, the crawler needs to be able to determine them. Some file formats may present challenges when it comes to this, which we will discuss in the following section.

Custom classifiers

Certain file formats have particularities that don't allow for *one-size-fits-all* parsing. For instance, the CSV format has many variations of the character used to separate fields (**tab-separated values (TSV)**, **pipe-separated values (PSV)**, and many custom ones). When crawling nested formats such as XML or JSON, the user might want to parse only a subset of the tree rather than the whole structure.

In order to support these variations, users can create custom classifiers that allow them to modify the behavior of the default built-in classifiers. Custom classifiers are available for the following file formats:

- **CSV**: Allows for the configuration of column delimiters, quote symbols, file headers, and crawler behavior when encountering abnormalities.

- **JSON**: Allows you to specify a path in dot or bracket notation to only parse parts of each record.

- **XML**: Allows you to specify a root tag so that only information below it is parsed.

- **Grok**: Custom Grok expressions can be provided to parse log files not directly supported by Glue. Grok classifiers can also be used to parse custom text files without a strong format or that are not supported by other classifiers.

A crawler can have several custom classifiers attached to it, and they will be used in the same order they were attached when classifying files.

Maintaining metadata

There's rarely a scenario in which Glue Data Catalog tables will be static entities defined once and never updated again. Whether your tables use partitioning and they need to be updated with new partition values, or you have a changing stream of incoming data that adds or modifies data types, you'll want to keep updating and refining your Data Catalog entities.

Glue provides several mechanisms to do so automatically without user interaction, although any of the methods described before can be used to update tables or partitions manually. Metadata can be automatically updated using crawlers or **extract, transform, load** (**ETL**) jobs, which we will discuss in this section.

Glue crawlers

Similar to how crawlers can define tables and partitions in the Data Catalog, they can also update them. Any subsequent runs of a successfully completed crawler will update objects the crawler initially defined as per the configuration options selected. There are several aspects to consider when using crawlers to maintain metadata.

Crawler behavior when re-crawling

For S3 targets, when a crawler is executed a second time (or any subsequent times after that), it will try to avoid re-analyzing all the contents of the S3 path, which saves both time and costs for the user. This is achieved by checking the start time of the last successful execution of the crawler and comparing that against the last modification time of all files within the S3 target path. Only files created or modified after the last start time will be crawled.

Scheduling

Crawlers can be set to execute on a regular schedule, allowing users to refresh their Data Catalog entities periodically. The crawler's schedule is configured as part of its properties and can be specified via either the web console or the API. Even though the web console offers preconfigured options (such as weekly, hourly, or daily), crawler schedules are always expressed in cron notation, and choosing a preconfigured option will automatically generate a cron expression (https://en.wikipedia.org/wiki/Cron).

The schedule is configured under the Schedule section of the Crawler object, as follows:

```
"Schedule": {
    "ScheduleExpression": "cron(08 11 ? * MON *)",
    "State": "SCHEDULED"
}
```

Even though schedules can be useful, some users prefer to update their Data Catalog definitions right after data has been pushed to the data lake, reducing the time it takes for new columns or changes to be updated. A common setup is to run crawlers right after ingestion has been completed, enabling new partitions within minutes. The following section will describe how to achieve that.

Automation

Crawlers can be automated in a variety of ways, and really any kind of custom automation can be developed thanks to the REST API. This section will discuss some of the most common automation options, typically based on other AWS services, as outlined here:

- **Glue workflows**: The Glue service itself offers the workflows feature (`https://docs.aws.amazon.com/glue/latest/dg/workflows_overview.html`), which allows you to create complex step-based automations involving not just crawlers, but also ETL jobs and custom conditions. A very common setup is to run a Glue ETL job as an ingestion job, then run a crawler over the output location of the job if the ingestion was successful.

- **AWS Step Functions**: Similar to Glue workflows, AWS Step Functions (`https://aws.amazon.com/step-functions/`) is a visual workflow service based on state machines that enables automation for many AWS service components, not just Glue ones. Step Functions allows for more complex integrations, such as running a crawler after an EMR cluster has completed running a job or running a crawler over the resulting Linux kernel logs of an **Elastic Compute Cloud (EC2)** instance.

- **AWS Lambda**: The fact that crawlers can be started and managed through the AWS SDK allows developers to write their own automation code if the previous workflow solutions don't fit their use case. Lambda (`https://aws.amazon.com/lambda/`) is a serverless code service that can very easily run complex code-based workflows with a variety of conditions and **inputs and outputs (I/Os)**.

Once again, given that the API can be accessed programmatically through the AWS CLI or any of the SDKs, the possibilities here are limitless. That said, there's a particular scenario we wouldn't recommend: Glue ETL job code. Even though we've proposed code-based services as an automation solution, and that you could potentially start a crawler programmatically as part of the code of a Glue ETL job, this is typically not recommended. Decoupling code and responsibilities from isolated components will make your workflows safer to run and easier to troubleshoot.

Incremental crawling

Even if the crawler will not re-analyze all files for every subsequent execution, there are still situations in which it can take increasingly longer for every execution.

Take a streaming ingestion system, for instance: most data streaming platforms such as Amazon Kinesis (`https://aws.amazon.com/kinesis/`) will write files to S3 in periodical batches, often resulting in a large number of small files, which is not optimal for querying with most big data platforms. Most users would typically run a compaction system after files have been written, merging small files into larger ones; however, that means creating a new last modification timestamp that would cause the crawler to re-analyze the files, even if their schema has not changed.

In order to provide a way to avoid these situations, the incremental crawling feature was developed for crawlers. This feature will only crawl new folders within the target S3 path rather than checking modification timestamps, meaning already crawled partitions can be safely edited or compacted without affecting the crawler's execution time.

This option can be enabled by either of the following:

- Checking the **Crawl new folders only** option when editing a crawler's configuration

- Setting `RecrawlPolicy` to `CRAWL_NEW_FOLDERS_ONLY` instead of `CRAWL_EVERYTHING` when using the API

When enabling this feature, all crawler behavior options are changed to `LOG`, meaning the crawler will not alter schemas or delete objects automatically. Because of this, the crawler will also ignore any objects that have a schema sufficiently different from the already existing one. Thus, this feature is only recommended for stable schemas where variations are known to be rare.

S3 event-based crawling

If incremental crawling is not an option because of its limitations, there's still another feature to accelerate crawling S3 targets. S3 offers the Event Notifications feature (`https://docs.aws.amazon.com/AmazonS3/latest/userguide/NotificationHowTo.html`), which can trigger notifications upon a variety of events (such as creating a new S3 object). These notifications can then be configured to be sent to **Amazon Simple Notification Service** (**Amazon SNS**) topics (`https://aws.amazon.com/sns/`), **Amazon Simple Queue Service** (**Amazon SQS**) queues, or AWS Lambda functions (`https://aws.amazon.com/lambda/`), which essentially enables automation based on S3 changes.

When writing data to a new partition in a Glue table, enabling the S3 Event Notifications feature essentially creates a log of all new objects within the target location, which is essentially what the crawler needs to avoid re-crawling older files.

In order to enable this option, you will first need to create an SNS topic and an SQS queue to handle S3 event notifications, then enable the option via either the web UI or by adding the SQS queue **Amazon Resource Name** (**ARN**) to the target definition object, as follows:

```
"S3Targets": [
    {
```

```
        "Path": "s3://sample-path/", "Exclusions":
[], "EventQueueArn": "arn:aws:sqs:us-east-
1:123456789123:samplequeue"
    }
]
```

Crawlers are a viable option to automate metadata maintenance, but their execution needs to be started. If this is to happen right after an ETL job is executed, the ETL job itself can be used to update metadata without requiring a crawler execution. The following section will go over how to achieve that.

Updating Data Catalog tables from ETL jobs

When running Glue ETL jobs that write results to the Data Catalog, it is possible for them to not just write output data but also to update the catalog with its respective metadata. This means the job itself can add new partitions or modify the table's schema without the need to run a crawler afterward or update the table manually.

This option is limited to only updating metadata with changes present in the data that is being written. Imagine a scenario where a single Data Catalog table is being updated with new data by two entities: a Glue ETL job and an EMR cluster. The metadata would only be updated with what the ETL job writes, and any changes made by the EMR cluster would not be reflected. This means the option is only suitable when the ETL job is the only entity writing to the target; otherwise, a crawler or manual updates will still be necessary.

ETL jobs can update the Data Catalog with the following:

- New tables
- New partitions being written to a table
- Schema changes being made to a table

Full instructions on how to enable these features, alongside code samples, can be found in the public AWS documentation (https://docs.aws.amazon.com/glue/latest/dg/update-from-job.html).

Partition management

In the previous sections, we discussed how to automatically update and add partitions to tables. This means that with an easy setup, Glue is capable of adding partitions continuously as your dataset grows.

For very large data lakes, however, this setup can easily run into issues. Glue supports up to 10 million partitions per table by default; however, having such a large number of partitions will increasingly lower your query execution times without proper management.

Partition indexes

Let's take the example of a table storing product sales information. The table is partitioned by product category, and even though the business started small and we had only a handful of categories, as we expanded and added external sellers, we are now in the tens of thousands of categories.

Our business analysts want to query data based on product families, and so their Glue ETL queries usually include a WHERE CATEGORY= clause, filtering by category. Every time the query is executed, Glue will have to list all product categories and filter out those that don't match the filter. This means running GetPartitions API calls, which are paginated and can get expensive the more values we have to retrieve, slowing our queries down.

In order to avoid this, Glue introduced **partition indexes**. These indexes basically hold a list of partition values known to already exist in the table beforehand, speeding the filtering process up by a large margin since it won't be necessary to retrieve all partitions and filter them.

Every Glue table can have up to three indexes defined for it, and once an index is created, Glue will validate all new partition values to ensure they belong to the right data type. Once an index has been created, all GetPartitions API calls can include a filter expression that Glue will try to match against the index. Limitations and considerations when using partition indexes can be consulted in the AWS public documentation (https://docs.aws.amazon.com/glue/latest/dg/partition-indexes.html).

Versioning and rollback

The previous sections described automated and autonomous metadata management for Glue tables. This, however, can lead to unexpected changes in the Data Catalog that might break pipelines relying on it. Even when not relying on automated changes, a human error could also break a table definition by mistake. This section describes the versioning and rollback mechanisms in place in the Data Catalog, designed to avoid and recover from such scenarios.

Table versioning

The Glue Data Catalog has a versioning mechanism for tables. Every time an edit is made to the table (even if the table definition passed as the edit is the same as the already existing one), a new version will be created, identified by a monotonically increasing integer starting at 1.

Only one table version can be active at any time, and only the active version can be accessed—it is not possible to read from a table specifying a previous version, for instance. At any time, the user can choose to pick an active version from all versions of a table; however, this operation can only be done through the AWS web console as there's no API call to do it.

This mechanism allows for rollbacks in the case of an error, and also provides traceability for changes—something critical when having both automated and manual entities modifying the catalog.

Lake Formation-governed tables

AWS Lake Formation (`https://aws.amazon.com/lake-formation/`) is a managed data lake service that enables secure, row-level access security for tables defined in the Glue Data Catalog. Lake Formation has a wide variety of features, but for the purposes of this chapter, we will discuss governed tables.

Governed tables are S3 tables managed by Lake Formation that provide an additional set of features not available to regular Glue tables. Two main differences occur when a table is governed, as outlined here:

- **Transactions**: Any operation made against the table will be encapsulated within a Lake Formation transaction, which includes both data and metadata. These transactions can be canceled and reverted if necessary, providing an automatic rollback mechanism for failures.

- **Manifests**: Lake Formation will keep a manifest of all S3 objects that represent the current dataset in the table. This means the S3 path specified as the location of the table can contain objects that are not part of the table. These objects can be part of a currently ongoing transaction that is not yet committed or could be data that has been deleted from the table.

These two key differences enable advanced features that regular Glue tables cannot provide, as outlined here:

- **ACID transactions**: Enable security and atomicity when multiple users are querying and inserting data into Data Catalog tables.

- **Automatic data compaction**: As mentioned in previous sections, having large amounts of small objects can negatively impact the performance of query engines accessing Data Catalog tables. Lake Formation automatically compacts objects for governed tables to ensure proper performance.

- **Automatic garbage collection**: Lake Formation can automatically delete objects that are not part of the table to save on costs.

- **Time-travel queries**: Each governed table keeps a manifest of S3 objects that represent the data within it. This manifest is versioned and can be used to query previous versions of data within the table, without the need to load them back.

- **Rollback mechanism**: In the case of a failed transaction, both data and metadata can be rolled back to their previous state. If an ETL job failed in the middle of writing data, Lake Formation can automatically remove data that was written up until the failure. If a streaming job needs to add a column to a table to insert data but then fails before it finishes, the new column can be automatically removed.

In order for a table to be governed by Lake Formation, its data needs to be stored in S3, and the S3 location must be registered with Lake Formation. Once that is done, a table can become Lake Formation-governed if any of the following actions are performed:

- Enabling the option in the Lake Formation web console when browsing or creating a table

- Setting the `TableType` property to `GOVERNED` through the Glue API

- Adding the property within `TBLPROPERTIES` in Athena, as follows:

```
TBLPROPERTIES (
    'table_type'='LAKEFORMATION_GOVERNED',
    'classification'='parquet'
)
```

Once a table is governed, the ways to interact with it change, with several important considerations to make, as follows:

- S3 objects within it should be considered immutable. Even though through S3 you could potentially upload a new version of an object, this will not update the Lake Formation manifest and thus could potentially break the table's functionality.

- Whenever data is written to the table, the `UpdateTableObjects` API must be called to update the manifest with the new S3 objects.

- In order to read from the table, any of the Lake Formation querying API calls should be used rather than simply reading from the S3 location. This will ensure the right S3 objects are queried, as well as applying the security access models defined in Lake Formation.

When it comes to metadata management, all operations should be handled through Lake Formation transactions. Several of the Glue APIs (listed in the official documentation at `https://docs.aws.amazon.com/lake-formation/latest/dg/transactions-metadata-operations.html#trx-enabled-glue-apis`) have been updated to include a transaction ID parameter, the value of which can be obtained with the Lake Formation `StartTransaction` API call. After the operation has been completed, the Lake Formation `CommitTransaction` API should be called to end the transaction. For instance, when creating a table, the user should do the following:

1. Call `StartTransaction` to obtain a transaction ID.
2. Run the `CreateTable` operation, passing the transaction ID as a parameter.
3. If the operation is successful, call `CommitTransaction` to commit it. If the operation failed for whatever reason, `CancelTransaction` should be called to revert the changes.

Lake Formation is a very powerful tool for metadata and access management. We recommend considering enabling Lake Formation to manage your Data Catalog whenever possible.

Lineage

Data lineage is the process of visualizing and understanding the flow of data within your data lake. Lineage is critical for data engineers and analysts to understand how data is processed and transformed within the data lake. This section covers the tools Glue provides in regard to lineage.

Glue DataBrew

Glue DataBrew (`https://aws.amazon.com/glue/features/databrew/`) is a serverless data lineage tool integrated within the AWS Glue ecosystem. DataBrew provides a visual and interactive way of visualizing, transforming, and automating data processing within a Glue data lake.

There are a few key components of DataBrew, as outlined here:

- **Datasets**: In order to work with data in DataBrew, it must be registered as a dataset. This can be an S3 location, a JDBC database, or a Glue table.

- **Projects**: A project is a visualization environment that loads a sample of a dataset and allows you to apply transformations and see their results live. Once the user is happy with the results of the transformations, they can be written onto a recipe.

- **Recipes**: A recipe defines a set of transformations to be applied to a particular dataset.

- **Jobs**: DataBrew jobs apply recipes to a given dataset in an automated fashion. Jobs can be scheduled or automated in a way similar to that of Glue ETL jobs.

DataBrew also provides data discovery and analysis features that let users get additional insights into their datasets, as outlined here:

- Profile jobs collect statistics and summaries on a dataset, such as the distribution of unique values, or the number of `null` values in a column. These can be run periodically on a dataset, like regular jobs.

- Data quality rules are validation checks that can be attached to a profile job. These include factors such as duplicated rows, missing values, or outliers.

DataBrew enables easy data management and discovery for Glue users, which in combination with the features and utilities described in previous sections result in powerful metadata management.

Summary

In this chapter, we discussed all aspects of metadata management, such as Glue Data Catalog and how it stores metadata. We went over different methods of populating it both manually (such as with the AWS CLI or running DDL statements) and automatically (through crawlers and their schema discovery features). We also discussed metadata maintenance and how it can become an issue for large organizations. We went over different options to not just keep metadata up to date but also automate the process and decouple it from the logic of your ETL processes.

We talked about metadata versioning and how to roll back versions causing issues. We also discussed how Lake Formation can help with not just metadata rollbacks but also data ones, as well as the wide variety of features it offers. Finally, we talked about lineage and how Glue DataBrew can help you discover, analyze, and transform your datasets in a visual way.

With these concepts, you should be able to fully manage the metadata of your data lakes. However, as important as metadata is, a very crucial aspect of maintaining a data lake is keeping it secure. In recent years, many countries and organizations have passed laws mandating companies to be responsible for the data they gather and store. Because of this, the security of a data lake is a very important aspect to manage for any large enterprise. The following chapter will go over data security and all the options Glue offers to tackle it.

8

Data Security

At AWS, we like to say that security is "job zero," in that security is more important than even priority tasks. Glue has been built from the ground up with that tenet in mind, and that, together with all the security features of AWS services, makes data security an easy – but powerful – area to cover.

The Glue security model relies and builds upon concepts common to all AWS services, such as IAM roles, policies, and S3 encryption. Throughout this chapter, we'll cover different approaches and configurations to ensure the security of your data lake and data pipelines. This will include dealing with concepts such as encryption (both in transit and at rest), logging, and retention.

In this chapter, we will cover the following topics:

- Access control
- Encryption
- Network

Technical requirements

The code for this chapter can be found in this book's GitHub repository at `https://github.com/PacktPublishing/Serverless-ETL-and-Analytics-with-AWS-Glue`.

Access control

A large part of security is determining who can access data and in which ways. In this section, we will cover how to configure access control for all the components of a Glue data lake.

IAM permissions

Much like other AWS services, AWS Glue relies on IAM (`https://aws.amazon.com/iam/`) to provide access control for the service itself, meaning users need to be granted access for IAM to Glue operations to manage and retrieve elements of the data lake.

All IAM permissions depend on the resource's specifications, which in AWS are uniquely identified through an **Amazon Resource Name (ARN)**. Within Glue, only certain types of resources get ARN identifiers. Other resources, such as workflows, for instance, do not support the use of ARNs, so permissions cannot be granted on a resource-specific basis. For a complete list of resource ARNs, please refer to the following AWS documentation page: `https://docs.aws.amazon.com/glue/latest/dg/glue-specifying-resource-arns.html`.

For Data Catalog resources, all permissions that have been granted to objects that depend on parents also need permission to access the parents. For instance, granting `john` access to `glue:GetTable` on the `sales` table will also require giving `john` access to the database and Data Catalog that holds the table. Additionally, all delete operations require the opposite: the user must also have permission to access all child objects. For instance, if `john` wants to delete the `sales` table, they will also need permission to delete all table versions and partitions present in the table.

Glue dependencies on other AWS services

AWS Glue relies on capabilities provided by other services, such as VPC networking or CloudWatch for logging. When using the AWS Web UI to configure Glue resources, it will list and filter results, which means access will also have to be granted to them to fully manage a data lake. This includes the following:

- **IAM itself** to list and assign IAM roles to Glue resources
- **CloudWatch logs** to list and read the execution logs of Glue resources
- **VPC** to list and assign network resources such as VPCs, subnets, and security groups to Glue resources
- **S3** to list, read, and write buckets and objects
- **Redshift** to list and access clusters
- **RDS** to list and access databases

Without access to these permissions, the Web UI will often display error messages and incomplete results.

Resource-based versus identity-based policies

Within the AWS permissions model, IAM permissions policies can be attached to either a resource (an AWS component, such as an S3 bucket) or an identity (such as a user). With resource-based policies, the resource defines who can access or control it. Identity-based policies work the other way round: access to resources is defined by the permissions attached to a user or role.

Resource-based policies allow you to compact all access rules down to a single document, whereas identity-based policies offer more flexibility and allow for individual user management. Typically, and unless you are managing a small set of resources and entities, identity-based policies are preferred since it's easier to associate each user or IAM role with its permissions, rather than having to modify the permissions of all resources it has to access.

In the case of Glue, only the Data Catalog accepts policies – it is not possible to attach policies to Glue databases, tables, crawlers, or jobs. Let's say you wanted to grant john access to the payments table in the sales database. You could achieve this with either a resource-based policy attached to your catalog, or an identity-based policy attached to john. Let's compare how both are used in their JSON form:

- **Resource-based policy**: The following example showcases a JSON-formatted policy attached to a Glue Data Catalog. The policy grants john access to the glue:GetTable operation, but only against the payments table within the sales database:

```
{
    "Version": "2012-10-17",
    "Statement": [
      {
        "Effect": "Allow",
        "Action": [
          "glue:GetTable"
        ],
        "Principal": {"AWS": [
          "arn:aws:iam::account-id:user/john"
        ]},
        "Resource": [
          "arn:aws:glue:us-east-1:account-id:table/sales/
payments"
        ]
      }
    ]
}
```

- **Identity-based policy**: The following example showcases granting the same permissions but by attaching them to john rather than the Data Catalog itself:

```
{
    "Version": "2012-10-17",
    "Statement": [
```

```
        {
          "Sid": "AccessPayments",
          "Effect": "Allow",
          "Action": [
            "glue:GetTable"
          ],
          "Resource": "arn:aws:glue:us-east-1:account-
    id:table/sales/payments"
        }
      ]
    }
```

Managing access through a policy attached to your AWS account's Data Catalog comes with two main limitations:

- Only one policy can be attached to the Catalog.
- This policy is limited to 10 KB.

These limitations reinforce the fact that using resource-based policies is not recommended for large accounts or organizations, as the policy will be limited in size. There are additional limitations in the clauses that can be specified in the policy, which you can find in the AWS documentation at `https://docs.aws.amazon.com/glue/latest/dg/glue-resource-policies.html#overview-resource-policies`.

Cross-account access

A very common strategy in large multi-account AWS organizations is to centralize all table definitions into a single Data Catalog, then use other secondary accounts to process the data in them. Much like with other AWS services, cross-account access is possible and can be configured through IAM permissions, both with resource-based and identity-based policies.

Now, let's assume that the Data Catalog holding the `sales` database and the `payments` table is stored in one AWS account (account A) and that `john` is located in another (account B). The following resource-based policy will have to be attached to the Data Catalog in account A:

```
{
  "Version": "2012-10-17",
  "Statement": [
    {
      "Effect": "Allow",
      "Action": [
```

```
        "glue:GetTable"
      ],
      "Principal": {"AWS": [
        "arn:aws:iam::account-B:user/John"
      ]},
      "Resource": [
        "arn:aws:glue:us-east-1:account-A:catalog",
        "arn:aws:glue:us-east-1:account-A:table/sales/payments"
      ]
    }
  ]
}
```

On top of that, the administrator of account B will have to grant john permission to run glue:GetTable on account A, as follows:

```
{
  "Version": "2012-10-17",
  "Statement": [
    {
      "Effect": "Allow",
      "Action": [
        "glue:GetTable"
      ],
      "Resource": [
        "arn:aws:glue:us-east-1:account-A:catalog",
        "arn:aws:glue:us-east-1:account-A:table/sales/payments"
      ]
    }
  ]
}
```

For identity-based policies, the best way to achieve this is through IAM role assumption (https://docs.aws.amazon.com/STS/latest/APIReference/API_AssumeRole.html). This mechanism allows a user or IAM role to assume the credentials and permissions of another IAM role. Cross-account access is granted by the owner of account A by creating an IAM role and modifying its trust policy to be allowed by john in account B. The owner of account B will then have to give john permission to assume the role in account A, after which john should have access to the table.

Accessing cross-account Data Catalog resources is always done by giving a value to the `CatalogId` parameter, whether it is an API call, an AWS CLI command, or code in a Glue ETL job. Keep in mind that the ID of the Data Catalog is the same as the ID of the AWS account holding it.

Unlike with S3 objects, cross-account tables and databases must be owned by the account hosting the Data Catalog rather than whoever created them. In the example given earlier, the AWS account holding the `sales` database will be the owner of any tables or databases created by `john` and will have immediate access to them.

Note that cross-account access has certain limitations, the most notable of which is the inability to use Glue crawlers with cross-account setups. For a complete list of the limitations, check out the AWS documentation at `https://docs.aws.amazon.com/glue/latest/dg/cross-account-access.html#cross-account-limitations`.

Tag-based access control

IAM policies support the use of conditionals to determine which resources are affected by the permissions rule. A very common practice with AWS resources is to attach tags to them and make use of those tags to determine access and permissions. For instance, given an organization with two teams (sales and marketing), each team could tag their resources with a tag that specifies their team's name and, through that, restrict access to only themselves. Tags can also have other management purposes, such as separating billing into groups or for automated resource management.

Tags are always expressed in the form of a key/value pair. AWS Glue supports the use of tags for some of its resources, including the following:

- Connections
- Crawlers
- ETL jobs
- Development endpoints
- ML transformations
- Triggers
- Workflows

Tags can be added to any of these resource types at creation time, but they can also be added or removed for as long as the resource exists. The following is an example of an IAM policy that allows access to an ETL job's definition based on a tag with a `"team"` key and a `"marketing"` value:

```
{
    "Effect": "Allow",
    "Action": [
```

```
    "glue:GetJob"
    ],
  "Resource": "*",
  "Condition": {
    "ForAnyValue:StringEquals": {
      "aws:ResourceTag/team": "marketing"
    }
  }
}
}
```

This covers the Glue side of permissions management. In the next section, we'll discuss managing permissions in terms of S3.

S3 bucket policies

The previous sections described how to grant access to Glue resources. However, you will also need to restrict access to the data in your data lake. The process for this will vary, depending on where the data is stored. **Java Database Connectivity (JDBC)** databases can restrict access through user credentials and database permissions while DynamoDB tables can use IAM policies. In terms of S3 buckets, an effective way of restricting access is by using an S3 bucket policy.

S3 bucket policies are a form of resource-based access control where an IAM policy is attached to a bucket. This policy then determines what actions can be performed on objects within the bucket, and who can perform them. Only the bucket owner can attach a policy to the bucket, and the policy will only apply to objects owned by the bucket owner – not third accounts. For instance, the following is a bucket policy that's been designed to give read access to a third AWS account:

```
{
    "Version": "2012-10-17",
    "Statement": [
        {
            "Sid": "AddCannedAcl",
            "Effect": "Allow",
            "Principal": {
                "AWS": [
                    "arn:aws:iam::111122223333:root",
                    "arn:aws:iam::444455556666:root"
                ]
            },
            "Action": [
```

```
                    "s3:PutObject",
                    "s3:PutObjectAcl"
            ],
            "Resource": "arn:aws:s3:::DOC-EXAMPLE-BUCKET/*"
        }
    ]
}
```

In the context of Glue, using S3 bucket policies means specific access rules will need to be granted – not to the users using Glue, but to the IAM roles attached to crawlers, ETL jobs, and development endpoints.

S3 object ownership

When writing results with a Glue ETL job or development endpoint, the resulting objects will be owned by the account that owns the IAM role attached to said job or endpoint. In most scenarios, where your job or endpoint is writing to a bucket you own, this is meaningless as access will always be guaranteed. However, read access problems can arise when the destination S3 bucket is owned by a different AWS account.

When writing cross-account access, the objects will be in a bucket owned by a different account. However, each will have the writer account as its owner – resulting in access errors when they are read afterward. The best way to avoid this is by tackling the issue from both ends, as follows:

1. Set the right object owner when writing.

    Your ETL job or development endpoint can be configured to write objects that are owned by the same owner as the bucket containing them, avoiding the problem. To do so, a special configuration property must be passed onto the Hadoop configuration object, like so:

    ```
    sc = SparkContext()
    glueContext = GlueContext(sc)
    spark = glueContext.spark_session
    job = Job(glueContext)
    job.init(args['JOB_NAME'], args)

    glueContext._jsc.hadoopConfiguration().set("fs.s3.canned.
    acl", "BucketOwnerFullControl")
    ```

    Any subsequent write operations after this configuration change will address the issue.

2. Forbid non-bucket-owned writes.

You can also configure the S3 buckets in your data lake to reject any writes that don't set the object owner properly. This will not modify the owner of already-existing objects, but it will cause any future incorrect writes to fail, forcing the writer to set permissions properly and avoid situations where data must be rewritten or reassigned to a different owner.

Such a configuration can be achieved by configuring the S3 bucket policy of your buckets. The following example shows how this can be done:

```
{
    "Version":"2012-10-17",
    "Statement":[
      {
        "Sid":"OnlyAllowBucketOwnerFullControl",
        "Effect":"Allow",
        "Principal":{"AWS":"1234567890"},
        "Action":"s3:PutObject",
        "Resource":"arn:aws:s3:::my-bucket/*",
        "Condition": {
          "StringEquals": {"s3:x-amz-acl":"bucket-owner-
full-control"}
        }
      }
    ]
}
```

With that, we have discussed permissions management from both the Glue and S3 perspectives. However, permissions can only be granted to whole tables without any other granularity. While this works, recent legal requirements that have been imposed by regulations around the world have caused use cases where users only have access to parts of a table valid. In the next section, we will discuss Lake Formation, an AWS service that provides such capabilities.

Lake Formation permissions

AWS Lake Formation is a service that provides data lake capabilities on AWS resources. Even though it is separate from AWS Glue and can be used independently, Lake Formation and Glue share the same Data Catalog and are designed to work together from the ground up.

Lake Formation provides a wide array of features to support and manage data lakes. However, in this chapter, we are going to focus on permissions. Lake Formation permissions are an additional layer on top of IAM permissions that can be used to control access to both data and metadata. Lake Formation also provides fine-grained access control to not just tables, but also the rows and columns within those tables. This is particularly powerful for any company or organization going through compliance regulations.

When using Lake Formation, the data lake administrator decides which S3 locations and Data Catalog databases/tables are part of the data lake. For any request to any resource that is part of the data lake, the necessary permissions will have to be validated against both IAM and Lake Formation – otherwise, the request will fail.

The Lake Formation permissions management system is very similar to that of relational databases, where permissions are granted or removed using the GRANT or REVOKE statements, respectively. Now, let's discuss the different capabilities of Lake Formation and the permissions at different levels.

Data Catalog permissions

Data Catalog permissions refer to the ability to manage, create, and delete resources within the Data Catalog. These can be granted to either databases or tables, with the option of adding row/column granularity when granting access to a table. Permissions can either be granted to IAM principals in your AWS account or principals in other accounts, giving them access to your databases, tables, and their underlying data.

Permission management is done through the GrantPermissions and RevokePermissions API calls, which take in the following parameters:

- **Principal**: The IAM principal that the operation involves. This can be an IAM user, an IAM role, or an AWS organization.

- **Resource**: The Data Catalog resource (database, table, or table subset) that the operation grants/removes access to/from.

- **Permissions**: The list of operations being granted or revoked access. Lake Formation supports the following operations:

  - SELECT

  - ALTER

  - DROP

  - DELETE

  - INSERT

  - DESCRIBE

  - CREATE_DATABASE

  - CREATE_TABLE

  - DATA_LOCATION_ACCESS

  - CREATE_TAG

  - ALTER_TAG

- DELETE_TAG
- DESCRIBE_TAG
- ASSOCIATE_TAG

For instance, the following AWS CLI command grants SELECT permissions to john on the sales table in the payments database:

```
aws lakeformation grant-permissions --principal
DataLakePrincipalIdentifier=arn:aws:iam::1234567890:user/john
--resource '{
  "Table": {
    "CatalogId": "1234567890",
    "DatabaseName": "payments",
    "Name": "sales"
  }
}' --permissions DESCRIBE
```

In the next section, we'll discuss how to manage permissions for large groups of entities, typically found in large organizations.

Tag-based access control

Granting permissions to individual entities can quickly become tedious or repetitive to manage in organizations with large amounts of users and resources. This is a similar problem that happens when dealing with IAM permissions on large AWS accounts, and the typical recommendation is to group resources through tagging and then use those tags to determine access permissions.

Lake Formation offers a very similar approach with tag-based access control (or LF-TBAC). This feature allows you to manage permissions on a larger scale by granting permissions to tags and then attaching those tags to all the resources that fall under the same permissions model. For instance, if the sales department within your company has upwards of 1,000 tables, giving john the right access to all of them can become problematic and also consume a very large amount of API calls. With LF-TBAC, all these tables can be tagged under the department: sales key/value pair, and then john can be granted access to the tag. All tables with the tag will immediately inherit the permissions of the associated tag, reducing the amount of management overhead.

Keep in mind that Lake Formation tags are different than regular AWS resource tags. Lake Formation tags only exist within the domain of Lake Formation and only serve the purpose of managing Lake Formation permissions. Resources can still be attached regularly to AWS resource tags and their IAM access can be managed through those tags, regardless of their Lake Formation tags.

Granting permissions based on Lake Formation tags is similar to basing them on regular Lake Formation resources. For both the `GrantPermissions` and `RevokePermissions` API calls, the only difference is to specify a Lake Formation tag instead of a Lake Formation resource. For instance, to grant `john` select access on all tables with the `department: sales` tag, the following AWS CLI command can be executed:

```
aws lakeformation grant-permissions --principal
DataLakePrincipalIdentifier=arn:aws:iam::1234567890:user/john
--resource '{
    "LFTagPolicy": {
        "CatalogId":"1234567890",
        "ResourceType":"TABLE",
        "Expression": [{"TagKey": "department","TagValues":
["sales]}]'
--permissions SELECT
```

With this, we've covered all there is to Data Catalog permissions. The next section will go over data permissions.

Data – S3 permissions set

Lake Formation also requires data lake administrators to set permissions for their data locations in S3. A user with permissions for a data location will not just be able to read data from that location, but also create databases and tables that point to it. Therefore, unless a user has a very particular use case where only metadata access is needed, most users will need data access on top of metadata access – if it's not for data reading, it's at least to be able to create and define tables.

Granting and revoking permissions to/from an S3 data location is no different than doing so to/from a Data Catalog resource, with the only difference being the resource parameter will have to be a data location rather than a catalog resource. The following AWS CLI command shows how to grant `john` access to an S3 location defined by its ARN resource:

```
aws lakeformation grant-permissions --principal
DataLakePrincipalIdentifier=arn:aws:iam::1234567890:user/john
--resource '{
    "DataLocation": {
        "CatalogId":"1234567890",
        "ResourceArn":"arn:aws:s3:::bucket_name/key_name"'
--permissions DATA_LOCATION_ACCESS
```

Notice how the permission being granted here is `DATA_LOCATION_ACCESS` rather than the usual `SELECT` or `DESCRIBE`. This is a static value that must always be used with data location permissions.

Data location permissions can also be granted to a different account. The following code shows an example of the 1 2 3 4 5 6 7 8 9 0 account granting access to the 0987654321 account:

```
aws lakeformation grant-permissions
--principal DataLakePrincipalIdentifier=0987654321
--permissions "DATA_LOCATION_ACCESS"
--resource '{
    "DataLocation":{
"CatalogId":"1234567890",
"ResourceArn":"arn:aws:s3:::bucket_name/key_name "
}}
```

When granting cross-account access, the receiving account can also be permitted to grant access to others by itself. This can be done through the permissions-with-grant-option parameter of the API call, as shown here:

```
aws lakeformation grant-permissions
--principal DataLakePrincipalIdentifier=0987654321
--permissions-with-grant-option "DATA_LOCATION_ACCESS"
--permissions "DATA_LOCATION_ACCESS"
--resource '{
    "DataLocation":{
"CatalogId":"1234567890",
"ResourceArn":"arn:aws:s3:::bucket_name/key_name "
}}
```

This concludes all Lake Formation features related to data security. In the next section, we'll talk about the different aspects of encryption, and how they can be configured in Glue.

Encryption

Encryption is the basis of all data security policies, as it ensures critical data cannot fall into the hands of potential attackers. In recent years, encryption has also taken increased importance because of compliance and personal data protection regulations. AWS Glue offers several features to support encrypting your data both at rest and in transit. This section will cover all encryption options and features while providing examples and best practices.

Encryption at rest

When it comes to encryption at rest in Glue, it can happen at three different levels:

- Encrypting the metadata that defines your data lake, which is handled by Glue itself

- Encrypting the data auxiliary to executing Glue resources

- Encrypting the data within your data lake

In this section, we will go through each level. For encryption, Glue relies on **AWS Key Management Service (KMS)**, an AWS service that provides serverless hosting and management of encryption keys. All encryption features support the use of KMS keys. However, Glue only supports symmetric ones – keys that are used to both encrypt and decrypt data. When specifying KMS keys for any of the encryption features, make sure you enter the ARN of a symmetric key as Glue will not validate whether it is symmetric or not before attempting to encrypt or decrypt, resulting in potential failures down the line.

Metadata encryption

Glue is capable of encrypting all metadata in your Data Catalog using a KMS key. This covers the following catalog objects:

- Databases

- Tables

- Table versions

- Partitions

- Connections

- User-defined functions

Metadata encryption works as a toggle (either it is enabled or not). Despite that, encryption only takes effect for objects created *after* it has been enabled and doesn't happen retroactively. Let's say the following happens:

1. john creates a Glue table (table A) in the Data Catalog.
2. The AWS account administrator enables Glue metadata encryption.
3. john creates another Glue table (table B) in the Data Catalog.
4. The AWS account administrator disables Glue metadata encryption.

In this scenario, table A would not be encrypted (even after the administrator has enabled encryption) and table B would be encrypted (even after the administrator has disabled encryption).

Additionally, Glue can encrypt passwords that have been used for Glue connections using a KMS key. This can be different from the one used for the Data Catalog. This will ensure connection passwords are encrypted when stored in AWS, and that any entity requesting them must also have IAM permissions to run kms:Decrypt on the KMS key used to encrypt the data.

Data Catalog and connection password encryption can be enabled using either the AWS Web Console or the SDK/CLI through the PutDataCatalogEncryptionSettings API call. This call takes parameters in the following structure:

```
{
  "EncryptionAtRest": {
    "CatalogEncryptionMode": "DISABLED"|"SSE-KMS",
    "SseAwsKmsKeyId": "string"
  },
  "ConnectionPasswordEncryption": {
    "ReturnConnectionPasswordEncrypted": true|false,
    "AwsKmsKeyId": "string"
  }
}
```

If no KMS key is specified for either of the encryption options, Glue will use the service's default encryption key (aws/glue). To access any encrypted objects, the requesting entity (whether it is an IAM user or an IAM role) will need to have IAM permissions to use the kms:Decrypt, kms:Encrypt, and kms:GenerateDataKey API calls, allowing access to the KMS key that was used for encryption.

If a non-default key was configured to encrypt the Data Catalog and it is deleted from the AWS account, *all objects encrypted by it will become non-decryptable permanently*. Always make sure to manage KMS keys properly.

Auxiliary data encryption

When running Glue resources such as crawlers or ETL jobs, data is generated in the form of execution logs and job bookmarks. Even though this data may seem harmless at first, more often than not, it will contain critical information such as table column names or data samples, which can represent data leaks. Glue also supports encrypting these data sources so that your data lake is properly secured and fully compliant with regulations.

Encrypting Glue resources is always handled through Glue security configurations. A security configuration is a set of defined encryption rules that can be attached to a Glue crawler, a Glue ETL job, or a Glue development endpoint, determining how logs and bookmarks are encrypted for them.

Security configurations can be created through the `CreateSecurityConfiguration` API call, which takes parameters in the following structure:

```
{
  "S3Encryption": [
    {
      "S3EncryptionMode": "DISABLED"|"SSE-KMS"|"SSE-S3",
      "KmsKeyArn": "string"
    }
    ...
  ],
  "CloudWatchEncryption": {
    "CloudWatchEncryptionMode": "DISABLED"|"SSE-KMS",
    "KmsKeyArn": "string"
  },
  "JobBookmarksEncryption": {
    "JobBookmarksEncryptionMode": "DISABLED"|"CSE-KMS",
    "KmsKeyArn": "string"
  }
}
```

For the specified KMS keys to be used, the account administrator must grant AWS KMS IAM permissions to the roles used for Glue resources. This process is described in the KMS documentation at `https://docs.aws.amazon.com/AmazonCloudWatch/latest/logs/encrypt-log-data-kms.html`.

Data encryption

The process of encrypting the data that resides in your data lake will be a task shared between all silos or services involved: your RDS-backed tables will have to use RDS encryption (`https://docs.aws.amazon.com/AmazonRDS/latest/UserGuide/Overview.Encryption.html`). DynamoDB offers similar encryption-at-rest capabilities (`https://docs.aws.amazon.com/amazondynamodb/latest/developerguide/EncryptionAtRest.html`) and any S3 bucket can benefit from S3 encryption (`https://docs.aws.amazon.com/AmazonS3/latest/userguide/bucket-encryption.html`).

That said, Glue offers some additional features when writing data as part of the output of an ETL job. ETL jobs can be configured to write either S3-encrypted or KMS-encrypted output when the target is an S3 location, ensuring all the results of your jobs are protected, regardless of the configuration present at the storage layer.

ETL job data encryption can be enabled in two ways, depending on the type of encryption. Let's look at these two encryption types, as follows:

- S3-based encryption (SSE-S3) is configured by passing a property to the ETL job definition, either at the time of creation (CreateJob) or when editing it (UpdateJob). This property is defined inside the DefaultArguments property of the job:

```
"DefaultArguments": {
    "—TempDir": "s3://path/ ",
    "—encryption-type": "sse-s3",
    "—job-bookmark-option": "job-bookmark-disable",
    "—job-language": "python"
}
```

- KMS-based encryption (SSE-KMS) is configured by creating a security configuration and attaching it to the ETL job, similar to how log and bookmark encryption work.

  If both options are configured simultaneously, KMS encryption will be used over S3. For security configurations to take effect within an ETL job, the Job.init() statement must be executed within the job's code:

```
job = Job(glueContext)
job.init(args['JOB_NAME'], args)
```

This covers all the features and aspects of at-rest encryption. In the next section, we'll discuss encryption in transit.

Encryption in transit

Glue relies on **Secure Sockets Layer** (**SSL**) encryption for encryption in transit, which means connections to other AWS services (such as when reading or writing to S3 or DynamoDB) are made securely and are encrypted. For non-AWS connections (such as when connecting to a JDBC database), Glue supports enforcing SSL connections, which will cause the crawler or ETL job trying to use the connection to fail if connecting over SSL doesn't work.

Enforcing an SSL connection also allows you to configure the usage of custom SSL certificates to authenticate the connection, which allows users to connect securely to JDBC databases using a proprietary certificate that hasn't been publicly validated. The connection can also be configured to pass values to the SSL_SERVER_CERT_DN (for Oracle databases) or hostNameInCertificate (for SQL Server databases) parameters of the target database, which allows you to configure custom distinguished names and domain names for the database server, respectively.

FIPS encryption

AWS offers service endpoints that use cryptographic modules compliant with **Federal Information Processing Standards (FIPS)** rather than standard SSL for communication. If the purposes of your Glue usage must meet such a standard, Glue offers FIPS-compliant endpoints for all North American regions, including GovCloud ones.

Development endpoint connections

Glue offers development endpoints, (`https://docs.aws.amazon.com/glue/latest/dg/dev-endpoint.html`), which can be used to create a static development environment in the cloud that users can log into and use to develop and test scripts for ETL jobs. Development endpoints can be accessed via SSH and do not support authentication through a user/password combination – only SSH keys are supported. The use of SSH for communication also ensures all traffic between your local computer and the development endpoint is encrypted.

When creating a development endpoint, you must provide one or more public keys. These will be used to authenticate users logging in. If the development endpoint is going to be shared among several users, it is within best practices to create individual key pairs for each one and pass all public keys to the development endpoint, thus avoiding having to share SSH keys between users.

Once the endpoint is up and running, the `UpdateDevEndpoint` API call allows you to add new keys and delete unused ones. Reviewing and rotating SSH keys is a good practice that will prevent unwanted access to the development endpoint.

With this, we've covered all aspects of encryption in AWS Glue. In the next section, we'll discuss network security, which handles the security of all communications happening between Glue resources and external ones.

Network

Even though AWS Glue is a serverless service, understanding its network infrastructure and how it connects to resources is a critical part of guaranteeing your data's security and your organization's compliance. By default, Glue will always attempt to use the less public route to direct network traffic. However, it is crucial to understand how this routing works to avoid public calls that could compromise your information.

Glue network architecture

Much like with other AWS services, all AWS Glue resources are stored and executed in internal AWS accounts that are not accessible or part of any public infrastructure. This includes your Data Catalog, crawlers, ETL jobs, development endpoints, triggers, and workflows. This is shown in the following diagram:

Figure 8.1 – AWS resources within the AWS cloud

If a Glue resource needs access to an S3 location, this communication happens privately and internally through the AWS infrastructure, as shown in the following diagram:

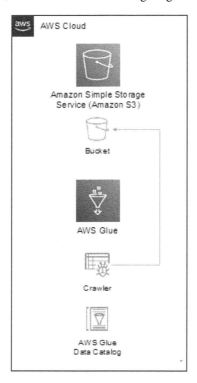

Figure 8.2 – AWS resources communicating through the AWS cloud

However, connecting to any other resource will require Glue to set up a bridge between its internal infrastructure and your AWS VPC. To make this happen, whenever a Glue resource is in execution, Glue will create **Elastic Network Interfaces (ENIs)** in your VPC and attach them to the nodes running your Glue resource – a process known as **requester-managed network interfaces** (https://docs. aws.amazon.com/AWSEC2/latest/UserGuide/requester-managed-eni.html).

Let's say you want to run a crawler to automatically detect the schema of your MySQL database, which is running in an EC2 instance in your account. Since you followed security best practices, this EC2 instance is running in a private subnet within your VPC, which means it is not accessible over the public internet. When you run your Glue crawler, Glue will create ENIs in your VPC, assign private IP addresses to them, and attach them to the nodes that execute the crawler process in the internal AWS infrastructure. Once the crawler finishes running, the ENIs will be detached and deleted, and their IP addresses will be released. The following diagram shows how this works:

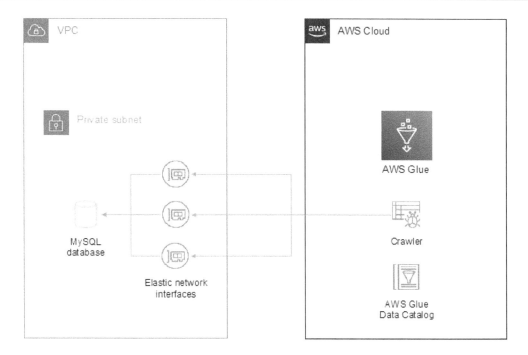

Figure 8.3 – AWS resources communicating with resources in a user's VPC

This process allows Glue to securely and privately connect to other resources in your AWS account without ever leaving the internal AWS infrastructure. To carry it out, though, Glue will need several parameters, such as the location of your database (VPC and subnet), proper security clearance to access it (security groups), and a way to authenticate within your database. All these parameters are supplied as part of a Glue connection.

Glue connections

A Glue connection is a set of configuration parameters that define the location and way of accessing an external resource so that Glue can automate its access. These parameters include the following:

- VPC and subnet combination.
- One or more security groups.
- If authentication is required, all the necessary parameters for it. These include the following:
  - The JDBC URL, which can include parameters to be passed onto the database
  - Username and password combination

When creating a Glue connection, you must also specify its type. There are as many types as there are supported connection targets:

- **JBDC**: Connects to a relational database that supports JBDC

- **MongoDB**: Connects to a MongoDB or DocumentDB cluster

- **Kafka**: Connects to an Apache Kafka cluster

In addition to the previous types, there are also three special connection types:

- **Network**: This connection will simply specify a VPC and subnet without any other parameters. This is designed to route connections through a VPC rather than connecting to a specific resource within it.

- **Marketplace**: This connection specifies parameters for connectors that have been obtained through the AWS marketplace.

- **Custom**: This connection specifies parameters for custom connectors that have been created by you.

Attaching a connection to a Glue resource will cause the resource to automatically infer its properties when it is being executed. For instance, a crawler will automatically know which subnet and database to connect to and will have the right security groups to do so.

Network configuration requirements and limitations

For connections to work properly, certain requirements must be met. Let's look at a few of these requirements, as follows:

- At least one of the security groups that's attached to the connection must include a self-referencing inbound rule that allows all traffic. Even though such a wide permission may seem like a security issue, permission will only be granted to incoming – not outgoing – connections, and will only take effect between resources that have the security group attached. This rule is necessary to allow proper communication between all Glue resources.

- When creating ENIs to attach to Glue resources, only private IP addresses will be granted to them to guarantee their security. If your resources need to connect to endpoints over the public internet, the lack of public IP addresses will make it impossible.

- When creating a development endpoint, any attached security groups will need to include access to TCP port 22 to allow for SSH logins – otherwise, the endpoint will be inaccessible.

- Connections to databases will require the involved security groups to allow the necessary traffic. For instance, if you're connecting to a MySQL database, you will need to allow traffic on TCP port 3306.

In the next section, we'll discuss the requirements and considerations to connect to resources on the public internet.

Connecting to resources on the public internet

As mentioned in the previous section, Glue resources will only get private IP addresses, which makes them unable to communicate with a resource on the public internet. Although this can be a benefit in terms of security, there are situations in which you might be interested in connecting over the public internet, such as trying to reach a resource in your on-premise network or reading data from a publicly-accessible API. There are two ways to make this possible:

- **VPC endpoints**: If this public communication is necessary for reaching an AWS service (for instance, making an API call to AWS Secrets Manager in your ETL job code to retrieve credentials), you can use VPC endpoints (`https://docs.aws.amazon.com/vpc/latest/privatelink/vpc-endpoints.html`) to route it through AWS infrastructure instead of the public internet. This is shown in the following diagram:

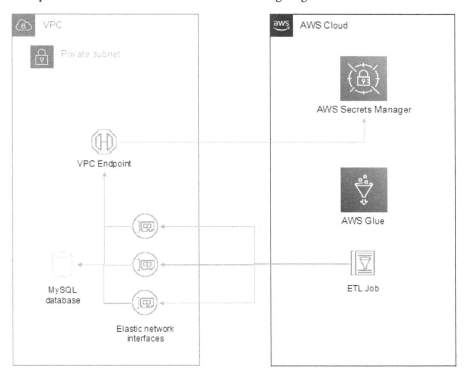

Figure 8.4 – Connecting over a VPC endpoint

Deploying a VPC endpoint in a subnet and updating its associated route table will direct traffic internally, allowing you to communicate with AWS services securely.

- **NAT Gateways**: If, instead of an AWS service, you are trying to reach a public resource over the internet, the only solution is to grant your resources a public IP address to communicate. VPC NAT Gateways (https://docs.aws.amazon.com/vpc/latest/userguide/vpc-nat-gateway.html) are NAT translation resources offered by AWS VPC that multiplex private IP addresses behind a public one, assigned to the gateway itself. The following diagram shows how such a connection happens:

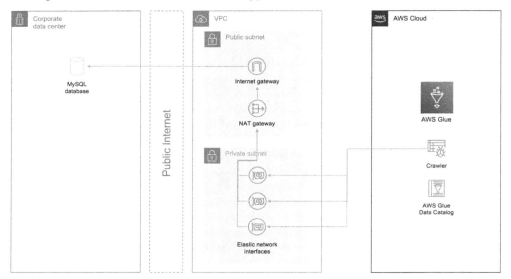

Figure 8.5 – Connecting through a NAT Gateway

When using NAT Gateways, resources can initiate connections to the public internet. However, the opposite can't happen, which means they are still protected in your private subnet.

This covers all the ways of connecting to resources through the public internet. Next, we'll discuss other ways to connect to resources offered by AWS VPC.

VPC peering

VPC peering (https://docs.aws.amazon.com/vpc/latest/peering/what-is-vpc-peering.html) is a VPC feature that allows traffic between two VPCs by simply adding routes between them as if they were part of the same network. This feature allows you to solve a variety of challenges that can affect your Glue connectivity.

Managing IP address pools

As mentioned previously, each ENI that's created by Glue will be assigned an IP address from the subnet it resides in. The amount of IP addresses is directly proportional to the number of nodes that are part of the resource – for instance, in the case of ETL jobs, it will depend on the number of DPUs or workers you assigned to the job.

There are certain situations in which there may not be enough IP addresses for the resource to run properly. For instance, your VPC subnet could have a small range, or it could already have a large number of resources running within it. Alternatively, you may want to run a large ETL job that requires a significant number of addresses.

VPC peering allows you to solve this challenge by creating a new VPC subnet dedicated to Glue resources receiving IP addresses. Since communication between the VPCs is as if they were part of the same, Glue will be able to work without issues and the original VPC or its resources won't have to be modified.

Connecting to cross-account resources

VPC peering also allows you to create a peering relationship between VPCs in different accounts. This allows for easy cross-account, private connections where a Glue resource can connect to a database owned by a different account in the same organization, for instance.

Connecting to cross-region resources

VPC peering can also bridge two VPCs placed in different regions, allowing for private connections through the AWS infrastructure and avoiding complicated setups to connect to resources in other regions.

AWS PrivateLink

AWS PrivateLink (`https://docs.aws.amazon.com/whitepapers/latest/aws-vpc-connectivity-options/aws-privatelink.html`) is a VPC service that allows you to publish an endpoint into a VPC, ensuring that traffic between clients on the VPC and the endpoint is always routed through the internal AWS infrastructure and never goes through the public internet. PrivateLink can be used in Glue setups to, for example, publish an endpoint to a JDBC database in the VPC where Glue resources run. PrivateLink endpoints can be published cross-account and cross-region, enabling solutions for complex setups.

Connecting to resources in your on-premise network

Glue is also capable of reaching resources in your local network, allowing, for instance, you to crawl your self-hosted JDBC databases. Just like with any other public resource, Glue can connect through the public internet via a public endpoint; however, this is not a good approach in terms of security. There are several AWS services and products that can help tackle this issue:

- AWS Direct Connect (`https://aws.amazon.com/directconnect/`) is an AWS service that can establish a direct link between your on-premise data center and the AWS infrastructure. This is a benefit not just in terms of security, but also that it can provide greatly increased speeds and lower latency, which makes it easier to transfer datasets, for instance.

- AWS Site-to-Site VPN (`https://docs.aws.amazon.com/vpn/latest/s2svpn/VPC_VPN.html`) allows you to create VPN connections between your on-premise network and your AWS VPC. This will still route traffic over the public internet, but it will be encrypted and protected as per the specifications of the VPN software of your choice.

- AWS Managed VPN (`https://docs.aws.amazon.com/whitepapers/latest/aws-vpc-connectivity-options/aws-managed-vpn.html`) is similar to Site-To-Site VPN in that it uses a VPN solution to encrypt traffic, but this software is managed and deployed by AWS. This may reduce or eliminate the technical overhead of managing such a solution.

- Finally, the AWS Snow Family (`https://aws.amazon.com/snow/`) is an alternative solution to establishing network links. These are hardware products that can be delivered to your premises and allow you to copy and deliver your datasets to AWS, who will then upload them to your account. This is a more effective solution if you are intending to upload your data to AWS and stop using your on-premises network.

This covers all the options and features for network security. Now, let's summarize this chapter.

Summary

In this chapter, we discussed all the aspects of security within AWS Glue. We talked about limiting access through IAM permissions on both Glue and S3 and how to extend this through different AWS accounts. We also talked about fine-grained access permissions through AWS Lake Formation.

We discussed how encryption works and how Glue relies on AWS KMS keys to encrypt and decrypt data. We also discussed all the entities within Glue that can be encrypted. We saw different options for auditing access to Glue resources.

Finally, we discussed how Glue works in terms of networking and discussed the different architectures and AWS services that can be used to access resources over networks, including best practices when it comes to connecting over the public internet.

This covers all aspects of security in terms of Glue within your AWS account. The next chapter will also be related to security and permissions to some degree, as it will talk about data sharing and best practices to let others access your Glue resources.

9
Data Sharing

When you build a cloud-native data platform at scale on AWS, you may want to share your data with multiple stakeholders under governance. Today, data sharing is one of the key topics in data democratization for making business decisions driven by data and driving business. Typically, the data platform is used by different users, such as data engineers, business analysts, and data scientists.

For example, data engineers own the data platform and maintain it, business analysts generate a daily report that represents business revenue and end user activities, and data scientists may want to unveil complex data patterns and build a data model for their applications. In such situations, these users can belong to different business units and organizations. For enterprise data platforms, democratizing and sharing data with different organizations under data governance securely is a high-demand requirement.

In this chapter, you will learn about three common data sharing strategies and characteristics, and how you can share your data on AWS using AWS Glue and AWS Lake Formation through a step-by-step tutorial with sample data. After completing this chapter, you will be able to design a data sharing model by choosing a strategy that fits your use case. You will also gain some hands-on skills to build a data sharing mechanism for your data platform.

In this chapter, we will cover the following topics:

- Overview of data sharing strategies
- Sharing data with multiple AWS accounts using S3 bucket policies and Glue catalog policies
- Sharing data with multiple AWS accounts using AWS Lake Formation permissions

Technical requirements

For this chapter, you need the following resources:

- An AWS account
- An AWS IAM role
- The AWS CLI

Overview of data sharing strategies

At the time of writing, depending on the organizations and use cases, there are different ways to share data. There are three typical strategies for sharing data:

- Single tenant
- Hub and spoke
- Data mesh

In this section, you will learn about each of these strategies and discuss their backgrounds, challenges, and benefits.

Single tenant

Data lakes have become a popular approach for people who want to store and query data in a centralized repository. It allows you to store all the structured data, semi-structured data, and unstructured data at any scale. Here, cloud storage such as Amazon S3 fits well with data lakes because there are no data size limits. You do not need to convert your data into a predefined fixed schema in advance. Instead, you can just ingest data as-is. When you want to analyze the data, you can easily convert the data into your preferred schema on the fly, then analyze it on top of the data lake.

The simplest use case is a single-tenant data platform. In this model, you will have all the components in a single AWS account that is, an Amazon S3 bucket for data lake storage, AWS Glue Data Catalog as a metadata store, Amazon Athena as the query engine, and more. To achieve data governance, you can just focus on IAM permissions; you do not need to think about ways to share data across multiple AWS accounts. This is simple and good for getting started, or for small use cases where you only have a few stakeholders in your organization.

However, in real-world use cases, you may have multiple AWS accounts. This is because AWS best practices recommend that you segregate your resources and workloads into multiple AWS accounts to isolate resources and ownership, categorize workloads, and reduce the blast radius when things go wrong. For such use cases, you will need to think about how organizations can collaborate through the data, and how you can share the data across different AWS accounts.

The following screenshot shows how the single-tenant model works with two consumer applications in the same account.

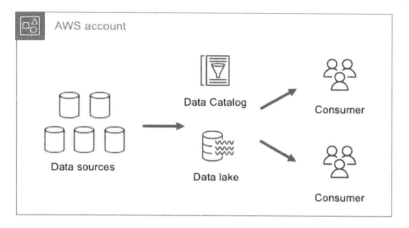

Figure 9.1 – Single-tenant model

In the single-tenant model, you must use one AWS account for all the components – that is, ingesting data from data sources, storing data in data lakes, cataloging data, and consuming/analyzing data.

Hub and spoke

The hub-and-spoke model was introduced to achieve a cross-organization data platform. In this model, a central "hub" account hosts all the data and metadata and shares it with multiple consumer accounts. Consumer accounts receive the shared data and metadata and run analytic workloads on their compute resources, such as Amazon Athena, Amazon Redshift, and so on. This centralized hub model is intended to simplify both data engineering operations and end user experiences. For data engineers who manage the data platform, the operational cost is not significant when they need to manage a single data platform as a "hub." For end users, all the data and metadata is stored in a single hub so that end users have good visibility of the data. It won't require deep technical expertise just to consume the data, so it can also reduce training costs.

Typically, there are different stakeholders in the data platform. There can be multiple data sources owned by different teams. There can also be different consumers who analyze the data and make decisions. The central data engineering team is responsible for managing the data lake in the following ways:

- Collecting data from the different data sources
- Enriching data to meet business requirements
- Ingesting the data into the data lakes
- Orchestrating components to extract, transform, and load data

- Maintaining the end-to-end data flow

- Ensuring that the data platform meets business SLAs, such as data freshness, data accuracy, cost, and so on

However, the central team has the problem of managing data through this kind of central data platform. Since the data pipeline is owned by the central team but data sources are owned by other teams, it is hard for the central team to understand the specific needs of a data domain. This can cause issues in terms of ownership and accountability. In addition, there is the challenge of scaling. For example, you may need to transport data into the hub account, even though it is already on cloud storage. Furthermore, you may also require intervention from the central data engineering team when you want to add more datasets or change the way you enrich and validate the data.

The following diagram shows how the hub-and-spoke model works with a single hub account and two consumer accounts:

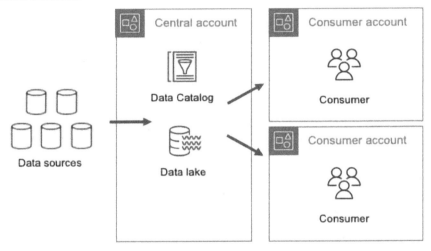

Figure 9.2 – Hub-and-spoke model

Data mesh

A data mesh is a design pattern that addresses the challenges of scaling, ownership, and accountability that the hub-and-spoke model often faces by introducing the data-as-a-product paradigm. The data mesh strategy is designed to overcome these challenges by allowing the data owner teams to build and publish the data as a product and making the teams accountable for the data.

A data mesh defines how you organize and deliver data as a product. The data is published by the data owner and shared with the consumers. A data mesh also provides federated access across consumers in different teams and organizations through a central catalog in the mesh account. Each organization will be a data owner who is responsible for maintaining the end-to-end data flow by building, operating, and serving the data products. They are also responsible for maintaining the data quality by monitoring and resolving any data. Data accountability lies with the data owner.

Data product owner teams are responsible for maintaining the data catalog regularly so that it's up-to-date and keeping the data discoverable and searchable on the catalog. These are the domain experts of the datasets in both the content and the data platform. When usage increases, the consumers of the data product may report some data issues, such as increased data latency and missing records. The data product team is the only team that can solve these data issues because they understand the context of the data, know the architecture of the data processing pipeline, and can identify the procedure to fix the issues. This reduces the overall friction for the data flow, where the data product team is responsible for the datasets and is accountable for the consumers, although the central data engineering team tends to be responsible for the dataset and accountable for the consumers in the traditional hub-and-spoke model. With the data mesh model, it is natural for them to keep the reliability of the data flow and the quality of the data, and improve the end-to-end data flow.

However, the data mesh model may not be the right pattern for your use case, and sometimes, it can be overkill since it brings more complexity than the hub-and-spoke model. You need to carefully validate whether your use case fits the data mesh pattern or not.

The following diagram shows how the data mesh model works with multiple accounts:

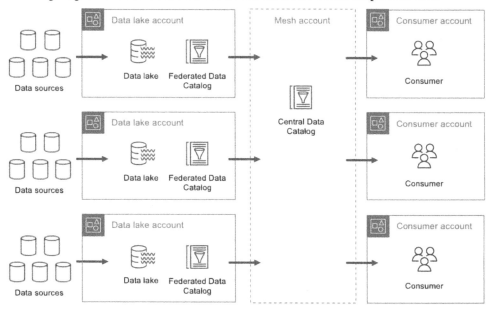

Figure 9.3 – Data mesh model

You can find more background information and reference architectures in the *Design a data mesh architecture using AWS Lake Formation and AWS Glue* blog (`https://aws.amazon.com/blogs/big-data/design-a-data-mesh-architecture-using-aws-lake-formation-and-aws-glue/`).

Sharing data with multiple AWS accounts using S3 bucket policies and Glue catalog policies

In this section, you will learn how to share your data with multiple AWS accounts using an S3 bucket policy and a Glue catalog policy.

When your use case is simple, and you want to share your data with a small number of accounts, it is possible to grant data access in S3 bucket policies (`https://docs.aws.amazon.com/AmazonS3/latest/userguide/bucket-policies.html`) and metadata access in Glue catalog resource policies (`https://docs.aws.amazon.com/glue/latest/dg/glue-resource-policies.html`). You will set these up in the following sections.

Scenario 1 – sharing data from one account with another using S3 bucket policies and Glue catalog policies

In the following scenario, there are two accounts – the producer account and the consumer account. Here, the producer account wants to share its table with the consumer account, and the consumer account wants to run `SELECT` queries against the shared table on Amazon Athena.

Prerequisite – S3

Let's look at the prerequisite for setting up the S3 resources. Follow these steps in the producer account:

1. Create a sample **JSON Line (JSONL)** file called `product_customer_sales.json`:

```
{"product_name":"Introduction to
Cloud","category":"Ebooks","price":15,"customer_
name":"Barbara Gordon","email":"gordon@example.
com","phone":"117.835.2584","purchased_at":"2022-04-
21T11:40:00Z"}
{"product_name":"Best practices on data
lakes","category":"Ebooks","price":25,"customer_name":"Tanya
Fowler","email":"tanya@example.net","phone":"(067)150-
0263","purchased_at":"2022-04-28T08:20:00Z"}
{"product_name":"Data Quest","category":"Video
games","price":30,"customer_name":"Rebecca
Thompson","email":"thompson@example.net","phone":"001-469-964-
```

3897x9041","purchased_at":"2022-03-30T01:30:00Z"}
{"product_name":"Final Shooting","category":"Video
games","price":20,"customer_name":"Rachel
Gilbert","email":"gilbert@example.com","phone":"001-510-198-
4613x23986","purchased_at":"2022-04-01T02:00:00Z"}

2. Create a `simple-datalake-<your-producer-account-id>` S3 bucket in your preferred region using the AWS CLI (replace the `<your-producer-account-id>` placeholder with your AWS account ID):

```
$ BUCKET_NAME="simple-datalake-<your-producer-account-id>"
$ aws s3api create-bucket --bucket ${BUCKET_NAME} --create-
bucket-configuration LocationConstraint=us-west-2
```

If you choose us-east-1, please remove the `--create-bucket-configuration` parameter.

3. Upload files to the S3 bucket by copying the sample data to your bucket:

```
$ aws s3 cp product_customer_sales.json s3://${BUCKET_NAME}/
simple_datalake/pcs/
```

With that, you have copied the sample data to your S3 bucket.

Prerequisite – Glue

Let's look at the prerequisite for setting up the Glue resources. Follow these steps in the producer account:

1. Create a database called `simple_datalake` in Glue Data Catalog by running the CREATE DATABASE DDL on Athena:

```
CREATE DATABASE simple_datalake
```

2. Create a `pcs` table in Glue Data Catalog by running the CREATE TABLE DDL on Athena (replace the `<your-producer-account-id>` placeholder with your AWS account ID):

```
CREATE EXTERNAL TABLE simple_datalake.pcs(
  product_name string,
  category string,
  price int,
  purchased_at string,
  customer_name string,
  email string,
  phone string)
```

```
ROW FORMAT SERDE 'org.openx.data.jsonserde.JsonSerDe'
STORED AS INPUTFORMAT 'org.apache.hadoop.mapred.
TextInputFormat'
OUTPUTFORMAT 'org.apache.hadoop.hive.ql.io.
HiveIgnoreKeyTextOutputFormat'
LOCATION 's3://simple-datalake-<your-producer-account-id>/
simple_datalake/pcs/'
TBLPROPERTIES ('classification'='json')
```

Please note that databases, tables, and partitions can be created in different ways. This time, we chose to run DDL on Athena to simplify the scenario. Of course, you can use the following as well:

- The Glue Data Catalog API

- A Glue crawler

- A Glue job

- DDL

Now, there is the new `pcs` table in the `simple_datalake` database. You can query the table like so:

```
SELECT * FROM simple_datalake.pcs
```

You will see four sample records in the result set:

Figure 9.4 – The SELECT query's result in the pcs table

Configuring S3 bucket policies and Glue Catalog resource policies

Follow these steps to configure S3 bucket policies and Glue catalog resource policies so that you can share data from one account to another:

1. [Producer] Grant permission on the Glue Catalog resource policy.

   The producer will need to share the table on AWS Glue Data Catalog by introducing the following resource policy (replace the `<your-producer-account-id>` and `<your-consumer-account-id>` placeholders with your AWS account IDs):

```
{
    "Version": "2012-10-17",
    "Statement": [
        {
            "Effect": "Allow",
            "Action": [
                "glue:GetDatabase",
                "glue:GetDatabases",
                "glue:GetTable",
                "glue:GetTables",
                "glue:GetTableVersion",
                "glue:GetTableVersions",
                "glue:GetPartition",
                "glue:GetPartitions",
                "glue:BatchGetPartition",
                "glue:SearchTables"
            ],
            "Principal": {
                "AWS": [
                    "arn:aws:iam::<your-consumer-account-
id>:root"
                ]
            },
            "Resource": [
                "arn:aws:glue:us-west-2:<your-producer-account-
id>:table/simple_datalake/pcs",
                "arn:aws:glue:us-west-2:<your-producer-account-
id>:database/simple_datalake",
                "arn:aws:glue:us-west-2:<your-producer-account-
```

```
id>:catalog"
            ]
        }
    ]
}
```

Save the preceding JSON as `catalog-policy.json` and run the following command to put the resource policy in your Glue Catalog:

```
$ aws glue put-resource-policy --policy-in-json file://./
catalog-policy.json --enable-hybrid TRUE --region us-west-2
```

2. [Producer] Grant permission on the S3 bucket policy.

    If the producer account wants to grant read-only access to your `pcs` table, which is located at `s3://simple-datalake-<your-producer-account-id>/simple_datalake/pcs/`, to the consumer account, the S3 bucket will need to be configured with the following S3 bucket policy (replace the `<your-producer-account-id>` and `<your-consumer-account-id>` placeholders with your AWS account IDs):

```
{
    "Version": "2012-10-17",
    "Statement": [
        {
            "Effect": "Allow",
            "Principal": {
                "AWS": [
                    "arn:aws:iam::<your-consumer-account-
id>:root"
                ]
            },
            "Action": [
                "s3:GetObject"
            ],
            "Resource": "arn:aws:s3:::simple-datalake-<your-
producer-account-id>/simple_datalake/pcs/*"
        },
        {
            "Effect": "Allow",
            "Principal": {
                "AWS": [
```

```
                      "arn:aws:iam::<your-consumer-account-
id>:root"
                ]
            },
            "Action": [
                "s3:ListBucket"
            ],
            "Resource": "arn:aws:s3:::simple-datalake-<your-
producer-account-id>"
        }
    ]
}
```

Save the preceding JSON as bucket-policy.json and run the following command to put the bucket policy in your S3 bucket (replace the <your-producer-account-id> placeholder with your AWS account ID):

```
$ aws s3api put-bucket-policy --bucket simple-datalake-<your-
producer-account-id> --policy file://./bucket-policy.json
```

3. [Consumer] Connect to the Glue Data Catalog shared by the producer:

    I.   Open the Athena console.

    II.   Click **Data sources**.

    III. At the top right, click **Connect data source**.

    IV.   In the **Data source selection** section, click **S3 - AWS Glue Data Catalog**, then **Next**.

    V.   In the **AWS Glue Data Catalog** section, click **AWS Glue Data Catalog in another account**.

    VI.   For **Data source details**, enter the following information:

    • **Data source name**: Enter producer_catalog

    • **Catalog ID**: Enter the AWS account ID of the producer account ID

    VII. Click **Next**, then **Create data source**.

Now. you can select the new `producer_catalog` data source instead of the default of `AwsDataCatalog` in the Athena query editor in the consumer account. Run Athena from the consumer account, as follows:

Figure 9.5 – The SELECT query's result in the consumer account

To summarize, we configured the producer account to grant permissions on the Glue Catalog resource policy and grant permissions on the S3 bucket policy. After that, we configured the consumer account to register the new data source in Athena so that it points to the producer's Glue Data Catalog. You will notice that there was no need to create/update any of the Glue catalog resources on the consumer account side. All the changes in the producer account will be visible and accessible without you needing to perform any manual operations in the consumer account.

This model works in simple use cases and is easy to understand. However, there are some challenges, as follows:

- First, you need to maintain both S3 bucket policies and Glue Data Catalog resource policies every time you want to grant or revoke access.

- Second, you need to manage permissions at the S3 object (file) level, even if your daily operation may be SQL style. When you know only tables and run only SQLs, you may not know about the underlying files you are touching in your queries. However, when you manage permissions in an S3 bucket policy, you need to find the underlying files under the target table and manage the relationship between the logical table and the physical files on S3. In addition, you cannot manage permissions at a more granular level, such as the column level or row level, since the S3 bucket policy can only be defined at the file level.

- Third, S3 bucket policies are limited to 20 KB in size, while Glue Data Catalog resource policies are limited to 10 KB in size. If you want to have a central place to manage all the permissions and have more flexibility in terms of granularity, you should try using Lake Formation permissions.

In the next section, you will learn about a scalable way to achieve cross-account data sharing using AWS Lake Formation permissions.

Sharing data with multiple AWS accounts using AWS Lake Formation permissions

In this section, you will learn how to share data with multiple AWS accounts using AWS Lake Formation permissions.

Lake Formation permission model

As you learned in the previous section, there are challenges in managing S3 bucket policies and Glue Data Catalog resource policies. AWS Lake Formation is the service that is designed to overcome those challenges and simplify data platform management. Lake Formation provides a central layer for defining, classifying, tagging, and managing fine-grained access control to the AWS Glue Data Catalog and Amazon S3 locations. The permission model is designed in an RDBMS-like style so that you can grant permissions on databases, tables, or columns instead of S3 objects. Once you have granted access to tables with Lake Formation permissions, Lake Formation automatically manages both data access and metadata access under the hood, so you don't need to manually take care of granting individual data access and metadata access.

Lake Formation cross-account sharing

The AWS Lake Formation permission model also simplifies cross-account configurations. With Lake Formation permissions, you can easily secure and manage data lakes across multiple AWS accounts at scale.

In terms of Lake Formation cross-account access control, there are two different approaches to sharing your databases and tables with another account:

- One approach is to use Lake Formation's named resource-based access control
- The other is to use Lake Formation's tag-based access control. This is a recommended approach.

Lake Formation tag-based access control is recommended because of its scalability and maintainability. We will look at these options in detail in the following sections.

Lake Formation named resource-based access control

Lake Formation named resource-based access control is a configuration option that manages permissions based on specific Data Catalog resources such as databases, tables, and columns. In this access control model, you can grant or revoke permissions on Lake Formation resources using the resource names. You can learn more by reading *Cross-Account Access: How It Works*: `https://docs.aws.amazon.com/lake-formation/latest/dg/crosss-account-how-works.html`.

We only recommend using named resource-based access control when you prefer granting permissions explicitly to individual resources. It works with a small number of resources, but if you have a large number of resources, then you should use Lake Format tag-based access control.

Lake Formation tag-based access control

Lake Formation tag-based access control is a configuration option that manages permissions based on logical attributes called **LF-tags**, instead of specific resources. It requires two separate configurations: LF-tag – Data Catalog resources (databases, tables, and columns), and LF-tag – Lake Formation principals (IAM users, roles, SAML users, and QuickSight users). First, LF-tags need to be configured on Data Catalog resources. Second, you must grant and revoke permission on the LF-tag (instead of specific Data Catalog resources) to Lake Formation principals. With these configurations, Lake Formation allows you to access those resources when the LF-tag that the principal has permission on matches the LF-tag that the resource has.

LF-tag-based access control is efficient and useful in environments that are growing rapidly. Imagine a scenario where there are five databases, and each database has 10 tables. There are three different organizations, and each organization has specific visibility per table. Today, a new employee joins your organization, and you need to grant the required permissions as a data lake administrator. Without LF-tag-based access control, you need to grant access to five individual databases and 50 tables for this user. With LF-tag-based access control, all you need to do is grant access to the LF-tag that matches the organization's permission for this user. LF-tag-based access control also helps with situations where resource-based policies become too complicated as you will need far fewer permission configurations than in traditional resource-based access control.

You can learn more about tag-based access control by reading *Easily manage your data lake at scale using AWS Lake Formation Tag-based access control*: `https://aws.amazon.com/blogs/big-data/easily-manage-your-data-lake-at-scale-using-tag-based-access-control-in-aws-lake-formation/`.

We recommend Lake Formation tag-based access control for the following use cases:

- You have a large number of Data Catalog resources (databases, tables, and columns) and principals (IAM users, roles, and more) that you need to grant access to
- You want to manage data access based on logical attributes or classifications of data
- You want to grant permissions dynamically, especially for new tables and principals

To learn more, please read *Securely share your data across AWS accounts using AWS Lake Formation*: `https://aws.amazon.com/blogs/big-data/securely-share-your-data-across-aws-accounts-using-aws-lake-formation/`.

Scenario 2 – sharing data from one account with another using Lake Formation Tag-based access control

In this scenario, we will use Lake Formation tag-based access control to share tables. There are two accounts: the producer account and the consumer account. Here, the producer account wants to share its table with the consumer account, and the consumer account wants to run `SELECT` queries on the shared table:

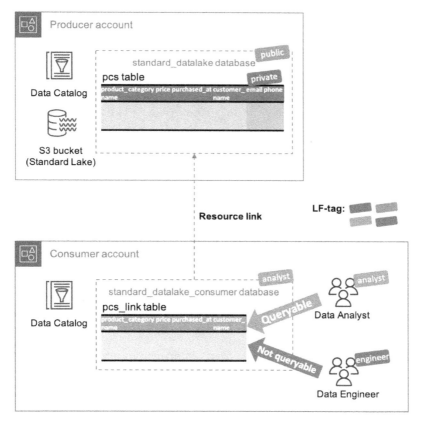

Figure 9.6 – Architecture of scenario 2

In addition to the preceding use case, there is a security requirement to hide specific columns from the consumer. The table has seven columns: product_name, category, price, purchased_at, customer_name, email, and phone. The producer wants to share this table with the consumer account for business reasons but does not want to share either the email or phone columns since they contain sensitive information. On the other hand, the consumer wants to grant access to this shared pcs table to people that belong to the analyst decision. You can achieve this by following the steps in the next section.

Prerequisite – S3

The following prerequisite is required to set up the S3 resources. Follow these steps in the producer account:

1. Create a sample JSONL file called product_customer_sales.json:

```
{"product_name":"Introduction to
Cloud","category":"Ebooks","price":15,"customer_
name":"Barbara Gordon","email":"gordon@example.
com","phone":"117.835.2584","purchased_at":"2022-04-
21T11:40:00Z"}
{"product_name":"Best practices on data
lakes","category":"Ebooks","price":25,"customer_name":"Tanya
Fowler","email":"tanya@example.net","phone":"(067)150-
0263","purchased_at":"2022-04-28T08:20:00Z"}
{"product_name":"Data Quest","category":"Video
games","price":30,"customer_name":"Rebecca
Thompson","email":"thompson@example.net","phone":"001-469-964-
3897x9041","purchased_at":"2022-03-30T01:30:00Z"}
{"product_name":"Final Shooting","category":"Video
games","price":20,"customer_name":"Rachel
Gilbert","email":"gilbert@example.com","phone":"001-510-198-
4613x23986","purchased_at":"2022-04-01T02:00:00Z"}
```

2. Create a standard-datalake-<your-producer-account-id> S3 bucket in your preferred region using the AWS CLI (replace the <your-producer-account-id> placeholder with your AWS account ID):

```
$ BUCKET_NAME="standard-datalake-<your-producer-account-id>"
$ aws s3api create-bucket --bucket ${BUCKET_NAME} --create-
bucket-configuration LocationConstraint=us-west-2
```

If you choose us-east-1, please remove the --create-bucket-configuration parameter.

3. Upload the files to the S3 bucket by copying the sample data to your bucket:

```
$ aws s3 cp product_customer_sales.json s3://${BUCKET_NAME}/
standard_datalake/pcs/
```

With that, you've copied the sample data to your S3 bucket. In the next section, you will set up a Glue table for this file on S3.

Prerequisite – Glue

The following prerequisite is required to set up various Glue resources – that is, the Glue database, Glue table, and its partitions – so that you can use them in the subsequent sections. Follow these steps in the producer account:

1. Create a `standard_datalake` database on Glue Data Catalog by running the `CREATE DATABASE` DDL on Athena:

```
CREATE DATABASE standard_datalake
```

Create a `pcs` table on Glue Data Catalog by running the `CREATE TABLE` DDL on Athena (replace the `<your-producer-account-id>` placeholder with your AWS account ID):

```
CREATE EXTERNAL TABLE standard_datalake.pcs(
    product_name string,
    category string,
    price int,
    purchased_at string,
    customer_name string,
    email string,
    phone string)
ROW FORMAT SERDE 'org.openx.data.jsonserde.JsonSerDe'
STORED AS INPUTFORMAT 'org.apache.hadoop.mapred.
TextInputFormat'
OUTPUTFORMAT 'org.apache.hadoop.hive.ql.io.
HiveIgnoreKeyTextOutputFormat'
LOCATION 's3://standard-datalake-<your-producer-account-id>/
standard_datalake/pcs/'
TBLPROPERTIES ('classification'='json')
```

With that, all the required Glue resources, the `standard_datalake` database, and the `pcs` table have been set up. You will use these resources in a sample dataset on your data lake.

Prerequisite – Lake Formation and IAM

The following prerequisite is required to set up baseline configurations for the Lake Formation resources. Follow these steps in both the producer account and the consumer account:

1. First, you must create a Data Lake Administrator if you do not have one. The Data Lake Administrator is an IAM user or an IAM role that has special privileges on Lake Formation resources. You will use this in the subsequent steps: `https://docs.aws.amazon.com/lake-formation/latest/dg/getting-started-setup.html#create-data-lake-admin`.

2. Next, you must update your default Lake Formation settings to migrate from traditional IAM-only access control to Lake Formation access control.

3. Sign in to the Lake Formation console using the Data Lake Administrator.

4. In the left menu, under the **Data catalog** category, click **Settings**. You will see the following settings:

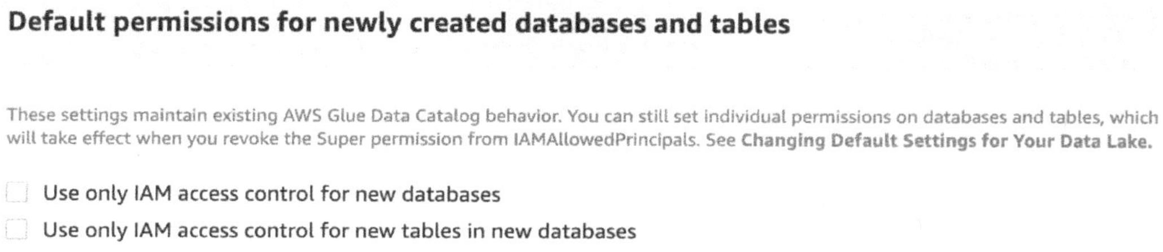

Default permissions for newly created databases and tables

These settings maintain existing AWS Glue Data Catalog behavior. You can still set individual permissions on databases and tables, which will take effect when you revoke the Super permission from IAMAllowedPrincipals. See **Changing Default Settings for Your Data Lake.**

☐ Use only IAM access control for new databases
☐ Use only IAM access control for new tables in new databases

Figure 9.7 – Updating the default permissions for Lake Formation resources

5. Deselect the **Use only IAM access control for new databases** and **Use only IAM access control for new tables in new databases** checkboxes.

6. Click the **Save** button.

Once you have done this, all your new databases and the new tables in those new databases will start following the Lake Formation access control model. Before updating this setting, special Lake Formation permissions are granted for `IAM_ALLOWED_PRINCIPAL` (any principals that are allowed through IAM authorization) on your databases and tables to keep backward compatibility. After updating the setting, the default permissions will be revoked, so you need to grant Lake Formation permission on those databases and tables expressly.

Next, to enable Lake Formation access control on the tables located in the `standard-datalake-<your-producer-account-id>` S3 bucket, follow these steps in the producer account:

1. From the left menu, under the **Register and ingest** category, click **Data lake locations**.

2. Click **Register location**. You will see the following output:

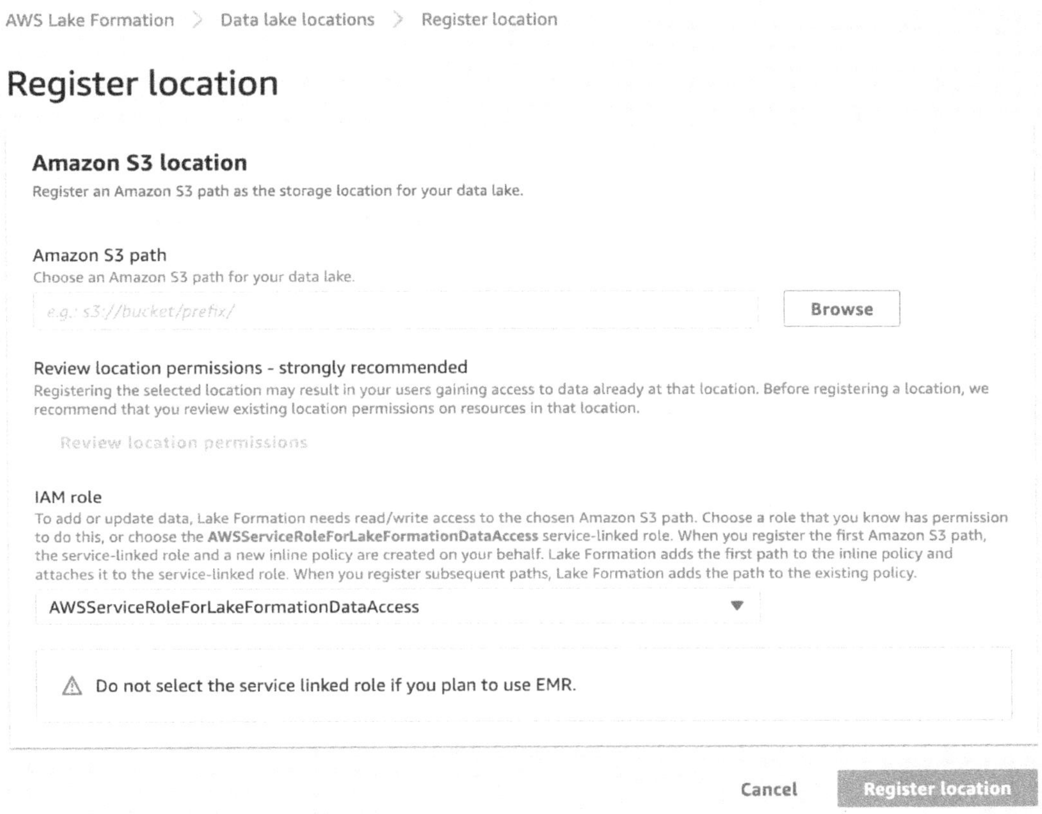

AWS Lake Formation > Data lake locations > Register location

Register location

Amazon S3 location

Register an Amazon S3 path as the storage location for your data lake.

Amazon S3 path
Choose an Amazon S3 path for your data lake.

| e.g.: s3://bucket/prefix/ | Browse |

Review location permissions - strongly recommended

Registering the selected location may result in your users gaining access to data already at that location. Before registering a location, we recommend that you review existing location permissions on resources in that location.

Review location permissions

IAM role

To add or update data, Lake Formation needs read/write access to the chosen Amazon S3 path. Choose a role that you know has permission to do this, or choose the **AWSServiceRoleForLakeFormationDataAccess** service-linked role. When you register the first Amazon S3 path, the service-linked role and a new inline policy are created on your behalf. Lake Formation adds the first path to the inline policy and attaches it to the service-linked role. When you register subsequent paths, Lake Formation adds the path to the existing policy.

| AWSServiceRoleForLakeFormationDataAccess ▼ |

⚠ Do not select the service linked role if you plan to use EMR.

Cancel   **Register location**

Figure 9.8 -- Register location

3. For **Amazon S3 path**, enter `s3://standard-datalake-<your-producer-account-id>/`.

4. Click the **Register location** button.

Now, all the Glue tables under this data lake location will start following Lake Formation access control.

Follow these steps in the consumer account:

1. Open the IAM console.

2. Create the `DataAnalyst` user by attaching the `AmazonAthenaFullAccess` AWS managed policy.

Now, all the IAM and Lake Formation resources have been successfully configured.

Step 1 – configuring Glue catalog policies

The producer needs to share the table on AWS Glue Data Catalog by introducing the following resource policy (replace the `<your-producer-account-id>` and `<your-consumer-account-id>` placeholders with your AWS account IDs):

```
{
    "Version": "2012-10-17",
    "Statement": [
        {
            "Effect": "Allow",
            "Action": [
                "glue:*"
            ],
            "Principal": {
                "AWS": [
                    "arn:aws:iam::<your-consumer-account-
id>:root"
                ]
            },
            "Resource": [
                "arn:aws:glue:us-west-2:<your-producer-account-
id>:table/*",
                "arn:aws:glue:us-west-2:<your-producer-account-
id>:database/*",
                "arn:aws:glue:us-west-2:<your-producer-account-
id>:catalog"
            ],
            "Condition": {
                "Bool": {
                    "glue:EvaluatedByLakeFormationTags": true
                }
            }
        }
    ]

}
```

You will notice that the preceding policy is a coarse-grained policy that allows glue:* actions for any databases and tables in the producer account. This lets Lake Formation manage fine-grained access control. This is still safe because the glue:EvaluatedByLakeFormationTags condition forces consumers to be authorized by Lake Formation permissions.

Save the preceding JSON as catalog-policy-lf.json and run the following command to put the resource policy in your Glue Catalog:

```
$ aws glue put-resource-policy --policy-in-json file://./
catalog-policy-lf.json --enable-hybrid TRUE --region us-west-2
```

Note that if you want to keep the existing policy you created in the previous section, you need to merge the catalog policies, as follows:

```
{
    "Version": "2012-10-17",
    "Statement": [
        {
            "Effect": "Allow",
            "Action": [
                "glue:GetDatabase",
                "glue:GetDatabases",
                "glue:GetTable",
                "glue:GetTables",
                "glue:GetTableVersion",
                "glue:GetTableVersions",
                "glue:GetPartition",
                "glue:GetPartitions",
                "glue:BatchGetPartition",
                "glue:SearchTables"
            ],
            "Principal": {
                "AWS": [
                    "arn:aws:iam::<your-consumer-account-
id>:root"
                ]
            },
            "Resource": [
                "arn:aws:glue:us-west-2:<your-producer-account-
```

```
id>:table/simple_datalake/pcs",
                "arn:aws:glue:us-west-2:<your-producer-account-
id>:database/simple_datalake",
                "arn:aws:glue:us-west-2:<your-producer-account-
id>:catalog"
            ]
        },
        {
            "Effect": "Allow",
            "Action": [
                "glue:*"
            ],
            "Principal": {
                "AWS": [
                    "arn:aws:iam::<your-consumer-account-
id>:root"
                ]
            },
            "Resource": [
                "arn:aws:glue:us-west-2:<your-producer-account-
id>:table/*",
                "arn:aws:glue:us-west-2:<your-producer-account-
id>:database/*",
                "arn:aws:glue:us-west-2:<your-producer-account-
id>:catalog"
            ],
            "Condition": {
                "Bool": {
                    "glue:EvaluatedByLakeFormationTags": true
                }
            }
        }
    ]
}
```

With that, your Glue Catalog policy has been successfully configured.

Step 2 – configuring Lake Formation permissions (producer)

Next, let's configure Lake Formation permissions using Lake Formation tags. This will allow you to publish your table from the producer account to the consumer account. Follow the steps provided in the following sections in the producer account.

Defining an LF-tag

Follow these steps to create a new LF-tag:

1. Sign in to the Lake Formation console using the Data Lake Administrator user.

2. From the left menu, under the **Permissions** category, click **LF-tags** under **Administrative roles and tasks**.

3. Click the **Add LF-tag** button. You will see the following output:

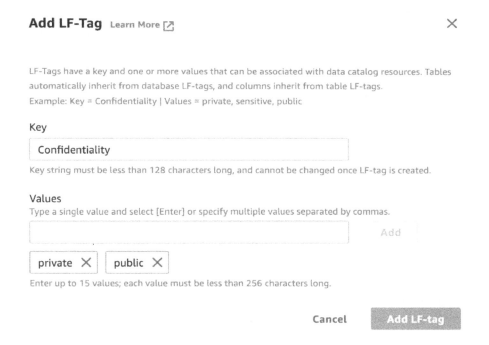

Add LF-Tag Learn More ↗   ✕

LF-Tags have a key and one or more values that can be associated with data catalog resources. Tables automatically inherit from database LF-tags, and columns inherit from table LF-tags.
Example: Key = Confidentiality | Values = private, sensitive, public

Key

Confidentiality

Key string must be less than 128 characters long, and cannot be changed once LF-tag is created.

Values
Type a single value and select [Enter] or specify multiple values separated by commas.

                                                                        Add

private ✕   public ✕

Enter up to 15 values; each value must be less than 256 characters long.

Cancel   Add LF-tag

Figure 9.9 – Add LF-Tag

4. For **Key**, enter Confidentiality, and for **Values**, enter private and public. Then, click **Add LF-tag**.

Now, you have a new LF-tag called Confidentiality that has two different values: private and public. We will use this LF-tag to manage access to the sample dataset.

Attaching an LF-tag

Attach the LF-tag that contains the public value to your standard_datalake database and update the value from public to private for the email and phone columns to indicate that this column contains sensitive data. Now, follow these steps:

1. From the left menu, under the **Data catalog** category, click **Databases**.

2. Select the standard_datalake database and, from the **Actions** menu, click **Edit LF-tags**.

3. Click **Assign new LF-Tag** to enter a new key and its value. You will see the following output:

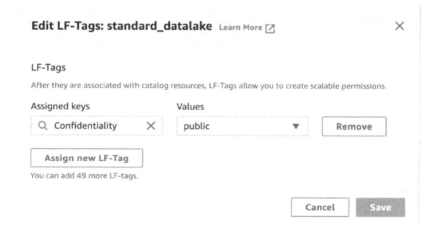

Figure 9.10 – Edit LF-Tags: standard_datalake

4. Add the Confidentiality key and the public value.

5. Click **Save**.

6. Then, select the standard_datalake database and click **View tables**.

7. Click the link to the pcs table.

8. Under **Schema**, click **Edit schema**.

9. Select the checkboxes for the email and phone columns and click **Edit tags**. You will see the following output:

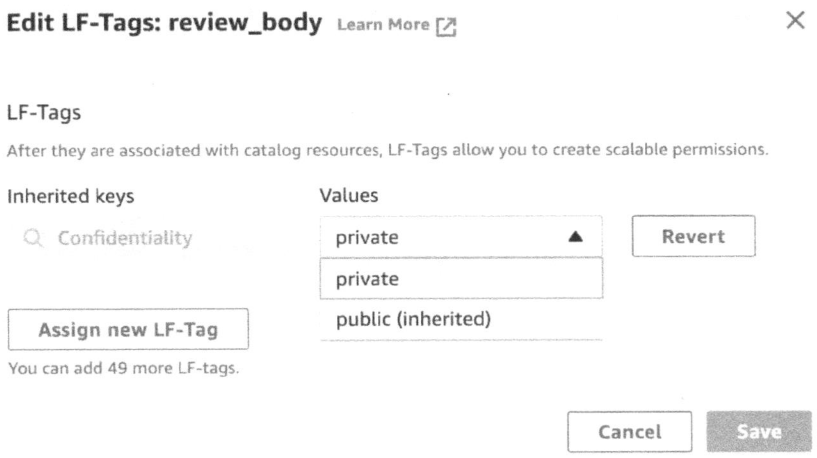

Figure 9.11 – Edit LF-Tags: review_body

10. Update the value of the `Confidentiality` key from `public` to `private`.

11. Click **Save**.

12. Click **Save as new version**.

The `Confidentiality` LF-tag and its `public` value have been configured to the `standard_datalake` database, and also recursively applied to the `pcs` table automatically. After that, the LF-tag's values were updated from `public` to `private` for the `email` and `phone` columns. This means that those who have `Confidentiality=public` LF-tag permissions can view all the columns except the `email` and `phone` columns.

Granting LF tag permission to the consumer account

Follow these steps to grant LF-tag permission to the consumer:

1. From the left menu, under the **Permissions** category, click **LF-tag permissions** under **Administrative roles and tasks**.

2. Click **Grant**. You will see the following output:

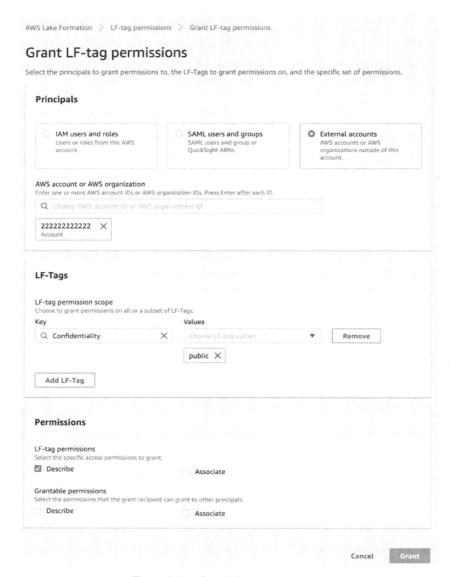

Figure 9.12 – Grant LF-tag permissions

3.   For **Principals**, select **External accounts**.

4.   For **AWS account or AWS organization**, enter the consumer account ID and press *Enter*.

5.   For **LF-Tag permission scope**, choose the `Confidentiality` key and the `public` value.

6.   For **LF-tag permissions**, select **Describe**.

7.   Click **Grant**.

Now, the `Confidentiality` LF-tag is visible from the consumer account and can be used to define data permissions that can share the data with the consumer account.

Granting data permission to the consumer account

Follow these steps to grant data permission using the `Confidentiality` LF-tag:

1. From the left menu, under **Permissions**, click **Data lake permissions**.

2. Click **Grant**. You will see the following output:

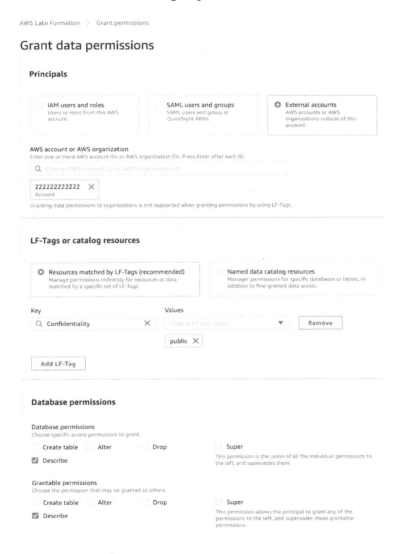

Figure 9.13 – Grant data permissions

3. For **Principals**, choose **External accounts**.

4. For **AWS account or AWS organization**, enter the consumer account ID and press *Enter*.

5. For **LF-tags or catalog resources**, select **Resources matched by LF-Tags (recommended)**.

6. Select `Confidentiality` as the key and `public` as the value.

7. For **Database permissions**, select **Describe** under **Database permissions** and **Describe** under **Grantable permissions**.

8. For **Table permissions**, select **Select** and **Describe** under **Table permissions** and select **Select** and **Describe** under **Grantable permissions**.

9. Click **Grant**.

With that, the data permission that uses the LF-tag has been configured.

Step 3 – configuring Lake Formation permissions (consumer)

Complete the following steps in the consumer account.

Creating a database

In this section, you will create a new database in the consumer account to add a resource link that points to the producer account. A resource link is a configuration that links to a local or shared database or table. It is required when you want to share your tables among multiple accounts. You can create a resource link with any preferred name to avoid name conflicts in a consumer account. Follow these steps:

1. Sign in to the Lake Formation console using the Data Lake Administrator user.

2. From the left menu, under **Data catalog**, click **Databases**.

3. Click **Create database**.

4. Select **Database**; do *not* select **Resource link** here.

5. For **Name**, enter `standard_datalake_consumer`.

6. Click **Create database**.

Creating a resource link under the database

Follow these steps to create a resource link pointing to the `pcs` table in the producer account:

1. From the left menu, under **Data catalog**, click **Databases**. You will see the `standard_datalake` database that was shared from the producer account.

2. Select the `standard_datalake` database and click **View tables**.

3. Select the `pcs` table and, in the **Actions** menu, click **Create resource link**. You will see the following output:

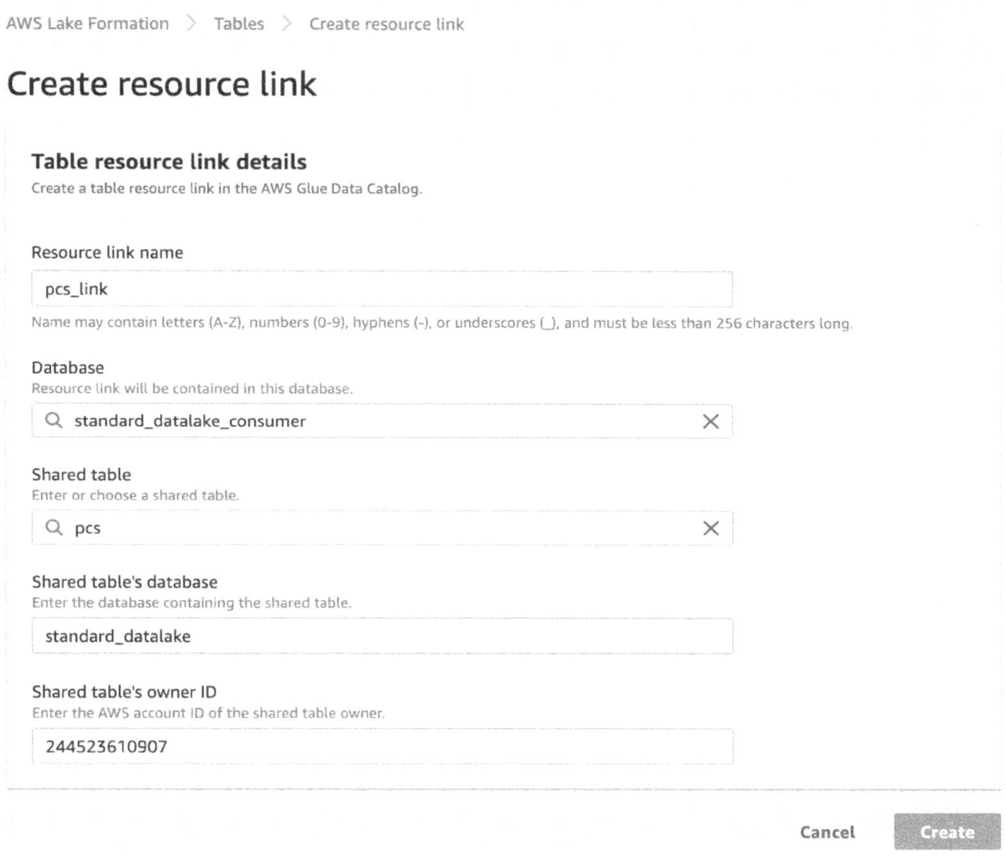

AWS Lake Formation > Tables > Create resource link

Create resource link

Table resource link details
Create a table resource link in the AWS Glue Data Catalog.

Resource link name

```
pcs_link
```

Name may contain letters (A-Z), numbers (0-9), hyphens (-), or underscores (_), and must be less than 256 characters long.

Database
Resource link will be contained in this database.

Q standard_datalake_consumer   ✕

Shared table
Enter or choose a shared table.

Q pcs   ✕

Shared table's database
Enter the database containing the shared table.

standard_datalake

Shared table's owner ID
Enter the AWS account ID of the shared table owner.

244523610907

Cancel   Create

Figure 9.14 – Create resource link

4. For **Resource link name**, enter `pcs_link`.

5. For **Database**, select `standard_datalake_consumer`.

6. Click **Create**.

Now, you can query the shared table in the consumer account using the Data Lake Administrator user. Open the Athena query editor and choose `AwsDataCatalog` under the **Data Source**. Run the following query on the Amazon Athena console:

```
SELECT * FROM standard_datalake_consumer.pcs_link
```

As shown in the following screenshot, there will be four sample records in the result set:

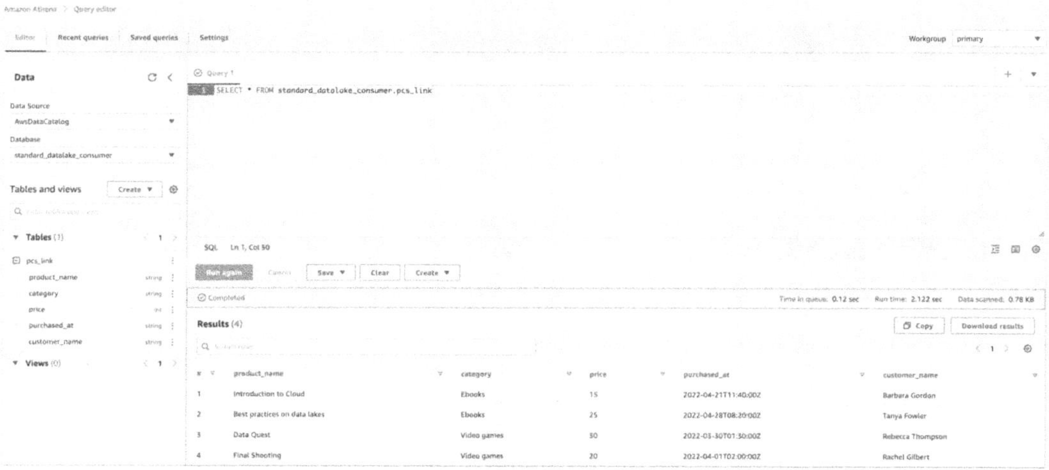

Figure 9.15 – The SELECT query's result executed by the Data Lake
Administrator user in the consumer account

You will notice that the records do not have either the email column or the phone column. This is because you marked the columns with an LF-tag where Confidentiality is private, and you only granted access to the consumer with an LF-tag where Confidentiality is public, not with an LF-tag where Confidentiality is private.

Defining an LF-tag

If you want to manage granular permissions for the IAM users and roles inside the consumer account, you can define a separate LF-tag and grant data permissions to the IAM users and roles using it. You will learn how to do this in the next few sections.

Note that Lake Formation tags are defined as resources in a single account, so LF-tags created in the producer account are not available to the consumer account. This means that the consumer account cannot use the producer account's LF-tag when granting access to the resource links. If you want to manage granular permissions for the IAM users and roles inside the consumer account, you need to create new LF-tags in the consumer account and grant separate permissions by using the new LF-tags on the resource links.

In this scenario, imagine that there are two different job roles in the consumer account – analyst and engineer – and you want to manage the visibility of the data based on these job roles.

Follow these steps to define a separate set of LF-tags:

1. Sign in to the Lake Formation console using the Data Lake Administrator user.

2. From the left menu, under the **Permissions** category, click **LF-tags** under **Administrative roles and tasks**.

3. Click **Add LF-tag**. You will see the following output:

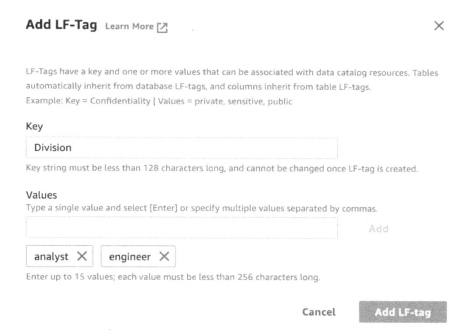

Figure 9.16 – Add LF-Tag

4. For **Key**, enter `Division`.

5. For **Values**, enter `analyst` and `engineer`.

6. Click **Add LF-tag**.

With that, you have defined a separate set of LF-tags: the key is `Division` and the values are `analyst` and `engineer`.

Attaching an LF-tag

Follow these steps to attach the separate `Division` LF-tag to the Data Catalog resources:

1. From the left menu, under **Data catalog**, click **Databases**.

2. Select the `standard_datalake_consumer` database and, from the **Actions** menu, click **Edit LF-tags.**

3. Click **Assign new LF-Tag**. You will see the following output:

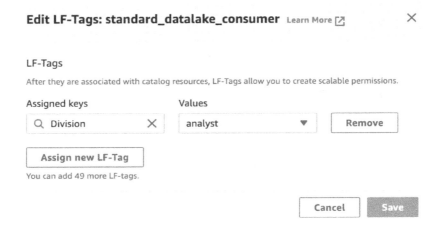

Figure 9.17 – Edit LF-Tags: standard_datalake_consumer

4. Add the `Division` key and the `analyst` value

5. Click **Save**.

With that, your `standard_datalake_consumer` database has been configured with the `Division` LF-tag.

Granting LF-tag permission to the IAM user in the consumer account

Follow these steps to grant LF-tag permission to the IAM users who reside in the consumer account to achieve granular access control:

1. From the left menu, under the **Permissions** category, click **LF-tag permissions** under **Administrative roles and tasks**.

2. Click **Grant**. You will see the following output:

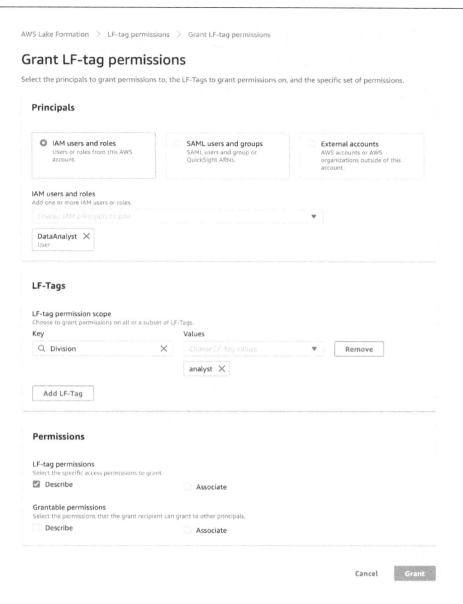

Figure 9.18 – Grant LF-tag permissions

3. For **Principals**, select **IAM users and roles**.

4. For **IAM users and roles**, select the DataAnalyst IAM user.

5. For **LF-Tag permission scope**, select the Division key and the analyst value.

6. For **LF-tag permissions**, select **Describe**. Then, click **Grant**.

Granting data permission to the IAM user in the consumer account

Follow these steps to grant data permission using the LF-tag to the IAM users in the consumer account:

1. From the left menu, under **Permissions**, click **Data lake permissions**.

2. Click **Grant**. You will see the following output:

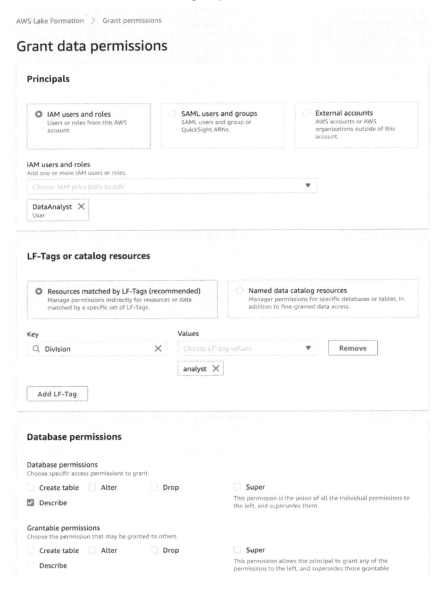

Figure 9.19 – Granting data permissions to DataAnalyst

3. For **Principals**, select **IAM users and roles**.

4. For **IAM users and roles**, select the `DataAnalyst` IAM user.

5. For **LF-tags or catalog resources**, select **Resources matched by LF-Tags (recommended)**.

6. Select the `Division` key and the `analyst` value.

7. For **Database permissions**, select **Describe** under **Database permissions** and **Describe** under **Grantable permissions**.

8. For **Table permissions**, select **Select** and **Describe** under **Table permissions** and **Select** and **Describe** under **Grantable permissions**.

9. Click **Grant**.

Now, `DataAnalyst` can query against the shared `pcs` table through the `pcs_link` resource link. As shown in the following screenshot, there will be four sample records in the result set:

Figure 9.20 – The SELECT query's result executed by DataAnalyst in the consumer account

As you can see, the records do not contain the `email` and `phone` columns. With that, you've configured cross-account Lake Formation permissions using Lake Formation LF-tags and confirmed that they work as expected on Athena queries.

Summary

In this chapter, you learned about three common data sharing strategies: single-tenant, hub-and-spoke, and data mesh. You also learned how to share data with different accounts using AWS Glue and AWS Lake Formation, as well as the benefits of doing so. At this point, you can design your data sharing model by choosing the strategy that fits your use case. You also gained hands-on skills in building a data sharing mechanism for your data platform.

In the next chapter, you will learn how to manage the data processing pipeline end to end.

10
Data Pipeline Management

Our data is composed of a lot of data types, such as IoT device logs, user logs, web server logs, and business reports. This data is generally stored in multiple data sources, such as relational databases, NoSQL databases, data warehouses, and data lakes, based on your applications, business needs, and rules. In this situation, there might be cases where you must obtain aggregated data results for user analysis, cost reports, and building machine learning models. To obtain the results, you may need to implement data processing flows to read data from multiple data sources by using a programming language, SQL, and so on. We usually call these flows data pipelines.

Recent pipeline flows consist of extracting data from data sources, transforming the data on computing engines, and loading the data into other data sources. This kind of pipeline is called an **extract, transform, and load** (ETL) pipeline, and it is used in a lot of cases. Additionally, the **extract, load, and transform** (ELT) and **extract, transformation, load, and transformation** (EtLT) patterns are used these days.

As you grow your data and data sources, the number of data pipelines increases. This can usually cause problems in scaling data pipelines, such as how you can build, operate, manage, and maintain pipelines. Therefore, effectively building and using data pipelines is one of the keys to effectively utilizing and operating your data for the growth of your company, organization, and team.

To tackle these problems, in this chapter, we'll look at data pipelines and the best practices to manage them. In particular, this chapter covers the following topics:

- What are data pipelines?
- Selecting the appropriate data processing services for your analysis
- Orchestrating your pipelines with workflow tools
- Automating how you provision your pipelines with provisioning tools
- Developing and maintaining your data pipelines

Technical requirements

For this chapter, if you wish to follow some of the walkthroughs, you will require the following:

- Internet access to GitHub, S3, and the AWS console (specifically the console for AWS Glue, Amazon Step Functions, Amazon Managed Workflows for Apache Airflow, AWS CloudFormation, and Amazon S3)

- A computer with Chrome, Firefox, Safari, or Microsoft Edge installed and the AWS Command Line Interface (AWS CLI)

> **Note**
> You can use not only the AWS CLI but also AWS CLI version 2. In this chapter, we have used the AWS CLI (not version 2). You can set up the AWS CLI (and version 2) by going to `https://docs.aws.amazon.com/cli/latest/userguide/cli-chap-getting-started.html`.

You will also need an AWS account and an accompanying IAM user (or IAM role) with sufficient privileges to complete this chapter's activities. We recommend using a minimally scoped IAM policy to avoid unnecessary usage and making operational mistakes. You can find the IAM policy for this chapter in this book's GitHub repository at `https://github.com/PacktPublishing/Serverless-ETL-and-Analytics-with-AWS-Glue/tree/main/Chapter10`. This IAM policy includes the following access:

- Permissions to create a list of IAM roles and policies for creating a service role for an AWS Glue ETL job.

- Permissions to read, list, and write access to an Amazon S3 bucket.

- Permissions to read and write access to Glue Data Catalog databases, tables, and partitions.

- Permissions to read, list and write access to Glue ETL Jobs, Crawlers, Triggers, Workflows and Blueprints.

- Permission to read, list and write access to **AWS Step Functions** resources.

- Permission to read, list and write access to **Amazon Managed Workflows for Apache Airflow (MWAA)** resources.

- Permissions to read, list and write access to **AWS CloudFormation** resources.

If you haven't set up the following resources, create or install necessary resources by following AWS documents:

- An S3 bucket for reading and writing data by AWS Glue. If you haven't had it yet, you can create one by going to the AWS console (`https://s3.console.aws.amazon.com/s3/home`) and choosing **Create bucket**. You can also create a bucket by running the `aws s3api create-bucket --bucket <your_bucket_name> --region us-east-1` AWS CLI command.

- The environment for Glue Blueprints. If you haven't set it up yet, you need to install the relevant modules and SDKs to use Glue Blueprints. Please refer to `https://docs.aws.amazon.com/glue/latest/dg/developing-blueprints-prereq.html`.

- The **MWAA** environment. If you haven't set it up yet, you need to create the environment from the MWAA console (`https://console.aws.amazon.com/mwaa/home#environments`). Please refer to `https://docs.aws.amazon.com/mwaa/latest/userguide/get-started.html`. At the time of writing, the latest Airflow version in MWAA is 2.2.2. This is the version we used.

What are data pipelines?

We generally use the word pipeline for a set of elements that are connected in a process, such as oil pipelines, gas pipelines, marketing pipelines, and so on. In particular, an element that is put into a pipeline is moved out via defined routes in a pipeline as output.

In computing, a data pipeline (or simply a pipeline) is referred to as a set of data processing elements that are connected in some series. Through a data pipeline, a set of elements are moved and transformed from various sources into destinations based on your implementation. A data pipeline usually consists of multiple tasks, such as data extraction, processing, validation, ingestion, pre-processing for machine learning use, and so on. Regarding the input and output of data pipelines, for example, the input is application logs, server logs, IoT device data, user data, and so on. The output of a data pipeline is analysis reports, a dataset for machine learning. The following diagram shows an example of a pipeline:

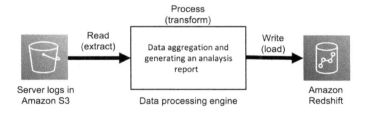

Figure 10.1 – A data pipeline that writes processed logs to an Amazon Redshift table

In this example, server logs are stored in S3 as raw data and are processed into an analysis report, then written to Amazon Redshift.

Usually, we run multiple pipelines as workflows by using scripts or automation tools. This creates various processed data to meet the various needs of multiple teams across multiple environments, such as multiple systems, programming languages, and so on.

Why do we need data pipelines?

We build and use data pipelines to process data and get results so that we can use the data further. Let's take a look at some popular use cases of data pipelines:

- **Data aggregation**: Through data pipelines, your data is processed and aggregated to generate a result that meets customer, team, and organization needs, such as analysis reports, cost usage reports, user activity reports, and so on. After processing the data via data pipelines, it's stored in various places, such as databases, data warehouses, and so on. If necessary, the aggregated data can be processed and combined with other aggregated data to generate a new report.

- **Data cleansing**: This use case is usually used for the raw data in your storage, such as application logs, user activity logs, server logs, IoT device data, and so on. Raw data often includes corrupted or garbage records. If you transform the raw data into data that other members such as analysts can process and visualize, you need to clean the raw data and also transform the data so that it matches your data source interface. For example, if analysts run analytic queries for a company's data warehouse, you need to transform the data into a new format so that it is compatible with the data warehouse schema.

- **Data anonymization**: Sensitive records in your data are masked and transformed as a password through data pipelines. This process aims to provide privacy protection. This type of data pipeline often consists of multiple tasks that process sensitive information based on various levels of privacy. For example, let's say that some data may include a user ID that must be masked for one team. However, another team needs that record, so the data doesn't need to be processed.

Now that we've looked at some data pipeline use cases, others are available. Data pipelines are widely used to process and transform data into a new form of data for future use.

How do we build and manage data pipelines?

So far, we've seen that a data pipeline is a set of data processing flows that consist of elements of data processing and data storage. We've also seen that data pipelines are used for data aggregation, cleansing, anonymization, and more.

To achieve this kind of data processing with pipelines, you need to design and build pipelines. Additionally, you need to update and maintain your pipelines based on your needs and data, such as organization/team updates, data schema changes, system updates, and so on. To effectively build and manage your data pipelines, you must understand the four main components of data pipeline management. We will cover these in the following sub-sections.

Selecting data processing services for your analysis

When you build a pipeline that extracts/writes data from/to your data storage, such as Amazon S3, relational databases, data warehouses, and so on, as a first step, you need to determine which data processing engines or services you use and how you process the data with them. To select data processing services, you need to consider things such as data usage, data format, data processing time, data size (which you try to process), and the relevant requirements such as the service latency, usability, flexibility, and so on. We'll cover the details of selecting data processing services in the *Selecting the appropriate data processing services for your analysis* section.

Orchestrating data pipelines with workflow tools

After building data pipelines combined with data processing services and your data sources, you may need to automate running your pipelines as a workflow to easily and safely run them without manual work. For example, you can create a scheduled-based workflow that automatically runs multiple pipelines, including multiple data processing jobs and multiple data sources, every morning. To run these pipelines, you don't need to manually run them one by one. You'll learn how to orchestrate your pipelines and workflow tools in the *Orchestrating your pipelines with workflow tools* section.

Automating how you provision your data pipelines and workflows

You can automatically run multiple data pipelines as a workflow with workflow tools. So, how can you build and manage multiple workflows if you have a lot of workflows? For example, let's assume you need to build hundreds of data pipelines that consist of the same data processing but various data sources. You can't imagine creating those pipelines each by one.

For this kind of use case, you can provision pipelines and workflows by using a template you define resources in with various provisioning tools, which we'll look at in the *Automating how you provision your pipelines with provisioning tools* section. Additionally, using provisioning tools, you can not only automate provisioning resources but also manage your resources via a template. By defining your pipeline resources with a template without manual operations in GUI applications, you can manage them with a versioning system and safely deploy them on your system by applying tests.

Developing and maintaining data pipelines

To build data pipelines and the relevant components, you also need to think about how you build them. In particular, you need to continuously update them without bugs based on company/organization/ team requirements, business needs, and so on. To achieve effective development cycles, a good solution is to apply the software practices of **continuous integration** (**CI**) and **continuous delivery** (**CD**) to your data pipeline development process. These concepts help with problem detection, productivity, release cycles, and so on. You learn how to utilize these concepts in your data pipelines development and management in the *Developing and maintaining your data pipelines* section. You'll learn how to develop Glue ETL jobs locally and how to deploy the ETL jobs and workflows in your environment in the section.

Next, we will cover four topics that we've looked at previously in terms of building and managing data pipelines using AWS Glue and combining it with other AWS services.

Selecting the appropriate data processing services for your analysis

One of the most important steps in using data processing pipelines is selecting the data processing services that meet the requirements for your data. In particular, you need to pay attention to the following:

- Whether your computing engine can process the data with the fastest speed you can allow
- Whether your computing engine can process all your data without any errors
- Whether you can easily implement data processing
- Whether the resource of your computing engine can easily be scaled as the amount of data increases (for example, you can scale it without making any changes to your code)

For example, if your data processing service doesn't have more memory capacity than your data, what does the computing engine do to your job? Having less memory capacity can cause **out-of-memory** (**OOM**) issues in your processing jobs and cause job failures. Even if you can process the data with that small memory capacity, it will slow down your data processing compared to processing the data in memory since you need to put some data aside in your disk to avoid issues. As another example, assuming that your job processes your data with a single node, what happens to your processing job in the future if the amount of data increases? You may need to scale up or scale out your computing resource for the engine as the job will need more time to process data as the amount of data increases. Then, when your computing engine reaches its limits in terms of its processing capabilities, you may need to select another computing engine that can process your data.

AWS provides multiple data processing services, such as AWS Lambda, AWS Glue, Amazon Athena, Amazon EMR, and more to match your environment's use cases and needs. In this section, we'll walk through each AWS-provided service for building data pipelines. Then, you'll learn how to choose the engine that satisfies your needs.

AWS Batch

AWS Batch is a fully managed service for running batch computing workloads based on your definition. Computing resources for AWS Batch are managed by AWS instead of customers. AWS Batch automatically provisions the resources and also optimizes your workload distribution based on workloads.

To run your batch computation, you must submit a unit of work, such as a shell script, a Linux executable, or a Docker container image, to AWS Batch. This definition is handled as a job. You can also flexibly define how jobs run – in particular, how many resources, such as CPU and memory, will be used, how many concurrency jobs will run, when AWS Batch executes jobs, and so on.

To use AWS Batch as a data processing service, you need to create a unit of work, a resource definition, and job scheduling. It runs on a single instance that you specify, so you need to care about resource limits such as memory, CPU, and so on. For more details about AWS Batch, please refer to `https://docs.aws.amazon.com/batch/latest/userguide/what-is-batch.html`.

Amazon ECS

Amazon Elastic Container Service (**ECS**) is a fully managed container orchestration service based on your container in a task definition. ECS also provides a serverless option, which is called AWS Fargate. Using Fargate, you don't need to manage resources, handle capacity planning, or isolate container workloads for security purposes.

Using ECS, all you need to do is build Docker images. After building these Docker images, you can deploy and run your images on ECS. You can also use this service as not only an application service but also as a data processing engine for big data. For example, you can deploy Apache Spark clusters, Kinesis Data Streams consumers, and Apache Kafka consumers by building Docker images.

Regarding container resources, ECS provides a wide variety of container instance types that are provided by Amazon EC2. Therefore, allocated resources such as memory and vCPUs are based on your instance images. Please refer to `https://docs.aws.amazon.com/AmazonECS/latest/developerguide/ECS_instances.html` regarding container instances.

AWS Lambda

AWS Lambda is a serverless computing service that runs your implemented code as Lambda functions on AWS-managed high-availability resources. All you need to do is write your code with a supported programming language, such as Python, Java, Node.js, Ruby, Go, or .NET, and a custom runtime.

Based on requests to Lambda, Lambda runs your defined Lambda functions with scaling automatically to respond to the requests. It can respond to up to 1,000 per second. You can use Lambda for a lot of use cases. The following are some examples:

- It can process batch-based data stored in S3

- It can process streaming-based data from streaming data sources such as DynamoDB Streams, Kinesis Data Streams, Managed Streaming Kafka, and others.

- It can work as an orchestrator of data pipelines to run data processing services such as AWS Glue, Amazon Athena, Amazon EMR, and others.

In addition to implementing the Lambda function code, you can set Lambda's resource configuration as follows:

- **Memory (MB)**: This determines the amount of memory that's available for your Lambda function. You can set this value between 128 MB and 10,240 MB. Regarding CPUs, they are linearly in proportion to the amount of memory that's been configured (at 1,769 MB, a function has the equivalent of 1 vCPU).

- **Timeout (seconds)**: This determines the Lambda execution timeout. If a function's execution exceeds this timeout, its execution is stopped. You can set this value to a maximum of 15 minutes.

Additionally, you can set asynchronous invocation, function concurrency, and so on.

As we've discussed, Lambda can be used in a lot of use cases and situations based on its implementation style. Therefore, it might be good to start using Lambda as a data processing service if you don't have a big data software environment such as Apache Hadoop, Apache Spark, and so on. Note that Lambda has memory limitations and that sometimes, duplicate invocation occurs.

Amazon Athena

Amazon Athena is a serverless query service. It allows you to run standard SQL queries for various data sources, such as CSV, JSON, Apache Parquet, Apache ORC, and so on, which are stored in your data stores, such as Amazon S3, JDBC/ODBC resources, and so on. Athena is based on Presto (`https://prestodb.io`), which provides a distributed SQL engine. This is useful for running ad hoc queries to obtain the analytical results of your data.

The Athena console provides an interactive view for users to easily run SQL queries, as shown in the following screenshot:

Figure 10.2 – Obtaining analytic data results by running a SQL query from the Athena console

In addition to the console, you can access Athena with APIs (`https://docs.aws.amazon.com/athena/latest/APIReference/Welcome.html`), SDKs (`https://aws.amazon.com/getting-started/tools-sdks/`), and more.

Athena can work with Glue Data Catalog as a Hive-compliant resource. Using Athena, you can create and read tables in/from the Data Catalog. If you need a data processing pipeline, you can build it with Athena. For example, you can build a simple pipeline so that Athena extracts data from S3 after creating a table with a Glue crawler, then writes the aggregated data to S3 using the access to Athena. This pipeline can be built by implementing a script that automates Athena queries and running the `StartQueryExecution` API (`https://docs.aws.amazon.com/athena/latest/APIReference/API_StartQueryExecution.html`) with AWS SDKs.

Athena charges your queries based on their data scanning size in terabytes. For more details about pricing, please refer to `https://aws.amazon.com/athena/pricing/`.

> **NOTE – Athena Service Quotas**
> When using Athena, you need to consider that Athena has default query quotas. For more information about service quotas, please refer to `https://docs.aws.amazon.com/athena/latest/ug/service-limits.html`.

AWS Glue ETL jobs

In AWS Glue ETL jobs, you can choose from **Spark**, **Spark Streaming**, and a **Python shell**. We'll look at these types here.

Spark

In terms of Spark, you can run Apache Spark applications as Glue jobs (hereafter, Glue Spark jobs) and process your data within Glue and Spark frameworks. To run Glue Spark jobs, you don't need to set up any resources for the computation. However, you need to implement scripts to process your data with Scala, Python (called PySpark), or SQL (called SparkSQL). Glue and Spark also provide many methods so that data processing can be enabled easily with a few pieces of code. The *Data ingestion from streaming a data source* section in *Chapter 3, Data Ingestion*, describes what Glue Spark is and how to use it. In Glue Spark jobs, you can choose a worker type that defines the memory, vCPUs, and disk size of each worker. **Worker type** is determined by your processing workloads, such as **Standard** for general use cases, **G.1X** for memory-intensive jobs, and **G.2X** for **machine learning** (**ML**) transform jobs.

Each worker type has a fixed allocated memory, vCPUs, and disk. At the time of writing, the details shown in the following table about these allocated resources are correct:

Allocated Resources per Worker (at the time of writing)	Standard	G.1X	G.2X
Memory	16 GB	16 GB	32 GB
vCPUs	4	4	8
Disk	50 GB	64 GB	128 GB

Figure 10.3 – Allocated resources of each worker type

In addition to the worker types, you need to set the number of workers, which defines how many workers with a specific worker type concurrently process your data.

The worker type and the number of workers define the capacity of the Glue computing resource (in other words, the Spark cluster) for your job. Specifically, they define how much memory and disk the job can use and how much concurrency the job processes. For example, when you set 10 G.1X workers to your Glue Spark job, the job can use a maximum of 160 GB memory, 40 vCPUs, and 640 GB disk for your entire Spark cluster.

> **Note – Data Processing Units (DPUs) and Maximum Capacity**
>
> The number of DPUs defines how many resources are allocated to your job. You are charged based on the DPUs you use in your job (please refer to `https://aws.amazon.com/glue/pricing/` for more information). A DPU has 4 vCPUs with compute capacity and 16 GB of memory.
>
> The maximum capacity is the same as the number of DPUs (for example, if you set 10 DPUs, the maximum capacity is also 10). When you choose Glue 1.0 and the Standard worker type, you need to set the **Maximum capacity** option instead of the **Number of workers** option.

Using Glue Spark jobs, you can use a distributed processing engine based on Spark, process your data with a lot of data processing methods, easily scale computing resources by changing the number of workers, and more.

Spark Streaming

Spark Streaming is one of the modules in Apache Spark for processing streaming data. This is different from Spark, which is typically used for batch jobs. Spark Streaming is used for streaming jobs for Glue (hereafter, Glue Streaming jobs). You can also implement Glue Streaming jobs with Scala, Python, or SQL, similar to Glue Spark jobs. The *Data ingestion from streaming a data source* section in *Chapter 3*, *Data Ingestion*, describes what Glue Streaming is and how to use it.

Regarding worker types and the number of workers for Glue Streaming jobs, you can configure them in the same way as you configure Glue Spark jobs. If you process the streaming data from streaming sources such as Amazon Kinesis Data Streams, Apache Kafka, and others, you can use this type. You are charged based on the DPUs per second you used in your job.

Python shell

If you select the Python shell type, you can run pure Python scripts, not PySpark, as Glue jobs (hereafter, Python shell jobs) on the Glue environment. Similar to the other Glue job types, you don't need to set up any resources for the computation. The *Data ingestion from the Amazon S3 object store* section in *Chapter 3*, *Data Ingestion*, describes what a Python shell is and how to use one.

Regarding worker types and the number of workers, you can only set the maximum capacity or DPUs for a Python shell job, not the worker types and number of workers. In particular, you can set the value to 0.0625 (the default DPU value) or 1. In addition to this, Python shell jobs can be integrated with other Glue components such as crawlers and Glue Spark jobs using a Glue workflow (which we'll see later in this chapter). You can also configure the job's timeout. The default is 48 hours.

When you don't need distributed processing via Spark jobs but you have a long-running job that, for example, simply checks multiple objects in S3 and deletes some objects based on a condition, you can use this type. You are charged based on the DPUs per second you selected (0.0625 or 1) in your job.

Amazon EMR

Amazon EMR (hereafter, EMR) provides a cluster management platform where you can run multiple big data-related applications such as Apache Hadoop, Apache Spark, Apache Hive, Presto/Trino, Apache HBase, Apache Flink, TensorFlow, and others in their latest versions. In addition to these applications, EMR also provides a lot of functionalities such as steps, bootstrap actions, and cluster configuration. We'll provide a summary of EMR here.

When you run multiple software applications, you don't always need to call each service API or log in each console/interactive shell. You can run these applications via **EMR Steps** (`https://docs.aws.amazon.com/emr/latest/ManagementGuide/emr-work-with-steps.html`), which runs applications on your behalf by adding your application implementation to EMR Steps.

You can also configure your cluster, such as its size, EC2 instance types, multiple versions of applications that match your needs, and so on. You can also add the software that you need to create an EMR cluster via the **EMR Bootstrap action** (`https://docs.aws.amazon.com/emr/latest/ManagementGuide/emr-plan-bootstrap.html`). This can be defined by implementing scripts and setting these scripts when creating the cluster. It's also possible to connect to AWS Glue Data Catalog.

Compared to AWS Glue, EMR enables you to provide various flexible options for selecting applications, cluster size, cluster scaling, cluster nodes, customizing the cluster node system, and so on. Furthermore, you can choose a cluster running environment from Amazon EC2 (EMR on EC2), Amazon EKS (EMR on EKS), AWS Outposts, and Serverless (this is a preview feature). However, note that EMR is not serverless except for EMR Serverless, so you need to manage clusters yourself.

Regarding EMR pricing, you are charged based on your running node type and running duration. For more details, please refer to `https://aws.amazon.com/emr/pricing/`.

Orchestrating your pipelines with workflow tools

After selecting the data processing services for your data, you must build data processing pipelines using these services. For example, you can build a pipeline similar to the one shown in the following diagram. In this pipeline, four Glue Spark jobs extract the data from four databases. Then, each job writes data to S3. In terms of the data stored in S3, the next Glue Spark job processes the four tables' data and generates an analytic report:

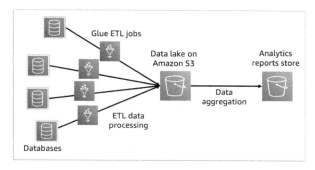

Figure 10.4 – A pipeline that extracts data from four databases, stores S3, and generates an analytic report by the aggregation job

So, after building a pipeline, how do you run each job? You can manually run multiple jobs to extract multiple databases. Once this has happened, you can run the job to generate a report. However, this can cause problems. One such problem is not getting a result if you run the generating report job before all the extracting jobs are completed. Another problem is that it will take a long time to generate a report if one of the extracting jobs takes a lot of time.

To avoid these problems, you can orchestrate pipelines with workflow tools such as AWS Glue workflows, AWS Step Functions, Apache Airflow, and others. Workflow tools for big data pipelines generally orchestrate not only multiple jobs but also multiple pipelines.

Recent modern workflow tools, such as the ones mentioned previously, represent the flow of jobs and the dependencies of jobs in a pipeline as a graph – in particular, a **directed acyclic graph** (**DAG**). A DAG has direction for each edge, but no directed cycles. In a cycle graph, the first and last edges are equal. The following diagram shows a DAG that represents the workflow example from earlier in this section, which involved generating a report pipeline:

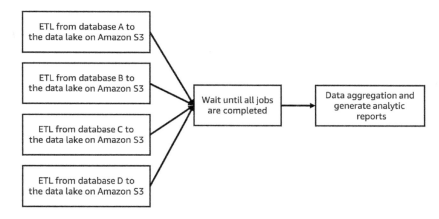

Figure 10.5 – A DAG workflow for generating a report pipeline

Using workflow tools, you can manage multiple jobs and pipelines as one workflow. Regarding the example of generating a report, a workflow tool can run each job, which may include extracting data from multiple databases, waiting for each job to complete, and generating a report. Thus, you don't need to run each job manually.

In this section, we'll walk through the workflow tools that AWS provides and learn how to combine them with the data processing services we looked at in the *Selecting the appropriate data processing services for your analysis* section:

- **AWS Glue workflows**
- **AWS Step Functions**
- **Amazon Managed Workflows for Apache Airflow (MWAA)**

First, we'll look at AWS Glue workflows.

Using AWS Glue workflows

AWS Glue workflows allow you to create workflows that combine dependent Glue functionalities such as crawlers and ETL jobs as an orchestrator. In particular, Glue workflows execute crawlers and ETL jobs using Glue Trigger, which triggers crawlers and ETL jobs based on your configuration, such as on-demand, scheduled, or conditional, or via an EventBridge trigger. More information was provided in the *Triggers* section of *Chapter 2, Introduction to Important AWS Glue Features*. In addition to the role of the orchestrator, Glue workflows allow you to monitor each workflow component's status, such as the success of ETL jobs, the failure of crawler runs, and so on.

To learn how we can configure and run Glue workflows, let's orchestrate a simple data pipeline by building a pipeline and using Glue workflows.

Example – orchestrating the pipeline that extracts data and generates a report using Glue workflows

We'll create the data pipeline that generates a product sales report by processing sales records that you can download from https://github.com/PacktPublishing/Serverless-ETL-and-Analytics-with-AWS-Glue/blob/main/Chapter10/sales-data.json. The following data shows example records in the sales data:

```
{"product_name":"tomato juice","price":"2.00","customer_
id":1698,"order_id":"DRE8DLTFNX0MLCE8DLTFNX0MLC","dateti
me":"2020-07-18T02:20:58Z","category":"drink"}
{"product_name":"measuring cup","price":"1.00","customer_
id":1643,"order_id":"KI6OOLNOY76OOLNOY7","datetime":"2021-02-
16T02:56:33Z","category":"kitchen"}
```

Then, we'll run this pipeline by creating a workflow. This workflow will run the pipeline by doing the following:

1. The Glue workflow will trigger the crawler (`ch10_1_example_workflow_acr`), which analyzes a table schema of the sales data and populates a table in Glue Data Catalog.

2. After running the crawler, the workflow will trigger the ETL job (`ch10_1_example_workflow_gen_report`), which will generate a report by computing sales by each product category and year. Then, the job will populate the report table in the Data Catalog.

Let's start by creating the data pipeline.

Step 1 – creating a data pipeline with a Glue crawler and an ETL job

We'll download product sales data and create the Crawler which populates a table in the Data Catalog based on the table schema of the sales data. Follow these steps:

1. Download the product sales data (`sales-data.json`) on your local machine from `https://github.com/PacktPublishing/Serverless-ETL-and-Analytics-with-AWS-Glue/blob/main/Chapter10/sales-data.json` Once downloading is completed, upload the file to your Amazon S3 bucket using the command; `aws s3 cp sales-data.json s3://<your-bucket-and-path>/sales` or from the S3 console (`https://s3.console.aws.amazon.com/s3/buckets`)

2. Access **Crawlers** (`https://console.aws.amazon.com/glue/home?region=us-east-1#catalog:tab=crawlers`) on the AWS Glue console and choose **Add crawler**.

3. Type `ch10_1_example_workflow` as the crawler's name and click **Next**.

4. Choose **Data stores** for **Crawler source type** and **Crawl all folders** for **Repeat crawls of S3 data stores**. Then, click **Next**.

5. Choose **Specified path in my account** in the **Crawl data in** section and specify `s3://<your-bucket-and-path>/sales/` that is the data location of `sales-data.json` for **Include path**. Then, click **Next**.

6. Set **No** for **Add another data store**.

7. Choose your IAM role for this crawler. You can also create an IAM role by clicking **Create an IAM role**.

8. Set **Run on demand** for **Frequency**.

9. Choose your database to create the report table in and type `example_workflow_` in **Prefix added to tables (optional)** for the table.

10. Then, review your crawler's configuration. If everything is OK, click **Finish**.

> **Specification of table name created by Crawler**
>
> The table name that Crawler creates is determined as `<Prefix><The deepest path that you specified in Include path>`. For example, if you set `example_workflow_` to **Prefix**, and `s3://<your-bucket-and-path>/sales/` to **Include path**, Crawler creates the table with its name **example_workflow_sales**.

At this point, you will see the `ch10_1_example_workflow` crawler on the console.

Now, let's create an ETL job to process the dataset and create a report table. Follow these steps:

1. Download the Glue job script from this book's GitHub repository (`https://github.com/PacktPublishing/Serverless-ETL-and-Analytics-with-AWS-Glue/blob/main/Chapter10/workflow-tools/glue-workflows/ch10_1_example_workflow_gen_report.py`).

2. Open **Jobs** in the AWS Glue Studio console (`https://console.aws.amazon.com/gluestudio/home?region=us-east-1#/jobs`). Then, choose **Spark script editor** in the **Create job** section and **Upload and edit an existing script** in the **Options** section. Now, upload the job script by clicking **Choose file**:

Figure 10.6 – The view for creating a Glue job in AWS Glue Studio

3. After uploading the `ch10_1_example_workflow_gen_report.py`, click **Create**.

4. Type `ch10_1_example_workflow_gen_report` as the job's name and choose your **IAM Role** for running the Glue job.

5. Scroll down the page and set **Requested number of workers** to 3, **Job bookmark** to `Disable`, and **Number of retries** to `0`.

6. Then, set each of your S3 bucket paths, using the details shown in the following screenshot:

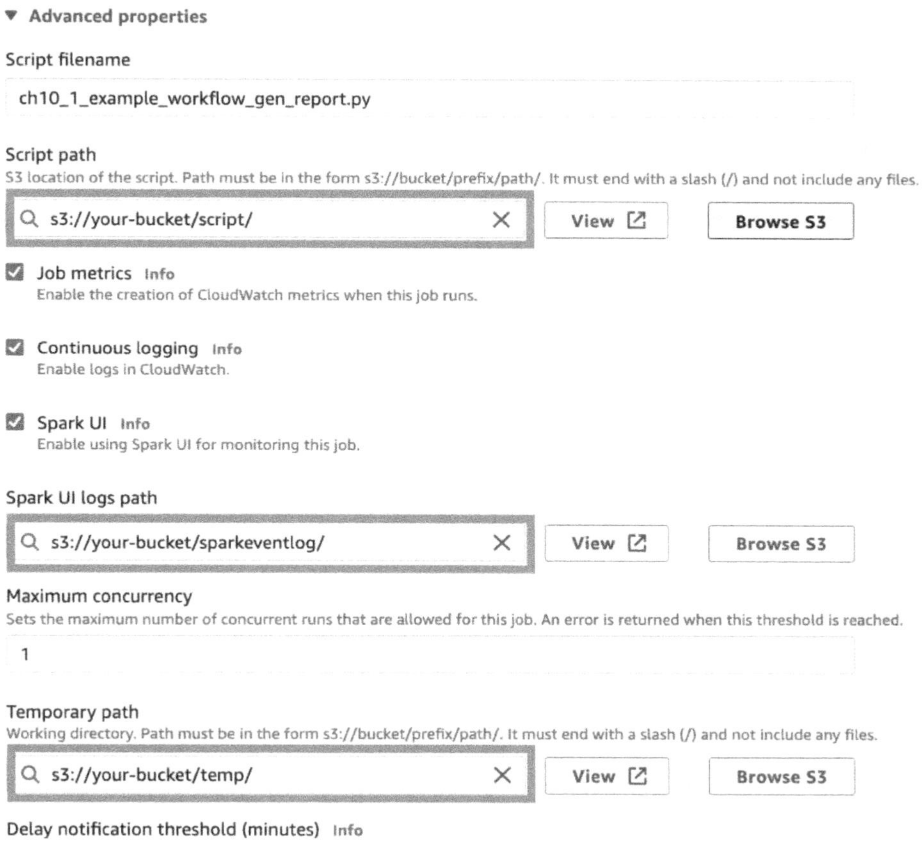

▼ **Advanced properties**

Script filename

ch10_1_example_workflow_gen_report.py

Script path
S3 location of the script. Path must be in the form s3://bucket/prefix/path/. It must end with a slash (/) and not include any files.

Q s3://your-bucket/script/   ×   View ⬈   Browse S3

☑ **Job metrics** Info
Enable the creation of CloudWatch metrics when this job runs.

☑ **Continuous logging** Info
Enable logs in CloudWatch.

☑ **Spark UI** Info
Enable using Spark UI for monitoring this job.

Spark UI logs path

Q s3://your-bucket/sparkeventlog/   ×   View ⬈   Browse S3

Maximum concurrency
Sets the maximum number of concurrent runs that are allowed for this job. An error is returned when this threshold is reached.

1

Temporary path
Working directory. Path must be in the form s3://bucket/prefix/path/. It must end with a slash (/) and not include any files.

Q s3://your-bucket/temp/   ×   View ⬈   Browse S3

Delay notification threshold (minutes) Info

Figure 10.7 – Setting a script, Spark event logs, and temporary locations

7. Scroll down the page and set `s3://crawler-public/json/serde/json-serde.jar` to **Dependent JARs path**.

8. Save the job.

Now that you've created the data pipeline, you will create a workflow by using the crawler and glue job you created.

Step 2 – creating a workflow

Let's create a workflow that will manage the crawler and ETL job that you created in *Step 1 – creating a data pipeline with a Glue crawler and an ETL job*. Follow these steps:

1. Open **Workflows** in the AWS Glue console (`https://console.aws.amazon.com/glue/home?region=us-east-1#etl:tab=workflows;workflowView=workflow-list`) and click **Add workflow.**

2. Set `ch10_1_example_workflow_gen_report` as the workflow's name and set the following workflow run properties:

    I.   **Key**: `datalake_location`, **Value**: `s3://<your-bucket-and-path>`; this is the report data S3 path.

    II.   **Key**: `database`, **Value**: `<the db name which you set to the Crawler you created in Step 1>`; this is the table of the Amazon Customer Review dataset.

    III. **Key**: `table`, **Value**: `example_workflow_sales`; this table is created by the crawler and its name is set to this value.

    IV.   **Key**: `report_year`, **Value**: `2021`; In this example, `2021` is set as the value.

3. Then, click **Add workflow** at the bottom of the page.

4. After adding the workflow, you can create a Glue trigger to run your workflow. Click **Add trigger**:

Figure 10.8 – Adding a trigger to the workflow

5. Go to the **Add new** tab, type `ch10_1_example_workflow_ondemand_start` as the workflow's name, and set **On demand** for **Trigger type**. Then, click **Add**.

6. At this point, you will be able to see the first trigger in the **Graph** tab. Let's add the **Crawler** first. Click **Add node**, as shown in the following screenshot:

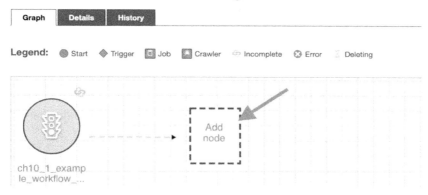

Figure 10.9 – Adding a node to the workflow

7. Go to the **Crawlers** tab, specify the `ch10_1_example_workflow` crawler, and click **Add**.

8. You will see the crawler in your workflow diagram. Now, create a new trigger to run the ETL job. Click **Add trigger** in the workflow diagram.

9. In the **Add new** tab, type `ch10_1_example_workflow_event_gen_report` as a new trigger name. Set **Event** as its trigger type and **Start after ALL watched event** as its trigger logic. Then, click **Add**.

10. The following screenshot shows the additional trigger that starts running ETL jobs. To set the job for this trigger, click **Add node**:

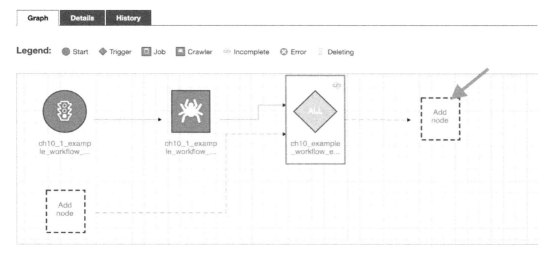

Figure 10.10 – Adding a new Glue job node to the workflow

11. Go to the **Jobs** tab, specify the `ch10_1_example_workflow_gen_report` job, and click **Add**.

Once you've configured the workflow, you will see the following diagram in the **Graph** tab:

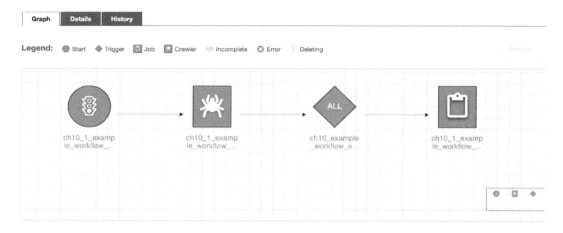

Figure 10.11 – The workflow diagram

Now, you're ready to run the workflow! This is what we'll do in the next step.

Step 3 – running the workflow

You can run the workflow via the Glue console. Follow these steps:

1. Go back to the workflow in the Glue console and choose your workflow (`ch10_1_example_workflow_gen_report`). Then, choose **Actions** and click **Run**.

2. After starting the workflow, you can see the workflow's running status by going to **View run details** in the **History** tab.

3. Once the workflow has finished running, you will see each node's status, as shown in the following diagram (this workflow run may take around 4 or 5 minutes):

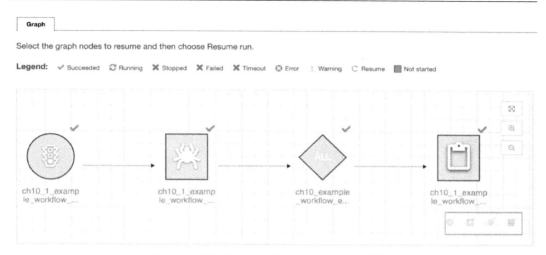

Figure 10.12 – The completed graphical workflow

The workflow run is now completed. Finally, let's check the result.

Step 4 – checking the result

By running this workflow, two tables were created by the crawler and the ETL job, and the reviews count report was provided as output in the S3 bucket you specified as `datalake_location` in *Step 2 – creating a workflow*. Let's have a look at these resources:

- The two tables that were created (you can see these tables in the Glue Data Catalog at `https://console.aws.amazon.com/glue/home#catalog:tab=tables`):

  - `example_workflow_sales`: This was created by the crawler; that is, `ch10_1_example_workflow`. This table contains the table schema of the sales data.

  - `example_workflow_sales_report`: This was created by the ETL job; that is, `ch10_1_example_workflow_gen_report`. This table has the table schema which includes, reported year as a partition key.

- The generated report data in the S3 bucket. The ETL job writes the report data in the S3 path as `s3://<your-specified-bucket-and-path>/serverless-etl-and-analysis-w-glue/chapter10/example-workflow/report/`. You can view the following bucket path and data by using the AWS CLI command:

  ```
  $ aws s3 ls s3://<your-bucket-path>/serverless-etl-
  and-analysis-w-glue/chapter10/example-workflow/report/
  --recursive
  ```

```
YYYY-MM-dd 01:23:45          799 <path>/serverless-etl-and-
analysis-w-glue/chapter10/example-workflow/report/report_
year=2021/run-xxxxxxxxxx-part-block-0-0-r-00113-snappy.
parquet
```

- The ETL job output in the CloudWatch logs. You can access the log link from the Glue Studio console by choosing **Output logs** in the **Runs** tab. The page will redirect you to the CloudWatch Logs console. Choose the Spark driver task ID that doesn't have an underscore (_) in the name of the Log stream; that is, jr_ea5565f6e248aa49dbbb....

You will see the following generated report. This report shows the product sales by each category in 2021:

```
+--------+-----------+-----------+
|category|total_sales|report_year|
+--------+-----------+-----------+
|   drink|        906|       2021|
| grocery|        983|       2021|
| kitchen|       4535|       2021|
|  office|       1316|       2021|
+--------+-----------+-----------+
```

Figure 10.13 – The Glue job's output in the Spark driver task log

In this section, we've done the following:

- Created a pipeline that is composed of a crawler and an ETL job:

  - The crawler populates a table in the Data Catalog

  - The ETL job generates a report by referring to the table data

- Created the workflow, which consists of two triggers for running the crawler and the ETL job. This workflow runs each component in the pipeline.

- Run the workflow and checked the result.

In this example, we learned that Glue workflows allow you to run data pipelines that consist of multiple crawlers and jobs. However, you may think that it's a bit hard to build multiple workflows that have multiple triggers/crawlers/ETL jobs because you need to set each component one by one. This can be solved by using provisioning tools such as AWS CloudFormation, Glue Blueprints, and so on. We'll look at these tools in the *Automating how you provision your pipelines with provisioning tools* section. Next, we'll look at another workflow tool: **AWS Step Functions**.

Using AWS Step Functions

AWS Step Functions is a serverless orchestration service that allows you to combine multiple AWS services such as AWS Lambda, AWS Glue, and so on. It can also be used to orchestrate and run multiple data pipelines, including multiple AWS data processing services and their related data storage. You can define workflows with Step Functions' graphical console, which visualizes your workflows.

Step Functions consists of **state machines** and **tasks**. Let's look at them in more detail:

- A **state machine** is a workflow.

- A **task** is a state (or a step) in a workflow. This state represents a single unit of work that's performed by a state machine.

To define a Step Functions workflow, you must create a state machine that has and combines multiple tasks, such as invoking a Lambda function, starting a Glue job run, running an Athena query, and so on.

Step Functions can handle AWS Glue APIs and you can create ETL workflows via Step Functions. Next, we'll orchestrate the same data pipeline that we built in the previous Glue workflows example by building a workflow with Step Functions.

Example – orchestrating the pipeline that extracts data and generates a report using Step Functions

In this example, we'll create the same data pipeline that we did in the *Example – orchestrating the pipeline that extracts data and generates a report using Glue workflows* section. Then, we'll orchestrate the pipeline with Step Functions' workflow. This pipeline will generate a product sales report by computing sales by each product category and year.

The Step Functions' workflow runs the pipeline by doing the following:

1. Step Functions' workflow triggers the crawler (`ch10_2_example_workflow_acr`), which analyzes a table schema of the sales data and populates a table in Glue Data Catalog.

2. After starting the crawler, the workflow polls the crawler's running status. If it confirms that the crawler has finished running, it triggers the ETL job (`ch10_2_example_workflow_gen_report`), which generates a report by computing sales by each product category and year. Then the job populates the report table in the Data Catalog.

First, let's create the pipeline.

Step 1 – creating a data pipeline with a Glue crawler and an ETL job

In this example, we'll create the crawler and the ETL job. These will have the same configuration as the crawler (`ch10_1_example_workflow`) and ETL job (`ch10_1_example_workflow_gen_report`) we created in the *Example – orchestrating the pipeline that extracts data and generates a report using Glue workflows* section. If you haven't created the crawler and ETL job, please refer to that section. Follow these steps:

1. Go to **Crawlers** (`https://console.aws.amazon.com/glue/home?region=us-east-1#catalog:tab=crawlers`) in the AWS Glue console and choose `ch10_1_example_workflow`. Then, choose **Duplicate crawler** from the **Action** tab.

2. Type `ch10_2_example_workflow` as the crawler's name and choose **Output** in the left pane.

3. In the crawler's **Output** view (in the left pane), type `example_workflow_sfn_` in **Prefix added to tables (optional)** for the table.

4. After reviewing the crawler's configuration, click **Finish**.

5. Next, you must create the Glue job. Before creating the job, download the Glue job script from this book's GitHub repository (`https://github.com/PacktPublishing/Serverless-ETL-and-Analytics-with-AWS-Glue/blob/main/Chapter10/workflow-tools/step-functions/ch10_2_example_workflow_gen_report.py`).

6. Open the job in the AWS Glue Studio console (`https://console.aws.amazon.com/gluestudio/home?region=us-east-1#/jobs`). Then, choose the **ch10_1_example_workflow_gen_report** job and choose **Clone job** from the **Actions** tab to take over the previous job configuration.

7. On the **Job details** tab, type `ch10_2_example_workflow_gen_report` as the job's name. Confirm that the script's filename is `ch10_2_example_workflow_gen_report.py`.

8. On the **Script** tab, copy the downloaded job script to the editor. Then, click **Save** to save the job.

Next, we'll create a Step Functions state machine by combining it with a Glue crawler and an ETL job.

Step 2 – creating a state machine

In this step, we'll create a step machine that orchestrates a Glue crawler and an ETL job:

1. Before creating the job, download the state machine definition from this book's GitHub repository (`https://github.com/PacktPublishing/Serverless-ETL-and-Analytics-with-AWS-Glue/blob/main/Chapter10/workflow-tools/step-functions/ch10_2_example_sfn.json`).

2. Open the AWS Step Functions console (`https://console.aws.amazon.com/states/home#/statemachines`) and click **Create state machine**.

3. On the **Define state machine** page, click **Write your workflow in code** and then **Standard** in the **Type** section. Then, copy the downloaded definition to the script editor (by clicking the **Reload** button, you can see the visualized workflow, as shown in the following screenshot). After copying the script, click **Next**:

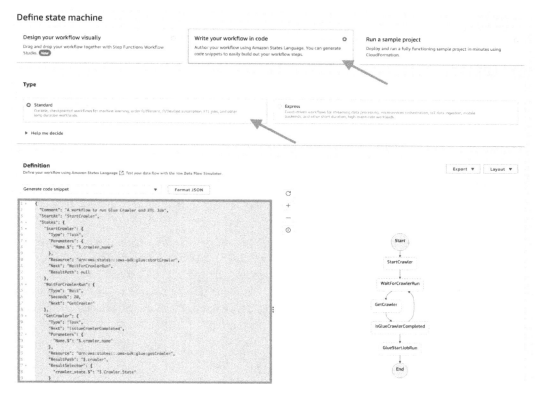

Figure 10.14 – Defining the state machine

As we've discussed, this state machine polls the crawler's running status periodically (every 20 seconds). After that, the state machine starts the ETL job.

4. On the **Specify details** page, type `ch10_2_example_workflow_sfn` as the state machine's name and click **Create new role** (the IAM Role that includes the necessary permission is created by AWS). Regarding the **Logging** section, by default, logging configuration is not enabled. If necessary, you can set any log level such as `ALL`, `ERROR`, and so on.

When you scroll down the page, you may see a notification about insufficient permissions that states "*Permissions for the following action(s) cannot be auto-generated ….*" After creating the state machine, we'll add these permissions to the IAM Role.

5. Click **Create state machine**. Upon doing this, the state machine you defined will be created.

6. To add the insufficient permission to the IAM Role for the state machine, open the IAM console (`https://console.aws.amazon.com/iamv2/home#/policies`) and create a new IAM policy. Copy the policy file in this bok's GitHub repository (`https://github.com/PacktPublishing/Serverless-ETL-and-Analytics-with-AWS-Glue/blob/main/Chapter10/workflow-tools/step-functions/ch10-2-sfn-additional-glue-policy.json`). After creating the policy, attach it to the IAM Role.

Now, you're ready to run the workflow. We'll do this in the next section via the Step Functions console.

Step 3 – running the state machine

Let's run the workflow. In this step, we'll run it manually from the Step Functions console. You can also invoke the state machine via the `StartExecution` API (`https://docs.aws.amazon.com/step-functions/latest/apireference/API_StartExecution.html`). Follow these steps:

1. Go back to the Step Functions console, choose the `ch10_2_example_workflow_sfn` state machine, and click **Start execution**.

2. Specify the following input to the state machine. You can copy the input from this book's GitHub repository: `https://github.com/PacktPublishing/Serverless-ETL-and-Analytics-with-AWS-Glue/blob/main/Chapter10/workflow-tools/step-functions/ch10_2_input.json`. Note that we need to replace the values of `--datalake_locaiton` and `--table`. These parameters are processed by the state machine and passed to the ETL job as job parameters:

Name - *optional*

```
ffdb5e40-9dc9-0025-f619-63cc56083822
```

Input - *optional*
Enter input values for this execution in JSON format

```
 1 ▾ {
 2        "crawler_name": "ch10_2_example_workflow",
 3        "etl_job_name": "ch10_2_example_workflow_gen_report",
 4 ▾      "etl_job_args": {
 5            "--datalake_location": "s3://<your-bucket-and-path>",
 6            "--database": "<your-database>",
 7            "--table": "example_workflow_sfn_sales",
 8            "--report_year": "2021"
 9        }
10    }
```

Figure 10.15 – The input to the state machine

3. After starting the execution, you will be able to see the running status of each task. Once the execution has finished, you will see the following diagram:

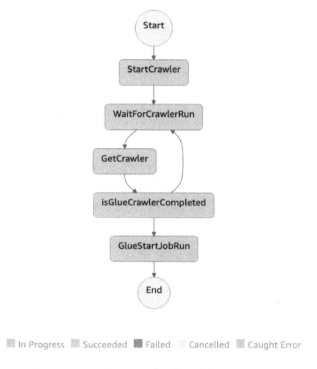

Figure 10.16 – The completed workflow diagram

Now, let's check out the result of executing the workflow.

Step 4 – checking the result

Here, we get the same result that we got in the *Orchestrating the pipeline that extracts data and generates a report by Glue workflows* section. Therefore, we won't look at the result in detail here, but we will look at the output:

- Two tables were created in the Data Catalog:

    - `example_workflow_sfn_sales`: This was created by `ch10_1_example_ workflow` crawler.

    - `example_workflow_sfn_sales_report`: This was created by the `ch10_2_ example_workflow_gen_report` job.

- The report data was generated in the S3 path as `s3://<your-specified-bucket-and-path>/serverless-etl-and-analysis-w-glue/chapter10/example-workflow-sfn/report/`.

- The ETL job's output shows the sales data of each category and year in CloudWatch Logs. You can access this log from the Glue job, as we've seen previously.

In this example, we learned that Step Functions also provides running data pipelines that consist of multiple crawlers and jobs, similar to what Glue workflows provide. Using Step Functions, you can manage your workflows using a JSON-like template. This can make it easier to build and manage workflows compared to manually creating workflows via a GUI application because all you need to do is manage your templates.

Step Functions supports not only AWS Glue but also other AWS services such as AWS Lambda, Amazon Athena, and others. By using Step Functions, you can create various workflows by combining multiple AWS services.

Now, let's look at **Amazon Managed Workflows for Apache Airflow (MWAA)** one of many available workflow tools.

Using Amazon Managed Workflows for Apache Airflow

MWAA is a distributed orchestration service that provides programmatic workflow management. MWAA is based on **Apache Airflow** (`https://airflow.apache.org`), whose resources are managed by AWS. Airflow runs workflows that are expressed as DAGs, as defined by Python. By defining workflows as DAGs, Airflow orchestrates and schedules your workflows. We won't explain the details of Airflow in this book, but you can refer to the public Airflow documentation if you want to learn more: `https://airflow.apache.org/docs/apache-airflow/stable/concepts/index.html`.

You can use MWAA to create workflows that combine not only AWS Glue but also other AWS services, such as Amazon Athena, Amazon EMR, and others. Next, we'll learn how to combine MWAA with AWS Glue by creating the same workflow that we created in the previous two examples.

Example – orchestrating the pipeline that extracts data and generates a report using MWAA

In this example, you'll learn how to use MWAA as a workflow tool for Glue by creating the same workflow and pipeline that you created for Glue workflows and Step Functions. In the workflow, MWAA runs a crawler. After completing the crawler run, it starts an ETL job. If you haven't set up the MWAA environment yet, please refer to `https://docs.aws.amazon.com/mwaa/latest/userguide/get-started.html` (this document link is also provided in the *Technical requirements* section).

Step 1 – creating a data pipeline with a Glue crawler and an ETL job

First, download the Glue job script from this book's GitHub repository at `https://github.com/PacktPublishing/Serverless-ETL-and-Analytics-with-AWS-Glue/blob/main/Chapter10/workflow-tools/mwaa/ch10_3_example_workflow_gen_report.py`. The crawler and the ETL job that you will create here will be the same ones that you created in *Step 1 – creating a data pipeline with a Glue crawler and an ETL job* in the *Example – orchestrating the pipeline that extracts data and generates a report using Glue workflows* section. You'll create the following resources with updating configuration:

- **Crawler** (`ch10_3_example_workflow`): Create this crawler by replicating `ch10_2_example_workflow_acr crawler`. Update the table prefix that the crawler creates from `example_workflow_sfn_` to `example_workflow_mwaa_`.

- **ETL job** (`ch10_3_example_workflow_gen_report`): Create this job by copying the `ch10_2_example_workflow_gen_report` job. Update the job script from `ch10_2_example_workflow_gen_report.py` to `ch10_3_example_workflow_gen_report.py` (this can be downloaded from the aforementioned GitHub repository).

Now that you've created the crawler and job, you must set up the workflow via MWAA.

Step 2 – creating a workflow with MWAA

To create and run the DAG, you need to upload the DAG file that's been written in Python to the S3 bucket that is specified for your MWAA environment. The DAG file (`ch10_3_example_workflow_dag.py`) can be downloaded from this book's GitHub repository at `https://github.com/PacktPublishing/Serverless-ETL-and-Analytics-with-AWS-Glue/blob/main/Chapter10/workflow-tools/mwaa/ch10_3_example_workflow_dag.py`. After downloading it, upload it to the DAG location in your S3 bucket.

After uploading the DAG file, you will see the `ch10_3_example_workflow_mwaa` workflow from **Airflow UI**. Now, you can trigger this workflow by using the **Trigger** button in the **Actions** column in the Airflow UI.

Step 3 – checking the result

After running the workflow, you will see the following DAG execution result from the Airflow UI. In particular, you will see if the DAG was successful or not, as well as concrete components such as the `sales_crawl` task (which is `ch10_3_example_workflow` crawler-run) and `gen_report` (which is `ch10_3_example_workflow_gen_report` job-run):

Figure 10.17 – The DAG's execution result in Airflow UI

You will also see each of the component's results, as follows:

- The `example_workflow_mwaa_sales` table is created by the `sales_crawl` task.
- The `example_workflow_mwaa_sales_report` table is created by the `gen_report` task.
- The `gen_report` task also writes the data in your specified S3 path.

By walking through this basic example, you've learned that you can also use MWAA as a workflow tool for Glue. Using MWAA, you can programmatically manage your workflows with Python. This can also make it easier to build and manage workflows compared to manually creating them. Additionally, you can provision workflows more safely by adding testing code steps (such as unit tests, integration tests, and so on) to your development life cycle.

As Step Functions does, MWAA supports not only Glue but also other AWS services, such as Amazon Athena, Amazon EMR, and others. You can find more examples of creating workflows, including Glue by MWAA in the AWS Glue public document and AWS big data blog posts. If you're interested in this example, please refer to the *Further reading* section at the end of this chapter.

As you've seen, several workflow tools, such as Glue workflows, Step Functions, and MWAA, can run your pipeline components step by step based on your workflow's definition, such as scheduling, on-demand, and so on. However, you need to create pipeline components before building and running workflows. If you need to create pipelines that consist of a lot of components, it's not easy to manually create, update, and replicate the pipelines, which you did in each of the preceding examples. To make these operations easy, you can use another tool that builds resources on your behalf. This tool is generally called provisioning tools. We'll look at this in the next section.

utomating how you provision your pipelines with provisioning tools

In the previous section, *Orchestrating your pipelines with workflow tools*, you learned how to orchestrate multiple pipelines and automate how they run with one tool. Using workflow tools for multiple pipelines can not only avoid human error but can also help you understand what pipelines do.

Note that as your system grows, you will build a lot of pipelines, and then you will build workflows to orchestrate them. If you have a lot of workflows as your system grows, you may need to consider how you should manage them. If you manually build several workflows and deploy them on your system, similar to how you would build and run pipelines manually, you may build some workflows that contain bugs. You can do this by specifying incorrect data sources, connecting incorrect pipeline jobs, and so on. As a result, this will corrupt your data and system, and pipeline job failures will occur due to broken workflows being deployed.

So, how can you avoid these kinds of errors when building workflows? One of the solutions involves using provisioning tools such as **AWS CloudFormation** (`https://aws.amazon.com/cloudformation/`), **AWS Glue Blueprints** (`https://docs.aws.amazon.com/glue/latest/dg/blueprints-overview.html`), **Terraform** (`https://www.terraform.io`), which is provided by Hashicorp, and others.

Provisioning tools generally deploy resources defined in the template, which you specify as JSON, YAML, and so on. Here's a simple example template of AWS CloudFormation, which creates the `glue_db` database and then the `glue_table` table in your Glue Data Catalog:

```
Resources:
  GlueDatabase:
    Type: AWS::Glue::Database
    Properties:
      CatalogId: !Ref AWS::AccountId
      DatabaseInput:
        Name: 'glue_db'
  GlueTable:
    Type: 'AWS::Glue::Table'
    DependsOn:
      - GlueDatabase
    Properties:
      CatalogId: !Ref AWS::AccountId
      DatabaseName: 'glue_db'
      TableInput:
        Name: 'glue_table'
```

Figure 10.18 – An example of a CloudFormation template

As mentioned previously, in this example, by using provisioning tools, you can manage your pipelines and workflows as a template that's in JSON, YAML format, and so on. In addition to this, there are provisioning tools that allow you to define and manage your pipelines and workflows as code. For example, you can define your data pipelines with popular programming languages, and you can also safely deploy them by running your resource definition code. AWS Glue provides this programmatic resource definition functionality via AWS Glue Blueprints. Other tools are provided by AWS for this purpose, such as **AWS Cloud Development Kit** (**AWS CDK**), which automatically creates CloudFormation templates based on your code.

In this section, you'll learn how to build and manage your workflows and pipelines with provisioning tools. Specifically, we'll focus on the following two services, which are provided by AWS:

- AWS CloudFormation
- AWS Glue Blueprints

First, we'll look at AWS CloudFormation.

Provisioning resources with AWS CloudFormation

AWS CloudFormation allows you to model and set up AWS resources with a template where you define the necessary resources. CloudFormation mainly provides the following features for users:

- **Simplifying your resource management**: All you need to do is create or update a template. Based on this template, CloudFormation sets up resources for your environment on your behalf.

- **Quickly replicating your resources**: Once you have defined a template, by reusing it, you can create or update your resources over and over.

- **Controlling and tracking changes in your resources**: By defining your resources as a text-based file (we've been calling this a **template**), you can control and track your resources.

You can define the resources that you want to deploy, and related resource properties in a template in JSON or YAML format. In CloudFormation, defined resources in a template are handled as a single unit. This unit is called a **stack**. If you want to change your running resources and update a stack, you can create sets of your proposed changes before making changes to them. These sets are called **change sets**. They allow you to see how your running resources change before you update them.

By using CloudFormation for your data pipelines, you can build data pipeline resources such as data processing services, workflows, and more with a template. Additionally, CloudFormation can track changes in your pipeline resources. Once you have defined data pipelines and workflows in a template, you don't need to manually create or update pipelines with GUI tools. Therefore, CloudFormation helps not only easily provision resources but also avoid human error, such as workflow misconfiguration and incorrectly setting data processing engines.

CloudFormation covers a lot of AWS services, including Glue. Through a template, you can set up Glue resources such as databases, tables, crawlers, jobs, and more. To learn more about the Glue resources that CloudFormation covers, please refer to `https://docs.aws.amazon.com/glue/latest/dg/populate-with-cloudformation-templates.html`.

Now, let's learn how to set up a schedule-based data pipeline that consists of Glue ETL jobs and Glue workflows by defining resources in a CloudFormation template.

Example – provisioning a Glue workflow using a CloudFormation template

In this example, you will extend the data pipeline that you created in the *Orchestrating your pipelines with workflow tools* section. In particular, you will provision the `ch10_4_example_cfn_` Glue workflow by CloudFormation (this workflow has been omitted in each component name in the following diagram). This workflow runs each component in the pipeline as follows:

1. The `ondemand_start` component triggers the `acr` crawler, which populates a table based on the sales data.

2. After `crawler-run` is completed, `event_run_partitioning` triggers the `partitioning` job. This job extracts the data from the Amazon Customer Reviews dataset and writes the data to the S3 path with year and month-based partitioning.

3. Once the `partitioning` job has finished running, `event_run_gen_report` triggers the `gen_report` job. This job generates the same report that the job in the *Orchestrating your pipelines with workflow tools* section did:

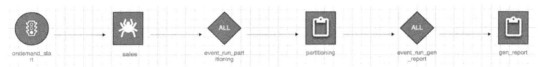

Figure 10.19 – The Glue workflow graph you'll create via CloudFormation

Let's create this workflow using CloudFormation.

Step 1 – putting ETL job scripts in your S3 bucket

Before provisioning the resources via CloudFormation, copy the necessary job scripts to your S3 bucket by using the S3 console or the `aws s3 cp <your_local_script_location> s3://<your-bucket-and-path>/` AWS CLI command. You can download these job scripts from the following GitHub repository links:

- `https://github.com/PacktPublishing/Serverless-ETL-and-Analytics-with-AWS-Glue/blob/main/Chapter10/provisioning-tools/cloudformation/ch10_4_example_cf_partitioning.py`

- `https://github.com/PacktPublishing/Serverless-ETL-and-Analytics-with-AWS-Glue/blob/main/Chapter10/provisioning-tools/cloudformation/ch10_4_example_cf_gen_report.py`

Next, you'll provision the crawler, ETL jobs, and workflow.

Step 2 – provisioning triggers, the crawler, ETL jobs, and the workflow via a CloudFormation template

You can provision the resources in this book's GitHub repository with a CloudFormation template (`https://github.com/PacktPublishing/Serverless-ETL-and-Analytics-with-AWS-Glue/blob/main/Chapter10/provisioning-tools/cloudformation/ch10_4_example_cf.yml`). Follow these steps:

1. Open the CloudFormation console (`https://console.aws.amazon.com/cloudformation/home`) and click **Create stack**, then **With new resources (standard)**, at the top right of the page.

2. Choose **Template is ready** and upload your downloaded YAML file (`ch10_4_example_cf.yml`).

3. Follow each description and type in the necessary information. Then, click **Next** so that you can provision the resources. It will take a few minutes to create resources via the CloudFormation stack:

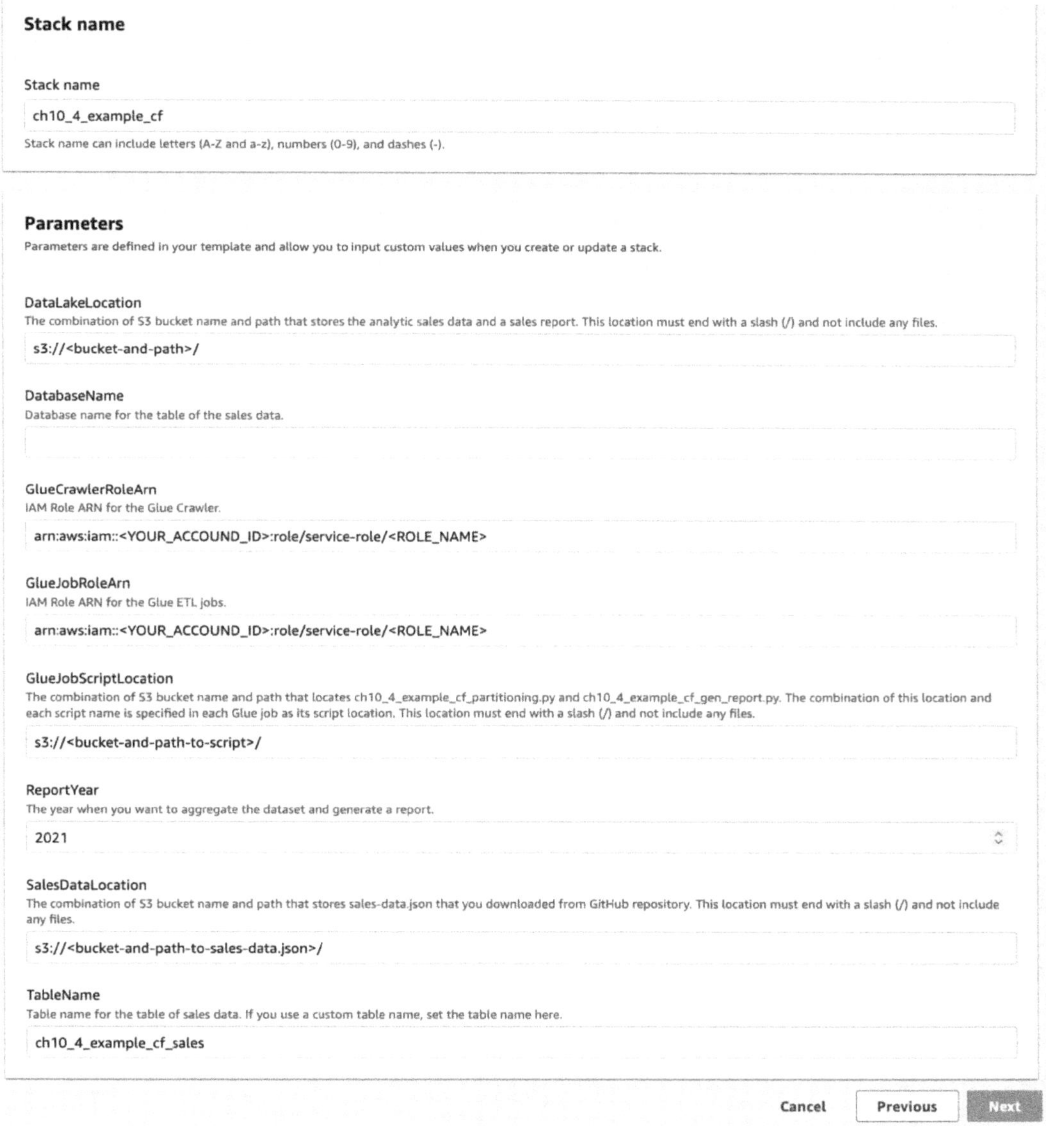

Stack name

Stack name

```
ch10_4_example_cf
```

Stack name can include letters (A-Z and a-z), numbers (0-9), and dashes (-).

Parameters

Parameters are defined in your template and allow you to input custom values when you create or update a stack.

DataLakeLocation

The combination of S3 bucket name and path that stores the analytic sales data and a sales report. This location must end with a slash (/) and not include any files.

```
s3://<bucket-and-path>/
```

DatabaseName

Database name for the table of the sales data.

GlueCrawlerRoleArn

IAM Role ARN for the Glue Crawler.

```
arn:aws:iam::<YOUR_ACCOUND_ID>:role/service-role/<ROLE_NAME>
```

GlueJobRoleArn

IAM Role ARN for the Glue ETL jobs.

```
arn:aws:iam::<YOUR_ACCOUND_ID>:role/service-role/<ROLE_NAME>
```

GlueJobScriptLocation

The combination of S3 bucket name and path that locates ch10_4_example_cf_partitioning.py and ch10_4_example_cf_gen_report.py. The combination of this location and each script name is specified in each Glue job as its script location. This location must end with a slash (/) and not include any files.

```
s3://<bucket-and-path-to-script>/
```

ReportYear

The year when you want to aggregate the dataset and generate a report.

```
2021
```

SalesDataLocation

The combination of S3 bucket name and path that stores sales-data.json that you downloaded from GitHub repository. This location must end with a slash (/) and not include any files.

```
s3://<bucket-and-path-to-sales-data.json>/
```

TableName

Table name for the table of sales data. If you use a custom table name, set the table name here.

```
ch10_4_example_cf_sales
```

Cancel   Previous   Next

Figure 10.20 – The AWS Management console view for filling in parameters

4. Once resource provisioning has been completed, the stack's status will appear as CREATE_ COMPLETE on the CloudFormation console.

Next, you will check the provisioned resources.

Step 3 – checking the provisioned resources

You will see the following resources that have been provisioned by the CloudFormation stack on the Glue console:

- Triggers:

  - ch10_4_example_cf_ondemand_start

  - ch10_4_example_cf_event_run_partitioning

  - ch10_4_example_cf_event_run_gen_report

- Crawler: ch10_4_example_cf

- ETL jobs:

  - ch10_4_example_cf_partitioning

  - ch10_4_example_cf_gen_report

- Workflow: ch10_r_example_cf

This workflow visualizes the same diagram as the one shown in *Figure 10.43*.

You can also run this workflow by choosing **Run** from the **Actions** menu in the Glue console (https:// console.aws.amazon.com/glue/home#etl:tab=workflows). In addition to the same generated reports that we got in the previous section, the pipeline also replicates the Amazon Customer Reviews dataset to the S3 bucket that you specified as the CloudFormation stack parameter. In particular, you will be able to see the replicated files by using the following AWS CLI command:

```
$ aws s3 ls s3://<your-bucket-and-path>/serverless-etl-and-
analysis-w-glue/chapter10/example-cf/data/ --recursive
YYYY-MM-dd 01:23:45          XXXX <path>/serverless-etl-and-
analysis-w-glue/chapter10/example-cf/data/category=grocery/
year=2021/month=6/run-xxxxxxxxxx-part-block-0-0-r-xxxxx-snappy.
parquet
```

In this example, you learned that CloudFormation helps with the resource provisioning process. If you create that workflow and pipeline on the AWS Glue console, you need to create and configure at least seven components – that is, three triggers, one crawler, two ETL jobs, and this workflow. Additionally, if you try to replicate this workflow too many times, the process will be difficult (for example, if you replicate this into 10 workflows, you need to set up at least 70 components). However, if you create a CloudFormation template and create resources using that template, it becomes easier to set up multiple workflows compared to setting up each workflow manually from the Glue console.

You can find more examples of Glue resource provisioning by CloudFormation in the AWS Glue public document and AWS big data blog posts. If you're interested in such examples, please refer to the *Further reading* section at the end of this chapter.

Provisioning AWS Glue workflows and resources with AWS Glue Blueprints

AWS Glue Blueprints allows you to create and share AWS Glue workflows by defining your workflow as a single blueprint, which is similar to using a template. In particular, you can build pipelines by specifying Glue ETL jobs, a crawler, and related parameters that are passed to your Glue jobs, crawlers, workflows, and so on in your blueprint. Based on a blueprint, Glue Blueprints automatically generate workflows. Therefore, you don't need to manually set up workflows from the AWS Glue console.

To create a blueprint, you need to define the following components and package them as a ZIP archive file:

- **A layout file implemented by Python**: You can define crawlers, ETL jobs, and the relevant workflow, including your pipeline logic, in this file. When the layout file is run by Glue, your defined workflows are returned and generated.

- **A configuration file**: You need to set the function name that returns workflows and is defined in the layout file. You can set relevant workflow components such as the workflow names, data types, user input properties, and so on.

- **ETL job scripts and the relevant files (optional)**: Here, you can specify the location of your ETL job scripts to create them and specify the relevant files in the layout to process them.

Let's look at a basic example of a blueprint that consists of a layout file (`layout.py`) and a configuration file (`blueprint.cfg`). By applying this blueprint for Glue, the workflow that contains an ETL job, `sample_etl_job_bp`, will be created. The job's configuration, such as the Glue job's script location, Glue job role, worker type, and so on, is set by the implementation in the `layout.py` file. Additionally, you can set any Glue job script location by parameterizing the script location that's defined in `ScriptLocation`, in `parameterSpec`, in `blueprint.cfg`.

The following code shows the Glue workflow and component definitions in `layout.py`:

```python
def generate_layout(user_params, system_params):
    etl_job = Job(
        Name=»sample_etl_job_bp",
        Command={
            «Name»: «glueetl",
            «ScriptLocation": user_params['ScriptLocation'],
            «PythonVersion": "3"},
        Role=»your_glue_job_role",
        WorkerType="G.1X",
        NumberOfWorkers=5,
        GlueVersion="3.0")
    return Workflow(Name="sample_worflow_bp",
Entities=Entities(Jobs=[etl_job]))
```

The following code shows the Glue workflow parameter configuration in `blueprint.cfg`:

```
{
    «layoutGenerator": "project.layout.generate_layout",
    «parameterSpec": {
        «ScriptLocation": {
            «type»: «S3Uri»,
            «collection»: false,
            «description»: «Specify the S3 path to store your
glue job script.»
        }
    }
}
```

After creating a workflow with this blueprint, you will be able to see the workflow in the AWS Glue console, as shown in the following screenshot:

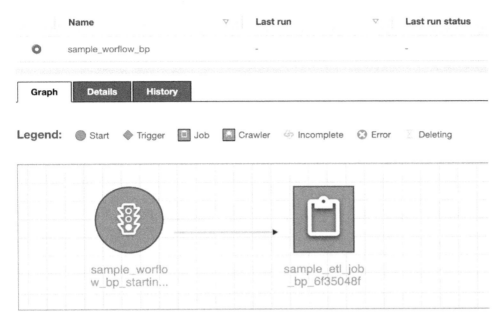

Figure 10.21 – A workflow that includes an ETL job generated by a blueprint

Using Glue Blueprints, you can easily create, replicate, and manage your workflow by implementing a layout file with Python and a configuration file with JSON. The AWS Glue public document (https://docs.aws.amazon.com/glue/latest/dg/blueprints-overview.html) shows what Glue Blueprints is, as well as what your job role needs to do based on three patterns of personas, such as Developer, Administrator, and Data Analyst. Next, you will set up the scheduled-based workflow that you tried to set up in the *Provisioning a Glue workflow using a CloudFormation template* section. You will do so by implementing a blueprint that includes a layout file and the necessary configuration.

Example – provisioning a Glue workflow using Glue Blueprints

In this example, by using Glue Blueprints, you will build the same workflow and pipeline that you did in the *Provisioning a Glue workflow using a CloudFormation template* section. In particular, the following resources will be provisioned via Glue Blueprints:

- **Workflow**: ch10_5_example_bp: This generates a report by running the necessary crawler and ETL jobs

- **Triggers**:

  - ch10_5_example_bp_ondemand_start: The entry point of the workflow. This triggers the ch10_5_example_bp crawler

- ch10_5_example_bp_event_run_partitioning: This triggers the ch10_5_ example_bp_partitioning job

- ch10_5_example_bp_event_run_gen_report: This triggers the ch10_5_ example_bp_gen_report job

- **Crawler**: ch10_5_example_bp: This populates a table based on the Amazon Customer Reviews dataset

- **ETL jobs**:

  - ch10_5_example_bp_partitioning: This extracts the dataset and writes the data to S3 with year and month-based partitioning

  - ch10_5_example_bp_gen_report: This generates the sales report

To create and provision those resources, complete the following steps.

Step 1 – downloading and uploading the blueprint package

Download the ZIP-archived package from this book's GitHub repository: https://github. com/PacktPublishing/Serverless-ETL-and-Analytics-with-AWS-Glue/ blob/main/Chapter10/provisioning-tools/blueprints/chapter10_5_ example_bp.zip. This package includes the following layout, configuration, and relevant job scripts. You can also view the content of each script in this book's GitHub repository (https:// github.com/PacktPublishing/Serverless-ETL-and-Analytics-with-AWS-Glue/tree/main/Chapter10/provisioning-tools/blueprints/scripts):

- layout.py
- blueprint.cfg
- ch10_5_example_bp_partitioning.py
- ch10_5_example_bp_gen_report.py

In this example, the two job scripts (ending with .py) are copied to the S3 location that you specify with layout.py. After downloading the ZIP package, upload it to your S3 bucket.

Step 2 – provisioning triggers, the crawler, ETL jobs, and the workflow via the blueprint

Now, you're ready to provision the resources. First, you need to set up the blueprint. Follow these steps:

1. Access **Blueprints** in the Glue console (https://console.aws.amazon.com/ glue/home#etl:tab=blueprints) and click **Add blueprint**.

2. Type ch10_5_example_bp as the blueprint's name and specify the S3 path where you uploaded the package. Then, click **Add blueprint**.

Once the blueprint's status is active, you must create the workflow. Follow these steps:

1. Click **Create workflow** on the **Blueprints** page.

2. Type in the necessary information, as shown in the following screenshot. Then, click **Next** so that you can provision the resources. After that, click **Submit**:

Create a workflow from ch10_5_example_bp

AWS Glue will run the blueprint to create a workflow.

WorkflowName
Name for the workflow.

```
ch10_5_example_bp
```

ScriptLocation
Specify the S3 path to store your glue job scripts.

```
s3://<your-bucket-and-path>
```
[Browse]

SalesDataLocation
Specify the S3 path to store the sales-data.json.

```
s3://<your-bucket-and-path>
```
[Browse]

DataLakeLocation
Specify the S3 path to store your sales data.

```
s3://<your-bucket-and-path>
```
[Browse]

GlueCrawlerRoleName
Choose an IAM role for Glue Crawler.

GlueJobRoleName
Choose an IAM role for Glue ETL Job.

DatabaseName
Specify a database name for the table of sales data.

ReportYear
Specify the year when you want to aggregate the dataset and generate a report.

```
2021
```

IAM role

Role assumed by AWS Glue with permission to create workflows and their AWS resources. For more information, see **Create an IAM Role for AWS Glue.**

[Cancel] [Submit]

Figure 10.22 – Workflow configuration

3. Once the blueprint successfully creates the `ch10_5_example_bp` workflow, go to **View** in the **Actions** menu in the Blueprints console. You will see the following output:

Figure 10.23 – Blueprint run status

Next, you will check the provisioned resources.

Step 3 – checking the provisioned resources

First, you can check the resources that have been created – that is, the workflow, triggers, the crawler, and the ETL jobs. The workflow visualizes the same graph as the one shown in *Figure 10.43*.

You can also run this workflow in the Glue console (`https://console.aws.amazon.com/glue/home#etl:tab=workflows`). Similar to what happened in the *Provisioning a Glue workflow using a CloudFormation template* section, the workflow replicates the Amazon Customer Reviews dataset to the specified S3 bucket and generates the report.

Blueprints also make provisioning resources easier than setting up resources manually from the Glue console. In addition to this basic example, you can try out more advanced examples by going to the GitHub repository provided by AWS: `https://github.com/awslabs/aws-glue-blueprint-libs/tree/master/samples`.

Developing and maintaining your data pipelines

Finally, let's learn how to grow and maintain data pipelines. Your requirements and demands for data are always changing based on your company's growth, market behaviors, business matters, technological shifts, and more. To meet the requirements and demands for data, you need to develop and update your data pipelines in a short period. Additionally, you need to care about the mechanism for detecting problems in your data pipeline implementations, safe pipeline deployment to avoid breaking your pipelines, and so on. For these considerations, you can apply the following system and concepts to your data pipeline development cycles. These are based on **DevOps** practices:

- **Version control systems** (**VCSs**): You can track changes, roll back code, trigger tests, and so on. Git is one of the most popular VCSs (more precisely, a distributed VCS).

- **Continuous integration** (**CI**): This is one of the software practices for building and testing all the changes on your system and integrating them only after successful tests.

- **Continuous delivery** (CD): This is similar to the concept of CI but is an extension of the concept. CI is usually for a single code base, while CD is for your systems. CD aims to continuously check if components, systems, and infrastructures have been prepared for production. The deployment usually needs explicit approvals. Sometimes, the deployment process is automated, which means that committed changes are instantly deployed on production after all tests are successfully passed. This automatic deployment is called **continuous deployment**.

There are a lot of references to deployment pipelines (NOT data pipelines), including the CI/CD process, such as about what CI/CD is, how to build CI/CD pipelines, and so on. Furthermore, actual deployment pipelines depend on company, organization, team, and system environments. Therefore, we won't cover the deployment process in this section. However, we will look at the basic development process of data pipelines by focusing on AWS Glue and the related tools we've seen so far:

- Developing AWS Glue ETL jobs locally

- Deploying your AWS Glue ETL jobs

- Deploying your workflows and pipelines using provisioning tools such as **Infrastructure as Code (IaC)**

First, you will learn how to develop Glue ETL jobs locally.

Developing AWS Glue ETL jobs locally

AWS Glue provides various local development environments for effectively coding Glue ETL job scripts. You can use various environments for your local development. Let's take a look at each module quickly:

- **AWS Glue ETL Library**: You can download the ETL library on your desktop and develop Glue ETL jobs using Python or Scala. The public documentation (`https://docs.aws.amazon.com/glue/latest/dg/aws-glue-programming-etl-libraries.html`) shows how to use the library.

- **Docker images for Glue ETL**: You can also use the ETL jobs with Docker images (`https://hub.docker.com/r/amazon/aws-glue-libs`) provided by AWS. At the time of writing, up to Glue 3.0 is supported. We won't cover the steps to develop Glue ETL jobs with a Docker image, but you can refer to the concrete steps that use PyCharm by going to `https://aws.amazon.com/blogs/big-data/developing-aws-glue-etl-jobs-locally-using-a-container/`.

- **Interactive Session**: This is one of the Glue functionalities that allows you to develop Glue ETL jobs easily. You can interactively develop your ETL job scripts on Jupyter Notebook by connecting the Glue ETL job system. In the Glue Studio console, you can set up Jupyter Notebook and use it for development purposes. Furthermore, AWS Glue provides a Python module so that you can connect from your local desktop to the Glue job system and use the interactive session. You can install the module via pip from `https://pypi.org/project/aws-glue-sessions/`. Please refer to the public document for details about the setup steps: `https://docs.aws.amazon.com/glue/latest/dg/interactive-sessions.html`.

> **Note – Local Development Restrictions**
>
> When you use the local library, at the time of writing, the **JobBookmarks**, **Glue parquet writer**, and **FillMissingValues/FindMatches** transforms in Glue ML are not supported. You need to use them within the Glue job system.

Regarding the Glue ETL job development cycle, Interactive Session is one of the ways to start checking how you process data, how you can implement Glue job scripts, and so on. If you already have Jupyter Notebook, you can use it on the Glue Studio console by uploading it to the console. You can also use Glue ETL Library and Docker images for your Glue ETL jobs development cycle to write tests, implement code, commit changes, build a package, and more.

Next, you will learn how to deploy your developed Glue ETL job code in the Glue job system.

Deploying AWS Glue ETL jobs

In this section, you'll learn how to deploy Glue ETL jobs by applying changes to your code base. When you initially create or update your ETL jobs, the following two styles are considered:

- **Update your job scripts and relevant packages in the S3 location**: In this style, you define the ETL jobs first. Then, you continuously update the scripts and packages in the S3 location that you specified as a script filename, Python library path, dependent Jars path, and/or reference files path in your ETL jobs.

- **Deploy your Glue jobs**: In addition to updating job scripts and packages, you can deploy your Glue job by using provisioning tools. We'll look at this in the next section.

For both styles, you can create a CI/CD pipeline and make it take on the following challenges while developing ETL jobs:

- Continuous development with unit and integration tests

- Continuous integration and build

- Testing ETL jobs for actual (small) datasets

- Testing the quality of datasets
- Delivering ETL jobs in test and production environments

For these challenges, AWS provides developer tools (`https://aws.amazon.com/tools/#DevOps_and_Automation`) such as **AWS CodeCommit**, **AWS CodePipeline**, **AWS CodeBuild**, and others. You can build CI/CD pipelines by combining these tools. There is a variety of AWS-provided tools to help with the development process, but we will cover the ones mentioned previously as they are often used in the ETL jobs development process to create a basic pipeline. Let's take a quick look at each tool and how to use it in the ETL jobs development process.

- **AWS CodeCommit**: This is an AWS-managed version control service. You can use it as a code repository to manage your job scripts using Git. CodeCommit can also integrate with other AWS tools such as AWS CodeBuild, AWS CodePipeline, AWS Lambda, and others.

- **AWS CodeBuild**: This is an AWS-managed build service. By using CodeBuild, you can compile your code, run tests, and create artifacts for deployment. CodeBuild covers various environments, such as operating systems (Amazon Linux 2, Ubuntu, and Windows Server 2019), programming language runtimes (Java and Python), build tools (Apache Maven and Gradle), and so on. You can also specify your custom image as a build environment. CodeBuild supports not only CodeCommit as a source provider but also Amazon S3, GitHub, BitBucket, and more. You can build, test, and create an updated ETL job script in this process.

- **AWS CodePipeline**: This is an AWS-managed continuous delivery service. By defining release pipelines, CodePipeline automates the pipelines, including build, test, and deploy. For CodePipeline, you define the source, build, and deploy stages. For the source stage, you can specify your code repository and its branch, such as AWS CodeCommit, Amazon ECS, Amazon S3, GitHub, and so on. For the build stage, you can select AWS CodeBuild or Jenkins. For the deploy stage, you can select a deployment provider, such as AWS CloudFormation, AWS ECS, or Amazon S3. For example, if you select Amazon S3 as your deployment provider, CodePipeline delivers your job scripts in your ETL job's S3 location. Then, you can run the updated job.

By using these tools, you can effectively develop Glue ETL jobs in a CI/CD pipeline.

> **Note – Data Quality Tests**
>
> AWS provides **Deequ** (`https://github.com/awslabs/deequ`), an open source data quality unit test tool. This tool checks whether your data is malformed or corrupted, and then computes quality metrics of your data. Please refer to the *Managing data quality* section in *Chapter 6, Data Management*, to learn how to use Deequ with Glue. If you wish to consider data quality tests for your data processing, please refer to the following blog post: `https://aws.amazon.com/blogs/big-data/test-data-quality-at-scale-with-deequ/`. This describes how to use it within Apache Spark.

Now, let's learn how to deploy workflows and pipelines.

Deploying workflows and pipelines using provisioning tools such as IaC

In this section, you'll apply the concept of the CI/CD pipeline for AWS Glue ETL jobs to the data pipelines and workflows you've developed. You can also manage the development process of your workflows, data pipelines, and relevant components such as Glue ETL jobs, Glue crawlers, and so on using CI/CD pipelines. In particular, you can use template-based workflows or provisioning tools to automatically deploy and manage your data processing infrastructure. This infrastructure management is based on IaC, which applies software development practices to infrastructure automation. By managing your infrastructure based on code, you can automate building or changing your infrastructure quickly and safely within CI, CD, and so on.

Regarding workflows and data pipelines, you can build, test, and deploy workflows and their relevant components in CI/CD pipelines by developing template-based files or provisioning tools such as AWS Step Functions, JSON templates, AWS CloudFormation, YAML templates, MWAA Python DAGs, Blueprint Python code, and more.

Let's take a quick look at the example from the *Provisioning AWS Glue workflows and resources with AWS Glue Blueprints* section. There, you defined workflows, a crawler, and Glue ETL jobs in the same repository and deployed each component. Blueprints allows you to programmatically manage workflows and the relevant components. Therefore, you can manage workflows, crawlers, and ETL jobs in the same repository as a data pipeline resource. You can also add tests for Blueprints, not just ETL job scripts. Then, you can build, test, and deploy the Blueprints code and ETL job scripts at the same time in a CI/CD pipeline that contains your data processing infrastructure. This can make your development process safer and faster compared to manually validating your infrastructure code.

Summary

In this chapter, you learned how to build, manage, and maintain data pipelines. As the first step of constructing data pipelines, you need to choose your data processing services based on your company/organization/team, supported software, cost, your data schema/size/numbers, your data processing resource limit (memory and CPU), and so on.

After choosing the data processing service, you can run data pipeline flows using workflow tools. AWS Glue provides AWS Glue workflows as workflow tools. Other tools you can use for this process include AWS Step Functions and Amazon Managed Workflows for Apache Airflow. We looked at each tool by covering examples.

Then, you learned how to automate provisioning workflows and data pipelines with provisioning tools such as CloudFormation and AWS Glue Blueprints.

Finally, you learned how to develop and maintain workflows and data pipelines based on CI and CD. To achieve this, AWS provides a variety of developer tools such as AWS CodeCommit, AWS CodeBuild, and AWS CodePipeline. You also learned how to safely deploy workflows and data pipelines based on IaC.

In the next chapter, you will learn to monitor your data platform and also learn about its specific components like AWS Glue.

Further reading

To learn more about what was covered in this chapter, take a look at the following resources:

- Examples of provisioning Glue resources by AWS CloudFormation:

  - `https://docs.aws.amazon.com/glue/latest/dg/populate-with-cloudformation-templates.html`

  - Build a serverless event-driven workflow with AWS Glue and Amazon Eventbridge: `https://aws.amazon.com/jp/blogs/big-data/build-a-serverless-event-driven-workflow-with-aws-glue-and-amazon-eventbridge/`

- An example of creating workflows using AWS Glue and MWAA: `https://aws.amazon.com/blogs/big-data/building-complex-workflows-with-amazon-mwaa-aws-step-functions-aws-glue-and-amazon-emr/`

Section 3 – Tuning, Monitoring, Data Lake Common Scenarios, and Interesting Edge Cases

Here, you will learn various ways to monitor and troubleshoot an AWS Glue job. You will also learn about different ways to consume data after it is processed by AWS Glue and apply the concepts introduced in this book to real-world data transformation scenarios.

This section includes the following chapters:

- *Chapter 11, Monitoring*
- *Chapter 12, Tuning, Debugging, and Troubleshooting*
- *Chapter 13, Data Analysis*
- *Chapter 14, ML Integration*
- *Chapter 15, Architecting Data Lakes for Real-World Scenarios and Edge Cases*

11
Monitoring

In the previous chapter, you learned how to build and manage your data pipeline with AWS Glue in detail. With that knowledge, you are now able to build a data platform powered by AWS Glue. Cool! But this is not the end of your work with the data platform. It is just the starting point.

Imagine that you have built your data platform using AWS Glue. If your data platform does not meet the predefined business requirements, end users will be confused and won't be able to make a reasonable decision based on the data. If your data platform gives outdated results, the decisions made based on the data will also be outdated. If your data platform is too slow, end users won't be able to make timely decisions and could lose business opportunities. If your data platform does not check data quality and accuracy, no one can use it for critical decisions. If no end users query your data platform due to a lack of knowledge, your data platform is meaningless.

To monitor the preceding situations, you need to have some visibility of what's going on, what situations need to be detected, and how to react to them. It is crucial to monitor your data platform to make and keep it valuable.

In this chapter, we will start with the bigger perspective of monitoring the entire data platform before diving deep into specific components such as AWS Glue. Through the topics discussed in this chapter, you will learn how to monitor your data platform and improve your data platform efficiently. Then, you will dive deep into how to monitor AWS Glue jobs, crawlers, and catalogs, and also learn how to monitor other services such as Amazon Athena, Amazon Redshift, and more.

In this chapter, we will cover the following topics:

- Defining a **service-level agreement (SLA)** for a data platform
- Monitoring the SLA of a data platform
- Managing the components of a data platform
- Analyzing usage

Defining an SLA for a data platform

When operating a data platform, it is essential to define a healthy state for the entire data platform and maintain that state. Think about what kind of state the data platform should be in. It would be good to define an SLA as an indicator of health. This SLA does not always need to be communicated to end users but is used as an internal indicator to measure whether your data platform is healthy or not.

The basic strategy is to maintain a certain data platform state where the SLA is met and then recover to the normal state when it fails. In other words, monitoring is performed to understand when the platform has deviated from a normal state to an abnormal state, and recovery is performed to return the data platform from an abnormal state to a normal state, as illustrated in the following diagram:

Figure 11.1 – The monitoring cycle

Now, I would like to look at an example of how to define the health of a data platform. First, there are a few key perspectives of a data platform to consider:

- The freshness of the data
- The accuracy of the data
- The performance of the queries
- The overall cost of the data platform

Regarding the normal state of the freshness of data, one approach to define the normal state is to determine a criterion, such as how long it can take from generating the data to the data being ready for queries. An example SLA is a one-hour threshold for the latency between the event timestamp and the timestamp that you can start querying from.

Another approach for defining the normal state of the freshness of data is to determine a deadline for data to be ready for queries. For example, let's say you have a business meeting at a fixed slot every week, and you need to create a report to use for that meeting. In this scenario, the normal state can be defined based on the fact that the data becomes available by the specified deadline. An example SLA is that data needs to be ready by 9:00 a.m. every Wednesday.

You can also think of health criteria and SLAs in terms of data accuracy, performance, cost, and more. You will need to organize your SLAs based on your use cases and your requirements.

In this section, you learned how to define a good SLA for your data platform. In the next section, we will learn how to monitor the defined SLA of your data platform.

Monitoring the SLA of a data platform

Let's think about the implementation of a mechanism to monitor the health of a data platform. There are two common strategies to identify the state of a data platform:

- **Fact-based approach**: Inspect the end user activities and retrieve the metrics.

- **Simulation-based approach**: Simulate the end user activities and measure the metrics.

To monitor performance and cost SLAs, you can inspect the end user activities from the metrics and log messages. For Amazon Athena, you will see a variety of metrics including query planning time and total execution time via Amazon CloudWatch (`https://docs.aws.amazon.com/athena/latest/ug/query-metrics-viewing.html`) or Amazon Athena's query history (`https://docs.aws.amazon.com/athena/latest/ug/querying.html#queries-viewing-history`). For Amazon Redshift, you can rely on system tables: `SVL_QUERY_SUMMARY` (`https://docs.aws.amazon.com/redshift/latest/dg/using-SVL-Query-Summary.html`) and `SVL_QUERY_REPORT` (`https://docs.aws.amazon.com/redshift/latest/dg/using-SVL-Query-Report.html`).

To monitor the SLA on the freshness and the accuracy of data, you can retrieve an end user's query results or simulate end user queries to get the latest status. Typically, a simulation-based approach is more useful because you can be flexible in terms of which value you rely on and how frequently you monitor the state. It is similar to synthetic monitoring (`https://en.wikipedia.org/wiki/Synthetic_monitoring`) for web applications and systems.

For example, you can run queries to select records that have been ordered by timestamp to extract the latest record to see how much latency you have in your data platform. Here's an example Athena query to extract the latest record using `ORDER BY` based on the `date` column:

Figure 11.2 – An Athena query example to monitor data freshness

As you learned in *Chapter 9, Data Sharing*, you can also monitor and manage data accuracy by defining data quality rules. Data quality rules allow you to populate data quality metrics and identify whether your data meets predefined criteria. You can use the result of data quality checks in your monitoring system for your data platform.

In this section, you learned how to monitor an overall data platform. Next, we will cover how to monitor each component of your data platform.

Monitoring the components of a data platform

Data platforms can consist of multiple components: data ingestion jobs, ETL jobs, data crawlers, data catalogs, ad hoc query engines, BI dashboards, and more. In order to detect potential issues that can affect an end user's experience, it is recommended that you monitor the individual components of your data platform. Here's a list of key topics to monitor AWS Glue and its related components:

- Monitoring overall statistics
- Monitoring state changes
- Monitoring delay
- Monitoring performance
- Monitoring common failures
- Monitoring log messages

In the following sub-sections, we will look at each of these key topics in detail.

Monitoring overall statistics

For AWS Glue jobs, Glue Studio gives you an aggregated view of the overall statistics, as shown in the following screenshot. This is useful for monitoring the trends of an entire AWS account/region:

Figure 11.3 – Monitoring with Glue Studio

Monitoring state changes

For AWS Glue jobs, crawlers, and data catalogs, you can configure the Amazon EventBridge rules (`https://docs.aws.amazon.com/glue/latest/dg/automating-awsglue-with-cloudwatch-events.html`) to monitor state changes, including job failures, crawler failures, and table partition updates. The rule also triggers an Amazon SNS topic, AWS Lambda function, and other supported services to perform actions for automation, notification, and recovery.

For Amazon Athena queries, you can configure the Amazon EventBridge rules (`https://docs.aws.amazon.com/athena/latest/ug/athena-cloudwatch-events.html`) to monitor query state changes including query failures.

Monitoring delay

For AWS Glue jobs, you can configure a timeout threshold for job duration, stop the job to avoid further charges, and trigger Amazon EventBridge rules for further actions.

For Amazon Athena queries, you can use CloudWatch metrics, such as `TotalExecutionTime` (`https://docs.aws.amazon.com/athena/latest/ug/query-metrics-viewing.html`), and configure a CloudWatch alarm for those metrics.

Additionally, you can configure scan size limits for your workgroup (`https://docs.aws.amazon.com/athena/latest/ug/workgroups-benefits.html`) to cancel queries that exceed the specified threshold to avoid any delay due to an unexpected amount of data.

Monitoring performance

For AWS Glue Spark jobs, you can monitor CloudWatch metrics. To monitor further details and tune the performance of the jobs, it is highly recommended that you enable Spark UI and use it:

Figure 11.4 – The event timeline on Spark UI

With Spark UI, you can identify how the Spark driver/executor works for your data, what Spark DAG and the physical plan look like, how much memory is consumed per executor, and more. It helps you to identify bottlenecks and optimize performance. You will learn more about performance tuning techniques in *Chapter 12, Tuning, Debugging, and Troubleshooting*.

Monitoring common failures

For AWS Glue Spark jobs, AWS Glue job run insights (https://docs.aws.amazon.com/glue/latest/dg/monitor-job-insights.html) that help you to troubleshoot and solve common job failures based on predefined rules extracted from common failure scenarios. It will give you the following insights:

- The line number of your job script
- Any exceptions
- Root cause analyses
- Recommended actions to solve the issue

It will help you solve common issues even if you do not have expertise in AWS Glue and Apache Spark.

To enable job insights, you need to select **Generate job insights** for your Glue Spark job in the Glue Studio console, or the API/SDK, before running the job:

☑ Generate job insights
AWS Glue will analyze your job runs and provide insights on how to optimize your jobs and the reasons for job failures.

Figure 11.5 – Generate job insights

When your job with job insights fails, you can see failure details such as the line number, the last Spark action executed, and concise time-ordered events from the Spark driver and executors in Amazon CloudWatch Logs.

Monitoring log messages

For AWS Glue jobs, log messages for `stdout`/`stderr` are written into Amazon CloudWatch Logs. If you enable **continuous logging** (`https://docs.aws.amazon.com/glue/latest/dg/monitor-continuous-logging.html`), Spark driver/executor logs are also written into Amazon CloudWatch Logs.

If you want to use an application-specific custom logger, you can retrieve the logger from `GlueContext` and use it in the Glue job script, as follows:

```
from awsglue.context import GlueContext
from pyspark.context import SparkContext

glueContext = GlueContext(SparkContext.getOrCreate())
logger = glueContext.get_logger()
logger.info("info log message")
logger.warn("warn log message")
logger.error("error log message")
```

Additionally, the custom logger writes into CloudWatch Logs via continuous logging.

You can also enable debug logging in Spark. This is useful for detailed troubleshooting. For `SparkContext sc`, you can set the log level using the following code:

```
sc.setLogLevel("DEBUG")
```

For `SparkSession spark`, you can set the log level using the following code:

```
spark.sparkContext.setLogLevel("DEBUG")
```

In this section, you learned how to monitor individual components (such as Glue, Athena, and more) on your data platform. Next, we will go over the general concept of analyzing end user activities on your data platform.

Analyzing usage

Due to the nature of a data platform, it is not practical to build it once and leave it as it is without any updates. This is because data volume, velocity, and variety increase day by day. Also, how the data is consumed and utilized can often vary. It is practical to build a platform based on the minimum requirement, start using it, measure end user activities, and continuously improve it based on end user feedback.

After you release the data platform to end users, you might see issues such as the following:

- Less usage than expected
- Less adoption in specific teams
- Too many escalations from end users

To make the data platform useful for your end users, you need to maintain and keep improving the platform by tracking and analyzing end user activities.

Let's look at how user activity can be measured for each type of activity. For example, if it is a simple data reference, it can be recorded and measured in the Amazon S3 server access logs, AWS CloudTrail, and more. From a query execution perspective, it's a good idea to look at the query log for each service. For Amazon Athena, 45 days of query history are recorded. From this evidence, you can gather the following insights:

- Common query patterns
- Popular tables/datasets
- Unique users
- Queries per user/team/organization

Other than that, with end user escalations, you can notice a lack of documentation.

For example, if you see too few unique users or too small a number of queries being made by a user/team/organization, it is possible that the stakeholders have not been notified correctly, the queries are not well documented, and more.

It is important to continuously evolve the data platform without leaving it as it is. Here is a diagram that shows you how analysis and improvement go hand-in-hand:

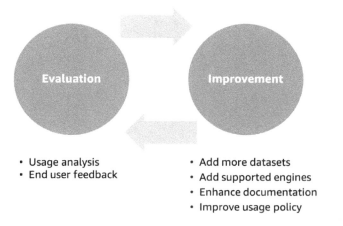

Figure 11.6 – Continuous improvement

As you can see in the preceding diagram, once you build a data platform, you should analyze usage and gather user feedback. Based on the usage and the feedback, you can take appropriate actions such as adding more datasets, increasing the number of analysis engines that are supported, enhancing documentation, promoting the data platform, improving usage policy, and more.

In this way, you can build a truly usable data platform by iteratively developing and operating it through repeated evaluation and improvement.

Summary

In this chapter, you learned how to monitor your entire data platform and your AWS Glue components and related services. Additionally, you learned how to analyze end user activities. Monitoring is essential to keep an SLA and also continuously improve a data platform. Now you should be able to define a reasonable SLA based on the requirement and implement a mechanism to monitor your data platform efficiently.

In the next chapter, you will learn more details about how to tune, debug, and troubleshoot issues when using AWS Glue.

12
Tuning, Debugging, and Troubleshooting

In the previous chapter, we explored some of the fundamental concepts involved in monitoring AWS Glue workloads, such as defining and monitoring **service-level agreements** (**SLAs**) of the data platform, as well as monitoring components of the data platform such as overall statistics, state changes, delay, performance, common failures, and log messages. We also explored how we can analyze usage using logs emitted by different AWS services based on the use case.

Based on the insights gained by monitoring AWS Glue workloads or downstream applications, we will be able to understand whether our workload is running optimally and whether we have over- or under-provisioned resources, and determine whether there is room for improvement. We can tune and enhance our workflows to obtain better performance and thereby save time and resources required by the components of our data integration workflow.

Upon completing this chapter, you will be able to explain how we can tune AWS Glue workloads to ensure we are taking full advantage of the resources we are allocating to our workloads. You will also be able to troubleshoot/debug some of the common issues we encounter in AWS Glue.

In this chapter, we will look at the following topics:

- Tuning AWS Glue workloads
- Troubleshooting and debugging common issues in AWS Glue **extract, transform, load** (**ETL**)

Now, let's explore some of the mechanisms we can use to tune our AWS Glue workloads based on the insights gathered by monitoring AWS Glue workloads or downstream applications and query engines.

Tuning AWS Glue workloads

Based on our discussions in the previous chapter, we already know that AWS Glue is a serverless data integration service wherein different components are bundled with a number of optimizations that cover *most* use cases—*most* being the operative word here. The optimizations already in place may not be the perfect fit for our use case, and they can be further improved to get the most out of the resources we are allocating.

It is still up to us to monitor workloads and implement optimizations where necessary to ensure that we are making use of resources efficiently. The performance of any Glue component is dependent on a number of factors such as input data, resources allocated, configuration, and the actual workflow itself.

Now, let's discuss some of the tuning mechanisms we can use to optimize different components of AWS Glue.

Tuning AWS Glue crawlers

As discussed in the previous section, the performance of a Glue component depends on factors such as input data, configuration, resources allocated, and the workflow itself. Similarly, for AWS Glue crawlers, the performance of the crawler run depends on a number of factors. Some key factors that influence the performance of an AWS Glue crawler run are noted here: type of input data store, number of items or objects to crawl/scan, and crawler configuration.

For instance, let's consider an AWS Glue crawler run where the crawler is crawling data in an **Amazon Simple Storage Service** (**Amazon S3**) location. In this case, if the directory structure is complex, the location contains a lot of small files/objects, and sampling/incremental crawl configurations are disabled, then the crawler would obviously be slower as it has to read a huge number of individual objects to infer a schema, build metadata, and populate the Data Catalog. In such cases, there are a number of optimizations we can implement to reduce the latency of a crawler run.

If the Amazon S3 location contains a large number of objects, we can specify the sample size parameter in the crawler configuration. It is important to note that while specifying sample size may improve the crawler runtime, this won't impact the time taken by the query engine to read source data. To improve the query runtime as well, we can run an ETL job to compact the data and reduce the number of files/objects. This can be achieved by using `coalesce()` or `repartition()` transforms in Apache Spark.

Similarly, if we are crawling Amazon DynamoDB, MongoDB, and Amazon DocumentDB data stores, we can implement sampling to reduce the amount of data scanned by using the **Enable data sampling** option.

If a dataset in an Amazon S3 data store is constantly growing and if the schema remains unchanged, the only reason to run a crawler on this dataset would be to register new partitions. In such cases, we can use the **Incremental Crawl** feature to only crawl new data that was written to an Amazon S3 location. When this feature is enabled, the Glue crawler keeps track of the `lastModifiedTimeStamp` value of Amazon S3 objects and determines whether the objects need to be crawled.

The compute infrastructure provisioned for crawlers is completely managed on the service side, and we do not have any control over the compute capacity allocated. So, the only optimizations we can apply for crawlers are configuration changes and input data optimization.

Now that we know how to improve crawler runtime, let's take a look at how we can tune AWS Glue ETL job performance.

Tuning the performance of AWS Glue Spark ETL jobs

Based on our discussions in the previous chapters, it is clear to us that we can monitor AWS Glue ETL jobs and gather job execution insights using a number of avenues—AWS CloudWatch metrics, logs written by AWS Glue ETL jobs, Spark UI, and AWS Glue job insights. Each of these tools/utilities provides different types of insights into job execution. We can use insights gathered from different tools/utilities to tune and optimize the job to make sure we are utilizing the resources efficiently.

The bottlenecks in an AWS Glue ETL job could be because of a number of reasons—for instance, there could be demanding stages or straggling tasks that are impeding the performance of the entire job. We can monitor ETL job metrics to identify such bottlenecks and implement optimizations. That being said, it is important to note that there is no one-size-fits-all approach to optimizing ETL jobs, and the series of steps required to optimize a particular job may be different from the ones required to optimize another job. Let's consider a few example scenarios to understand this better.

Optimizing ETL jobs with a straggler task

Consider an ETL job that has a non-uniform workload distribution that may be caused by a data skew (also known as a *hot partition issue*), and one of the tasks is processing a huge portion of the dataset. This doesn't just mean that the job is slow because one task in a particular executor is processing most of the data while the other executors are idle; there are chances that the jobs might fail with **out-of-memory (OOM)** or disk space issues if the amount of data being processed exceeds the resource allocation of the executor node.

Now that we know the importance of addressing the straggler task issue, how do we identify whether our ETL job run is experiencing this issue? We can monitor the driver and executor memory and **central processing unit (CPU)** CloudWatch metrics emitted by the job run to see whether all executors are busy.

If we notice that just one executor is busy and the rest of the executors are idle, then we have a bottleneck in our ETL job that needs to be addressed. Since the metrics emitted to CloudWatch have timestamps, we can use these timestamps to check the timeline in Spark event history logs (Spark UI) to identify operations being carried out by the ETL job around that time and focus on optimizing that particular section of our ETL. If event history logs are not available, we can check the Spark driver logs available in the `/aws-glue/jobs/error` log group in AWS CloudWatch Logs.

Let's say that in our example scenario here, we have a straggler task because of data skew and one task is processing most of the dataset, and we have identified that issue was happening during the `JOIN` operation. One way to solve this issue is to redistribute the workload across all executors by repartitioning the dataset based on the join key before performing the `JOIN` operation, as follows:

```
repartitionedDF = dataframe0.repartition(100, "JOIN_Key")
```

Here, we have identified `dataframe0` to contain a data skew that is affecting `JOIN` performance. To mitigate the issue, we are repartitioning the dataset into `100` partitions and distributing the dataset using the `JOIN_Key` key. Now, when we perform the `JOIN` operation in the next step, the operation will be distributed across different executors and not handled by just one executor.

Data skew is just one of the scenarios that can cause straggler task issues. There may be other use cases where this kind of issue can occur. The idea here is to identify such bottlenecks and make sure the workload is distributed and the resources allocated to the job are being used efficiently.

Optimizing ETL jobs with too many tasks

In the previous section, we discussed the issue where one or a few tasks were slowing down the job by processing a large volume of data. In this section, we will be looking at the other side of the coin where we have too many tasks for the job to execute and not enough resources.

Truth be told, this is one of the most common issues we face while executing a Glue ETL job. There are a number of use cases where a Glue ETL job can end up with too many tasks in the **directed acyclic graph (DAG)** and takes a long time to finish executing.

For instance, based on our discussions in the previous chapters, we know that by default, Spark uses a 1:1 mapping with the number of input partitions and the number of files/objects in the data source if the size of the file is less than the block size for the file format. If the file format and compression codec combination used is splittable, then the number of partitions created is equivalent to the number of splits generated. If the file format and the compression codec used cannot be split (for example, **JavaScript Object Notation (JSON)** data compressed in `gzip` format), then we have a much bigger problem as individual files have to be read by the ETL job and decompressed in memory before any operation can be performed.

If our data source has a large number of small files, Spark will eventually create a large number of tasks to read the data store, and the number of tasks that can be run in parallel is restricted by the number of CPU cores available. Having a large number of input files not only slows down the job due to the **input/output (I/O)** effort involved but can also potentially cause the job to fail as the metadata of the files in the data store is tracked by the driver and stored in driver memory before it is ingested into the ETL job. If the data store has too many files, there is a possibility that the driver memory gets filled up with file metadata and there is no memory available to actually execute the ETL job.

The solution to this problem is to reduce the number of tasks created by reading the input files in groups, perform compaction on the source data to reduce the number of files, or use predicate pushdown filtering to only read the data relevant to our ETL job.

We have discussed the option of reading input files in larger groups in *Chapter 3, Data Ingestion*, using the **Grouping** feature in AWS Glue ETL. This option essentially overrides Spark's default behavior by reading multiple files in the same input partition, thereby reducing the number of tasks created. While this is a very useful feature to optimize our ETL job, there are still limitations to this feature, the main one being that this feature does not support **Optimized Row Columnar (ORC)**, Parquet, and Avro file formats. In such cases, we can resort to the option of compacting our source data.

Now, compaction can be done in a number of different ways. The most common approach to compact a dataset is to use another ETL job and output a lower number of partitions. While Apache Spark is the preferred choice by a majority of data engineers, source data can also be compacted using a number of other tools and frameworks, such as Apache Hive, Presto, and `s3-dist-cp`.

If the dataset is in Parquet format and a catalog table for the dataset is registered in AWS Lake Formation as a GOVERNED table, we can use the built-in feature of Lake Formation to perform compaction. You can follow the steps available in the AWS Lake Formation documentation (`https://docs.aws.amazon.com/lake-formation/latest/dg/data-compaction.html`) to enable data compaction on a partitioned Parquet dataset registered as a governed table in AWS Lake Formation.

If grouping or compaction is not an option, we can use predicate filtering to filter out unnecessary data when a dynamic frame is being created. We can use the `push_down_predicate` parameter to perform predicate filtering. This parameter will be evaluated when the data is being read and only Amazon S3 objects matching the predicate expression are used in the ETL job. This is quite a powerful feature in optimizing an ETL job. Consider a use case where you are analyzing a dataset of almost a decade to identify sales trends for a particular month. There is no need to read the entire dataset as we are focusing on a particular month. In such cases, we can use predicate pushdown to push the filter onto the storage level and only read the relevant dataset.

Now, let's assume that in the same ETL job, we are reading the dataset using a catalog table, and the dataset is partitioned based on `year`, `month`, `day`, and `hour`. Even though we have optimized our data read operation by implementing predicate pushdown, the metadata for the entire table is pulled into the ETL job before Spark filters out unnecessary partitions. To avoid this, we can specify an additional `catalogPartitionPredicate` parameter that offloads the filtering of catalog partitions to the AWS Glue service, and only the catalog partitions matching the predicate expression are returned. This will reduce the I/O effort required in performing a `getPartition()` API call and matching the predicate expression specified in the `push_down_predicate` parameter.

The catalog partition predicate feature uses the partition keys registered as an index in the table configuration to perform filtering at the data catalog level, and it is much quicker to perform filtering at the catalog level compared to fetching all partitions registered in the table and performing filtering within the ETL job. You can refer to the AWS Glue documentation (`https://docs.aws.amazon.com/glue/latest/dg/aws-glue-programming-etl-partitions.html#aws-glue-programming-etl-partitions-cat-predicates`) to learn more about pushdown predicates in AWS Glue ETL.

One of the other reasons why an AWS Glue ETL job would end up with a lot of tasks is that we are using a transformation that is generating a large number of tasks. In such cases, identify the operation causing the bottleneck, restrict the number of tasks created, and ensure we have enough compute resources to handle the tasks.

If the number of tasks created is way too high, we may end up exhausting the driver memory as the Spark driver is responsible for tracking tasks created, and an increase in the number of tasks increases driver memory consumption. This will eventually cause the ETL job to fail with OOM errors.

Optimizing JDBC- and MongoDB-based ETL jobs

In this section, we will take a look at different optimization techniques to improve the performance of ETL jobs when reading from **Java Database Connectivity (JDBC)** and MongoDB data stores.

One of the biggest selling points of Apache Spark is that it offers a framework to ingest and reshape data in a distributed environment. However, when reading from a JDBC data store, Spark relies on the user to provide a partitioning strategy to read the data in parallel. If the user does not specify the number of partitions or the column(s) to partition by, Spark uses a single JDBC connection to read the entire table, and this can slow down the entire ETL job significantly. We have discussed how we can address this specific issue using `hashpartitions` in the *Data ingestion from JDBC data stores* section of *Chapter 3, Data Ingestion*.

In the same chapter, we have also discussed how we can use the `fetchSize` parameter to fetch rows from the JDBC data store in batches instead of fetching the entire table in one round trip. When we set the `fetchSize` parameter, this will be passed down to JDBC `PreparedStatement` and informs the driver of the number of rows to fetch per round trip. This parameter is extremely helpful in tuning the amount of data transferred, thereby reducing the pressure on executor memory.

If we are not interested in the entire table when reading from a JDBC data store, we can pass a SELECT query with a predicate expression using the `query` parameter.

For example, if we are just interested in values less than 100 in the `id` column, we can do the following in our ETL script:

```
connection_postgres_options = {
    "url": "jdbc:postgresql://HOSTNAME:5432/gluetest",
    "query": "select * FROM test where id < 100",
    "dbtable": "test",
    "secretId":"glue/postgres_test_db",
    "ssl": "true",
    "sslmode": "verify-full",
    "customJdbcDriverS3Path": "s3://S3_BUCKET/ postgresql.jar",
    "customJdbcDriverClassName": "org.postgresql.Driver"}
datasource0 = glueContext.create_dynamic_frame.from_options(
    connection_type="postgresql",
    connection_options=connection_postgres_options
)
```

As you can see, we still fetch all the columns using SELECT * in our query, but we are reducing the data fetched by using a WHERE condition. If we want to filter out any of the columns, we can do so using AWS Glue/Apache Spark transforms.

Similarly, when reading a MongoDB/DocumentDB data store, by default AWS Glue will read the entire collection. We can define a JSON string that denotes MongoDB's aggregation pipeline, and this will ensure that the filtering and aggregation operations defined in the pipeline string are performed at a MongoDB level instead of through a Spark ETL job. You can see an illustration of this in the following code snippet:

```
pipelineJSON = "{'$match': {'type': 'peach'}}"
mongo_options = {
    "uri": "MONGO_CONN_STR",
    "database": "test",
    "pipeline": pipelineJSON,
    "collection": "fruits",
    "username": "mongodb_test",
    "password": "XXXXXXX"
}
dynamic_frame = glueContext.create_dynamic_frame.from_options(
```

```
            connection_type="mongodb",
            connection_options=mongo_options
    )
```

In the preceding example, we are filtering documents by the `type` column name with the value `"peach"`. We can perform more advanced operations using MongoDB aggregation pipelines—for example, filtering documents with fields that contain data of a specific type, as follows:

```
    pipelineJSON = "{'$match': {'creationDate': {'$type': 'date'},
    'uid': {'$type': 'string'}}}"
```

Here, we are filtering documents based on the data type of the `creationDate` and `uid` fields. We are essentially ignoring documents that don't have values matching the data type specified.

Pipeline aggregations are incredibly helpful, both in reducing the amount of data read and ensuring the data being read conforms to a specific schema. There are other optimizations we can apply while reading data from MongoDB data stores, such as defining a partitioner class and configuration options for the partitioner class selected. The values for these configuration options are to be selected based on the use case. If our use case does not require a specific partitioner class to be defined, we can let the connector use the default options.

We can find a list of partitioner classes and configuration options supported in the AWS Glue documentation (https://docs.aws.amazon.com/glue/latest/dg/aws-glue-programming-etl-connect.html#aws-glue-programming-etl-connect-mongodb).

In this section, we discussed how we can tune the performance of a Glue Spark ETL job in specific use cases. For our discussion, we explored two to three scenarios here. There are a number of other optimizations we haven't discussed related to performance tuning, as it is a vast topic and largely depends on the specific use case and the performance bottleneck we are trying to address.

A rule of thumb here is to ensure that we are using compute resources efficiently by distributing the workload evenly, we have enough compute resources allocated to complete our ETL job, and we do not have too many pending tasks/operations waiting for resources to be allocated or blocked by a certain action within the ETL job.

Now that we have seen different optimization techniques for some of the use cases related to AWS Glue Spark ETL jobs and AWS Glue crawlers, let's take a look at some common issues we face while executing our AWS Glue workloads and how we can solve them.

Troubleshooting and debugging common issues in AWS Glue ETL

While AWS Glue makes it easy to implement data integration workloads using different components/microservices, depending on the user configuration and use case we may encounter a number of issues. In this section, we will discuss some common issues we may encounter while working with AWS Glue and different methods to solve those specific issues one by one.

ETL job failures

A Glue ETL job can fail for a number of reasons. Most job failures can be attributed to issues with configuration or resource provisioning, depending on the use case. Let's explore some common issues we may come across while working with Glue ETL.

OOM errors

When working with a large volume of data, it is not uncommon for us to run into OOM errors. OOM errors can appear in both drivers and executors, depending on the use case. How we approach the issue largely depends on where exactly the issue is occurring, whether in the driver or the executor.

Driver OOM

The Apache Spark driver is responsible for a number of things: executing user code, translating it to a DAG, coordinating with the cluster manager, distributing the workload to executors, and coordinating with all executors to ensure tasks are scheduled and executed successfully. As you might have guessed by now, most of these operations are carried out in memory.

Some common reasons for the driver to run out of memory are listed here:

- A large number of input files
- A large number of dynamic frames and transformations being defined, causing driver stack space to overflow
- A large amount of data being brought into the driver
- Too many tasks being generated as part of ETL code

If the Spark driver OOM is caused by a large number of input files' metadata being tracked in the driver, we can avoid such situations by enabling the `useS3ListImplementation` option in AWS Glue ETL. This option will inform AWS Glue to cache file lists in batches instead of all file metadata being cached in memory all at once. *It is a best practice to use this option with job bookmarking enabled* to ensure we are not fetching metadata for files that are not necessary for our job run.

It is imperative to ensure that the source code is optimized as well and not just the input dataset. For instance, `collect()` or `count()` statements are widely used by users to print information to logs while authoring and debugging ETL scripts. However, it is important to make sure we remove these statements when we finish authoring such scripts. Methods such as `collect()` and `count()` collect results on the driver and consume memory, which eventually leads to Spark driver OOM issues. We need to focus especially on `collect()` calls as they are extremely notorious for causing driver OOM issues.

The logic behind this is simple—when we are working on a sample dataset, we are bringing in a few rows to the driver when we call `collect()`. However, when we are running the same code on a large dataset, we end up bringing a huge volume of data into the driver memory, and this leads to driver OOM.

If we are seeing stack overflow errors in the Spark driver, this means that we are adding too many operations into the Spark DAG. This can happen when we are creating new DynamicFrames in a loop and performing different transformations on each of these DynamicFrames. For example, if we are reading all JDBC tables in a database in the same ETL job and there are hundreds of tables, this will end up causing Spark to build a DAG so large that it can no longer fit into the stack memory space.

In such cases, the recommendation would be to author the script in such a way that the table names are read from job parameters and multiple instances of the same job are being executed concurrently. The `getResolvedOptions()` utility method in AWS Glue is extremely helpful in such use cases to read job parameters in our ETL script. You can refer to the AWS Glue documentation (`https://docs.aws.amazon.com/glue/latest/dg/aws-glue-api-crawler-pyspark-extensions-get-resolved-options.html`) to read more about this utility method and view instructions on how to use it.

If we are encountering driver OOM issues because too many tasks were created, the recommendation would be to replace the ETL code causing this issue with more optimized code. For example, this is known to happen when we use `reduceByKey()` in our ETL code. Here, the goal is to ensure that the number of tasks is not too high for the driver to keep track of.

Now that we have seen some use cases that can cause Spark driver OOM, let's explore some use cases that can cause executor OOM.

Executor OOM

Similar to driver OOM, executor OOM can cause job failures as well. Most executor OOM errors can be typically resolved by scaling (vertical or horizontal, depending on the use case) resources assigned to the AWS Glue ETL job. However, a more sensible approach would be to examine the root cause of the issue before we blindly allocate more resources to the ETL job.

Executor OOMs can occur for a number of reasons, and the first step in addressing these issues is to identify which operation in our ETL job is causing executor OOM errors. The Spark UI, CloudWatch metrics, and CloudWatch Logs are extremely helpful in addressing executor OOM issues.

One of the common causes of executor OOM is that the task is processing a large amount of data that cannot fit into executor memory. If this is the case, the solution would be to optimize data load and ensure it is being done in parallel. The approach used to achieve this differs based on the data store we are working with.

For Amazon S3 data stores, the number of partitions is determined by data layout. If it is possible to optimize data layout before running Glue ETL. If our data store has large, unsplittable files, no amount of horizontal scaling will help. The recommendation here would be to optimize the data layout before using Glue ETL to transform the data.

As discussed in the earlier sections of this chapter, JDBC reads using DynamicFrames can be parallelized using `hashpartitions` and `hashfields/hashexpressions`. If we are using Spark DataFrames instead of dynamic frames, we will have to use `numPartitions`, `partitionColumn`, `lowerBound`, and `upperBound` parameters for JDBC reads. You can read more about these parameters in the Apache Spark documentation (`https://spark.apache.org/docs/latest/sql-data-sources-jdbc.html`).

If an executor OOM issue is happening with the DynamoDB read, the number of partitions created during the read operation is defined by the `dynamodb.splits` parameter, and the solution would be to increase the number of splits. While increasing the number of splits reduces the amount of data read per split, we also need to make sure that we have allocated enough workers to our ETL job to avoid tasks being backlogged in a pending state.

Executor OOMs can happen after the data read during transformations as well, and not just during data reads. In such cases, the best approach would be to identify the operation that was being performed when the Spark executor ran out of memory. Both Spark UI and driver logs can be helpful in these situations. If the executor OOM occurs during `JOIN` operations, we can try converting Glue dynamic frames to Spark DataFrames. AWS Glue dynamic frames are based on **resilient distributed datasets (RDDs)**, and RDD joins may result in more data shuffling than a DataFrame join. We can also reduce shuffle operations by repartitioning based on the join key just before performing a `JOIN` operation. This will essentially reduce the amount of shuffling required to perform a `JOIN` operation. This method can be used to reduce shuffling when we are writing partitioned data as well, in which case we would be repartitioning based on partition keys instead of `JOIN` keys.

You can refer to the *AWS Big Data Blog* post titled *Optimize memory management in AWS Glue*, available at `https://aws.amazon.com/blogs/big-data/optimize-memory-management-in-aws-glue/`, for more detailed information on different OOM use cases and how we can mitigate issues.

The Apache Spark framework has a number of query optimization techniques built into the Spark SQL engine: **Catalyst Optimizer** (used for query plan optimization for Spark SQL queries), **Project Tungsten** (focuses on optimizing memory and CPU utilization by Spark), and **Adaptive Query Execution** (AQE—reoptimizes and adjusts query plans based on runtime metrics), to name a few. These optimizations are enabled by default in Spark depending on the Spark version being used.

AQE and AWS Glue

AQE was made available in the Apache Spark 3.0 release and is available for use in AWS Glue 3.0 (Apache Spark 3.1.1). We can use the `spark.sql.adaptive.enabled` configuration parameter to enable AQE in AWS Glue ETL 3.0.

In this section, we discussed different causes of OOM errors in both the Spark driver and the executor and how we can address issues in each of those scenarios. In the upcoming sections, let's explore a few other common reasons why an ETL job would fail.

Permission issues

An AWS Glue ETL job can fail for permission issues originating from different sources. For instance, a job could fail because it is missing permissions to call a specific API in the **Identity and Access Management (IAM)** policy or it might be missing permissions to the **AWS Key Management Service (AWS KMS)** encryption key, permissions to Amazon S3 data stores, and Lake Formation catalog permissions. The only way to correctly debug permission issues is to check ETL job driver logs and find the stack trace containing the error message and check the operation that failed and the originating service. For instance, let's consider the following error message:

```
botocore.exceptions.ClientError: An error occurred
(AccessDeniedException) when calling the GetAuthorizationToken
operation: User: arn:aws:sts:: xxxxxxxxxxxx:assumed-role/
AWSGlueServiceRole-roleName/GlueJobRunnerSession is not
authorized to perform: ecr:GetAuthorizationToken on resource: *
```

In the preceding error message, it is clear that the IAM role being used by the AWS Glue ETL job (`AWSGlueServiceRole-roleName`) does not have enough permissions to call the `ecr:GetAuthorizationToken` action. The solution here would be to grant permissions in the IAM policy for this action on resource *. This is known to happen when a Glue ETL job is getting an **Elastic Container Registry (ECR)** container image for a Marketplace connector.

Let's take a look at another error message here:

```
org.apache.hadoop.hive.ql.exec.DDLTask.
MetaException(message:Insufficient Lake Formation permission(s)
on s3://BUCKET/path (Service: AWSGlue; Status Code: 400; Error
Code: AccessDeniedException; Request ID: xxxxxxxx-xxxx-xxxx-
xxxx-xxxxxxxxxxxx; Proxy: null))
```

In this particular error message, we can see that the IAM role is missing permissions to the Amazon S3 path in AWS Lake Formation. For instance, if an Amazon S3 path is managed by Lake Formation and the IAM role used by the job hasn't been granted permission to access this path, we will run into such errors. The solution here would be to grant permissions to relevant Amazon S3 locations in AWS Lake Formation.

Similarly, let's take a look here at another error message:

```
The ciphertext refers to a customer master key that does not
exist, does not exist in this region, or you are not allowed
to access. (Service: AWSKMS; Status Code: 400; Error Code:
AccessDeniedException; Request ID: 336e2c35-88b7-4859-ba2a-
da4e2bb9f5c3; Proxy: null)
```

The issue here is with the AWS KMS key in use. The best approach, in this case, would be to check AWS CloudTrail event history for events from the kms.amazonaws.com event source and check the key and the API being called. Once you have these pieces of information, make sure the IAM role being used by AWS Glue has the necessary permissions in both the AWS IAM policy and the AWS KMS key policy to perform the action in question.

Now that we know how to identify and mitigate permission issues, let's take a look at other issues that can cause AWS Glue ETL job failures.

Disk space-related error – No space left on device

This error message relates to the local disk space usage on the executor node. Spark uses the local disk for a number of reasons, and one of the most common use cases is that when the Spark executor memory is full, it starts spilling the content to the disk, and this can cause the disk attached to the executor node to fill up and cause job failures.

If the job failed because of this reason, the first step is to identify the root cause of the issue. If the issue was caused because of memory spilled to the disk, we can try using a bigger worker type (try a G.1X or G.2X worker type).

If the issue is still occurring, we can try to increase the number of shuffle partitions by tuning the spark.sql.shuffle.partitions Spark configuration parameter—this will redistribute the workload better. We can try to use the AWS Glue S3 shuffle service feature to write shuffle data to the Amazon S3 location. There's a downside to this approach as well. Considering the S3 location is being used for shuffle spills, the I/O effort required to read and write shuffle data is significantly higher—Amazon S3 reads/writes are computationally more expensive than memory/disk reads and writes. You can refer to the *AWS Big Data Blog* post titled *Introducing Amazon S3 shuffle in AWS Glue* (https://aws.amazon.com/blogs/big-data/introducing-amazon-s3-shuffle-in-aws-glue/) for a detailed explanation of how the AWS Glue S3 shuffle service works.

If none of these options works, we can implement bounded execution to limit the amount of data processed within an ETL job run and process the data in multiple batches. You can refer to the *Workload partitioning with bounded execution for Amazon S3 data stores* section of *Chapter 3, Data Ingestion* for more information on the **Bounded execution** feature.

Amazon S3 503 Slow Down errors

This is one of the most common errors we may come across when working with large datasets. This issue happens when AWS Glue ETL sends a large amount of **application programming interface (API)** requests to Amazon S3 API servers and the requests get throttled.

Amazon S3 API servers impose the following API limits by default: 3500 PUT/COPY/POST/DELETE or 5500 GET/HEAD requests per second per prefix in a bucket (reference: https://docs.aws.amazon.com/AmazonS3/latest/userguide/optimizing-performance.html).

Now, to resolve this issue, we have to identify whether we are being throttled during read operations or write operations and try to reduce the number of API requests being made from our ETL job. This can be achieved by checking the stack trace captured in the AWS CloudWatch logs for the job run and looking for the S3 operation being performed (for example, listBucket, putObject, or getObject). Depending on the operation being performed, there are a number of approaches we can take—depending on the use case—to reduce the number of API calls.

For instance, we can limit the number of files being read and thereby reduce the number of API requests made by using predicate pushdown filters and bounded execution.

If we are experiencing this issue during write, some of the ways we can fix this issue are noted here: we can redistribute the workload across different prefixes by introducing a new partition key in the data target, or we can reduce Spark partitions by using coalesce() or repartition() before writing the dataset. For Parquet data writes, we can use an **EMR File System (EMRFS)** S3-optimized committer to perform writes using a multi-part upload strategy that uses a smaller number of Amazon S3 API calls.

Essentially, the idea here is to identify the operation that is getting throttled and reduce the number of API calls being made using different strategies, depending on the use case. You can refer to the *AWS Big Data Blog* article titled *Best practices to optimize data access performance from Amazon EMR and AWS Glue to Amazon S3* (https://aws.amazon.com/blogs/big-data/best-practices-to-optimize-data-access-performance-from-amazon-emr-and-aws-glue-to-amazon-s3/), which discusses the Amazon S3 503 SlowDown issue in detail and outlines possible solutions for different use cases.

Now that we know how to identify the root cause of some common issues in AWS Glue ETL and how to address these options, it is clear from our discussion that the procedure to address any of these issues has something in common—we start by investigating the root cause of the issue by looking at the metrics and logs from different sources and looking at the error message(s) available. Consider the source of the error message (APIs/AWS service errors, data plane errors—errors thrown by libraries within Glue components), the type of error (for example, permissions; throttling), and then identify the cause of the error message (usually outlined in the form of error codes or messages).

Once we have the root cause, we identify the offending component or configuration and replace it with an appropriate source code or configuration fix and test the application. A similar approach can be used to tune and troubleshoot any component of AWS Glue, and not just AWS Glue ETL jobs. As mentioned in our earlier discussions, developing and maintaining a data integration workflow is an iterative process and often requires a lot of retries before we can come up with a fully functioning workflow that is suitable for our use case.

Summary

In this chapter, we discussed some of the options available at our disposal to tune AWS Glue Spark ETL jobs and AWS Glue crawlers based on the use case and understood how the procedure to tune a Glue ETL job or Glue crawler depends on data layout (input data type, partitioning structure, compression codec), crawler/job configuration, and downstream application/query engines. During our discussion on ETL job tuning, we explored different use cases and learned how to identify ETL jobs with straggler tasks and demanding stages and how we can optimize performance. We also discussed how to optimize ETL jobs with too many tasks and JDBC-/MongoDB-based ETL jobs to ensure we are using the resources allocated to the job to run quite efficiently.

We also outlined some common issues we may come across while working with an AWS Glue Spark ETL job and discussed different methods or steps to take to identify and mitigate such issues. It is important to note that while we discussed different issues we may encounter while working with Glue ETL, this is not an exhaustive list, and we may run into other issues. That being said, the approach to debug or mitigate an issue remains the same for any issue encountered. AWS Support Engineering is known to publish Knowledge Center articles addressing specific issues for different AWS services based on common trends in support cases raised by AWS customers. A list of Knowledge Center articles addressing specific issues related to AWS Glue can be found at `https://aws.amazon.com/premiumsupport/knowledge-center/#AWS_Glue`.

In the next chapter, we will be discussing some of the concepts of data analysis, such as running ad hoc queries using Amazon Athena and Amazon Redshift Spectrum. We will be exploring how we can take advantage of AWS Lake Formation-governed tables and run time-travel queries, and how we can perform near-real-time analysis using AWS Glue streaming. We will also be exploring how we can visualize data using Amazon QuickSight and how we can use an elastic/OpenSearch stack to search our dataset. This will help us understand how we can efficiently use the data output from AWS Glue with different downstream applications and query engines.

13
Data Analysis

In the previous chapter, we looked at the various buckets of Glue job expectation messages, why they occur, and how to handle them.

We learned about the impact of data skewness, how that can adversely impact job execution, and the techniques you can use to fix it. Additionally, we looked at some of the common reasons for **Out-of-Memory** (**OOM**) errors and the out-of-the-box mechanisms that are available in AWS Glue to handle them. Some of these tools and techniques can be used to be more effective in resource utilization in a pay-as-you-go cloud-native world. These techniques can not only be used for efficient processing but also help you reduce the processing time in a world that increasingly needs answers as quickly as possible.

But the question is, why put in all this effort? Why process data? This brings us to our current topic. One of the reasons for processing data is to analyze it. You might want to analyze the data to look at the larger picture or perhaps visualize the data in a way that makes some vital information stand out. Alternatively, you might want to search for a specific piece of information from a large pile, or you might want to check out the journey of a certain data item as it morphs from one state into another as a result of various factors that influence it. Sometimes, data is also processed for feature engineering to enable better predictions from **Machine Learning** (**ML**) models.

Each of the possibilities of data analysis listed earlier requires a special kind of processing. For example, the processing required for feature engineering is going to be different from the processing required for creating BI visualizations. Similarly, a search requirement on unstructured data might be better fulfilled if the data is stored as a NoSQL object, and a BI report might work better if the data is stored in a **Relational Database Management System** (**RDBMS**) data warehouse in Kimball's star format.

In this chapter, we will learn how AWS Glue can be used for diverse transformations, each suited for a specific objective. We will start by creating a sample dataset. This dataset will be used across the sections of this chapter. Then, we will dive into the common tools used for data analysis in the world of AWS. AWS Glue is often used to write this data. Then, we will look into **Transactional Data Lakes** and see how we can leverage technologies such as **Apache Hudi** and **Delta Lake** to upsert data in a data lake. We will follow this up with the mechanism used to write data in AWS Lake Formation's governed tables. Then, we will venture into the streaming area and use native Glue's method to consume streaming data, be it from **Apache Kafka** or **Amazon Kinesis**. Additionally, we will look at how we can use Hudi's DeltaStreamer in Glue to consume data from Apache Kafka. Finally, we will try to insert data into an OpenSearch domain and query it through **OpenSearch Dashboards**.

In this chapter, we will be covering the following topics:

- Creating Marketplace connections
- Creating the CloudFormation stack
- The benefit of ad hoc analysis and how a data lake enables it
- Creating and updating Hudi tables using Glue
- Creating and updating Delta Lake tables using Glue
- Inserting data into Lake Formation's governed tables
- Consuming streaming data using Glue
- Glue's integration with OpenSearch
- Cleaning up

We will start by creating some Marketplace connections. These Marketplace connections will be used as input into the CloudFormation template. The CloudFormation template that is shipped with this chapter will create 12 Glue jobs, an Amazon Redshift cluster, an Amazon MSK cluster, and an Amazon OpenSearch domain. Additionally, we will use all of the network plumbing and any other resources that might be required to understand the chapter.

> **Note**
>
> While I have taken care to use the minimum number of resources required for the execution of the code shipped with this chapter, please use your judgment to implement the CloudFormation template. Please delete the stack as soon as you have understood the concepts, and please take care when changing the network setting of the CloudFormation template to suit the needs of your organization. The CloudFormation template shipped with this chapter is built with a general requirement in mind. These requirements might not align with the guidelines of your organization. The reader bears the responsibility for any issues resulting from the implementation of the CloudFormation template, such as network and security compliance issues or the cost implications of creating the CloudFormation stack.

Creating Marketplace connections

We are going to create Marketplace connections for the Glue Hudi connector, the Glue Delta Lake connector, and the OpenSearch connector. We will be using these connectors in our code samples, and the names of these connectors will be used as input to the CloudFormation stack.

Creating the Glue Hudi connection

Let's begin by creating the Glue Hudi connection:

1. Navigate to **AWS Marketplace** (`https://aws.amazon.com/marketplace/`), search for the `Apache Hudi Connector for AWS Glue` product, and click on **Continue to Subscribe**:

Figure 13.1 – Subscribe to Apache Hudi Connector for AWS Glue

2. Click on **Accept Terms**:

Figure 13.2 – Accept the terms

3. After some time, when your request has been processed, the **Continue to Configuration** button will be enabled. Click on it:

Figure 13.3 – The Continue to Configuration button

4. Select **Glue 3.0** as the **Fulfillment option** setting, select **0.9.0 (Feb 16, 2022)** as the **Software version** setting, and click on the **Continue to Launch** button that is present in the upper-right corner of the screen:

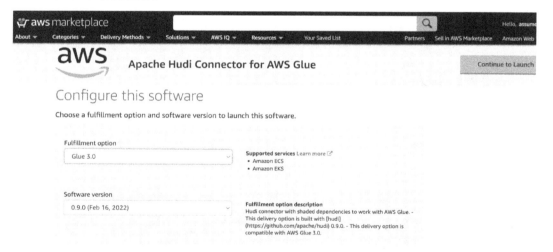

Figure 13.4 – Fill in the required options

5. Click on the **Usage instructions** link:

 Apache Hudi Connector for AWS Glue

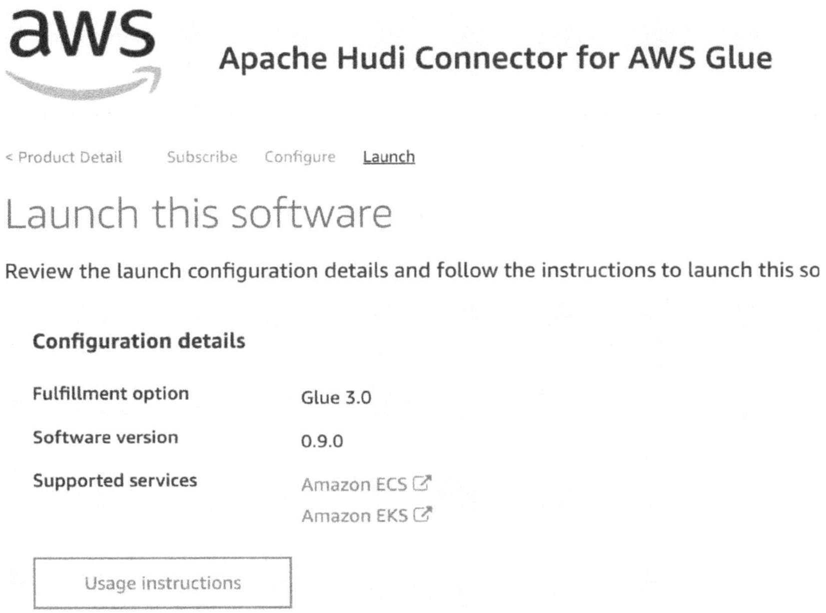

< Product Detail   Subscribe   Configure   Launch

Launch this software

Review the launch configuration details and follow the instructions to launch this so

Configuration details

Fulfillment option	Glue 3.0
Software version	0.9.0
Supported services	Amazon ECS ☑ Amazon EKS ☑

Usage instructions

Figure 13.5 – Launch the software

6. Click on the **Activate the Glue connector from AWS Glue Studio** link:

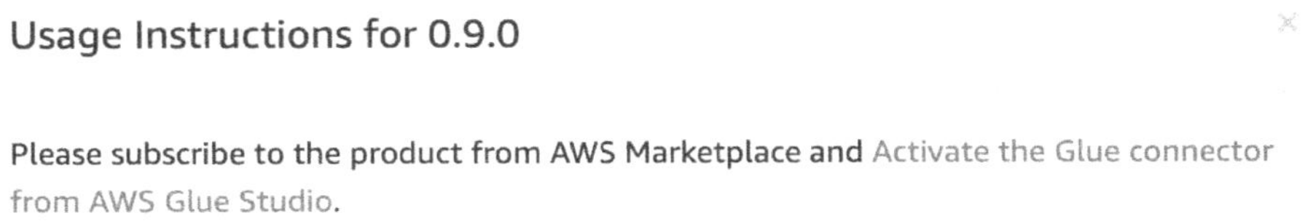

Usage Instructions for 0.9.0   ✕

Please subscribe to the product from AWS Marketplace and Activate the Glue connector from AWS Glue Studio.

Figure 13.6 – Activating the Glue connector

7. Give a name to the connection, and then click on the **Create connection and activate connector** button. Make a note of the name of the connection. This will be one of the inputs to the CloudFormation template:

Figure 13.7 – Create a connection

Now we will follow the same process for creating Delta Lake and Amazon OpenSearch connections.

Creating a Delta Lake connection

Search for `Delta Lake Connector for AWS Glue` in the Marketplace. We will use **1.0.0-2 (Feb 14, 2022)** as the **Software version** setting and **Glue 3.0** as the **Fulfilment option** setting. *Make a note of the name you give to the connection.* This name will be an input to the CloudFormation template.

Creating an OpenSearch connection

Search for `Elasticsearch Connector for AWS Glue` in the Marketplace. Use the one owned by **Amazon Web Services**. We will use **7.13.4-2 (Feb 14, 2022)** as the **Software version** setting and **Glue 3.0** as the **Fulfilment option** setting. *Make a note of the name you give to the connection.* This name will be an input to the CloudFormation template:

Figure 13.8 – Elasticsearch Connector for AWS Glue

Now we will be creating a CloudFormation stack. The stack will create all the network elements such as VPCs, subnets, and security groups along with Glue jobs and other resources such as a Redshift cluster, an OpenSearch cluster, and an MSK cluster. These resources will help you to successfully execute the Glue jobs associated with various sections of this chapter.

Creating the CloudFormation stack

First, let's go through the prerequisites for this section.

Prerequisites for creating the CloudFormation stack

Make sure that the Amazon OpenSearch, Delta Lake, and Apache Hudi connections have been created. Also, make sure that you have a KeyPair. This KeyPair will be used to connect to one of the EC2 instances created by the CloudFormation template.

The CloudFormation template will create IAM roles and policies, too. These roles and policies are required for the jobs to function. Please review the definition of these roles, policies, networks, and security groups, and ensure that they align with the standards of your organization. In the following sections, first, we will create the stack and then create the dataset.

Creating the stack

The CloudFormation stack creates 61 resources. These resources can be found in the **Resources** tab of the CloudFormation stack.

Import the template in CloudFormation and enter the name of the stack, the name of the Apache Hudi Marketplace connection, the name of the Delta Lake Marketplace connection, the name of the Amazon OpenSearch connection, the username and password for both the Redshift master user and the Amazon OpenSearch master user, the IP of your laptop, and the KeyPair that will be used to connect to the EC2 created by the **CloudFormation** (**CFn**). Keep the default settings for the rest of the parameters.

Please note that the password for Amazon OpenSearch master user must have at least 8 characters: one uppercase character, one lowercase character, one number, and one of the #$! special characters. The password for the Redshift master user must have at least 8 characters: one uppercase character, one lowercase character, and one number. Special characters are not allowed.

After the CloudFormation stack has been created, follow the next section to create a dataset.

Creating a dataset

Before we start looking at various techniques for data analysis, let's start by creating a basic dataset to work with.

Navigate to the **AWS Glue Studio** console (https://console.aws.amazon.com/gluestudio/home), check the checkbox next to the **01 - Seed data job for Data Analysis Chapter** job, and click on the **Run Job** button:

Figure 13.9 – The AWS Glue Studio console

The CloudFormation template shipped with this chapter will have created this job and the associated resources, such as the S3 bucket, the IAM roles and policies, and the AWS Glue Catalog database, that are required to run the job.

Now you can go to the AWS Glue Studio monitoring page (`https://console.aws.amazon.com/gluestudio/home?#/monitoring`) and check the status of the job. Note that you might see a lag for a few seconds for the job execution to be reflected on the AWS Glue Studio monitoring page (`https://console.aws.amazon.com/gluestudio/home?#/monitoring`):

Figure 13.10 – Checking the status of the jobs

The successful completion of this job will create an `employees` table in `chapter-data-analysis-glue-database`.

Now that we have some data, let's understand the past and current patterns of data analysis.

The benefit of ad hoc analysis and how a data lake enables it

Before the start of the data lake pattern, organizations used to offload their data into a data warehouse for analysis. This involved creating an **Extraction, Transformation, and Load (ETL)** pipe. Creating ETL pipes, moving the data into a warehouse, and creating reports take a substantial amount of time and resource investment. By the time all of this has finished, the requirements will have changed because of the change in the business over a period of time. Sometimes, business users discovered that they didn't get what they ordered and that there was a gap in requirement and implementation.

For example, a business user could request sales data, resulting in the IT team moving the sales data into the warehouse. However, the sales data in the warehouse might not be of the grain that the business user needs or does not include the sales data from all the sources of sales information. All of this involves a massive amount of rework.

With the advent of data lakes, organizations moved from code to configuration. Unlike a data warehouse, which requires the creation or modification of an ETL job, bringing a new source into the data lake usually involves adding a configuration to existing pipes. This is possible because the first layer of a data lake is generally the raw or the **bronze layer** and, usually, involves an extract and load job. Data is brought into this layer in a business-agnostic fashion. Since there is no transformation involved, the same jobs can be reused to bring in newer sources.

This hugely reduces the time required to make the data available, as bringing it from a new source to the data lake no longer requires any development effort and is, now, purely an operations ticket. However, this data in the raw/bronze layer is generally in the format of the source and is not standardized. This brings about the need for a semi-processed layer. This is generally called the **silver layer**.

Generally, the transformation between the bronze layer and the silver layer is also business agnostic. This is because the silver layer is considered the single source of truth for all downstream systems. We don't know what requirements we might have in the future. Hence, transforming the data in any way creates a possibility of not being able to transform it differently if we get such a requirement in the future.

However, the transformation from bronze to silver includes common sense operations such as partitioning, compression, the addition of audit columns, and creating derived fields. All of these operations are coded such that the jobs remain reusable for any new sources that we might have to bring in. The transformed data can be easily pulled by all the downstream systems that need it. Additionally, the transformations are designed to provide traceability to the ops team if they have to troubleshoot some data inconsistency in the downstream systems.

By now, we understand that the data is made available in the silver bucket using reusable code, but how do we access this data? That is where the central metadata catalog comes in. The AWS Glue Data Catalog can be the central repository of metadata, and the metadata can either be updated from within the Glue jobs or using AWS Glue crawlers. Other services, such as Amazon Athena and Amazon EMR, can also update the AWS Glue Data Catalog. The AWS Glue Data Catalog (`https://docs.aws.amazon.com/glue/latest/dg/components-overview.html#data-catalog-intro`) is also accessible from other AWS services such as Amazon EMR, Amazon RDS, Amazon Redshift Spectrum, Amazon Athena, and any application that is compatible with the Apache Hive metastore. Additionally, you can configure the AWS Glue Data Catalog of a different AWS account (`https://docs.aws.amazon.com/athena/latest/ug/data-sources-glue-cross-account.html`).

With this feature, business analysts do not have to wait for the creation of the ETL pipelines for the data to be available in the warehouse but can directly query the silver bucket using the AWS Glue Data Catalog. This enables them to do an analysis of the data and understand exactly which transformation has to be formalized and coded into the ETL pipelines and brought to the warehouse. This saves a lot of IT effort.

Now, that we understand the tangible benefit of ad hoc analysis and how a data lake enables it, let's look at the two primary means of computing in AWS that are used for ad hoc analysis. They are Amazon Athena and Amazon Redshift Spectrum.

Amazon Athena

Amazon Athena is a serverless interactive query service, based on the Presto platform, that can leverage the AWS Glue Data Catalog for getting the table metadata. Because Amazon Athena is serverless, there is no infrastructure to set up or manage.

While we will primarily use Amazon Athena for querying purposes, it can do a lot more than just that. We will spend the next few paragraphs learning about some of the most important features of Amazon Athena and what makes it so powerful. We discuss these features because Amazon Athena is one of the most important and widely used tools for data exploration and analysis in the AWS world. Having a good understanding of Amazon Athena is going to be important to be effective in data exploration on AWS.

Amazon Athena uses an asynchronous query arrangement. When a user submits a SQL query, Amazon Athena uses a hot cluster to execute the query and then writes the processed result into a temporary S3 location. Then, these results are read and returned to the client. You can use the AWS portal to use Amazon Athena, or you can also use the Athena JDBC driver (`https://docs.aws.amazon.com/athena/latest/ug/connect-with-jdbc.html`) in any application, such as SQL Workbench (`https://www.sql-workbench.eu/downloads.html`), that supports a JDBC connection. Additionally, you can use the identities stored in Okta for configuring federated access to Athena using JDBC and Lake Formation (`https://docs.aws.amazon.com/athena/latest/ug/security-athena-lake-formation-jdbc-okta-tutorial.html`). You can also use Microsoft's Azure **Active Directory** (**AD**) or Ping Identity's PingFederate for authentication. Additionally, you can choose to use the Amazon Athena ODBC drivers (`https://docs.aws.amazon.com/athena/latest/ug/connect-with-odbc.html`).

Recently, Amazon Athena upgraded to version 2 of the Athena engine, which is based on Presto 0.217. This brings new features and performance enhancements to the JOIN, ORDER BY, and AGGREGATE operations.

The integration with the AWS Glue Data Catalog allows the creation of a unified metadata repository across multiple AWS services. While the AWS Glue Data Catalog is generally used for the unified metadata store, you can also connect Athena to an external Hive metastore (`https://docs.aws.amazon.com/athena/latest/ug/connect-to-data-source-hive.html`).

In Amazon Athena, most results are delivered within seconds, and you are charged based on the amount of data scanned by the query (`https://aws.amazon.com/athena/pricing/`). Because you are charged for the amount of data scanned, you can greatly reduce your bills by following the best practices related to compression and partitioning that were introduced in *Chapter 5, Data Layout*. Also, you can use Amazon Athena workgroups to track costs, and control and set limits on each workgroup to control costs. You can also add tags to these workgroups and then use **Tag-Based IAM** access policies (`https://docs.aws.amazon.com/athena/latest/ug/tags-access-control.html`) to control permissions.

Amazon Athena query metrics can be published to CloudWatch. Then, these metrics can be used to create alarms that can trigger actions based on the alarms. Also, you can also use the Explain Analyze (`https://docs.aws.amazon.com/athena/latest/ug/athena-explain-statement.html`) statement in Amazon Athena to get the computational cost of each operation in a SQL query.

Additionally, Amazon Athena can use the fine-grained access control rules set up in your AWS Lake Formation. AWS Lake Formation allows administrators to configure column-, row-, and even cell-level permissions (`https://docs.aws.amazon.com/lake-formation/latest/dg/data-filtering.html`).

Amazon Athena also supports **Atomicity, Consistency, Isolation, and Durability** (**ACID**) transactions to allow for DML operations such as inserts, updates, and deletes along with the ability to time travel. This ACID transaction feature (`https://docs.aws.amazon.com/athena/latest/ug/acid-transactions.html`) is based on the open source Apache Iceberg (`https://iceberg.apache.org/`). Additionally, Amazon Athena supports read operations on AWS Lake Formation governed tables and Apache Hudi tables (`https://docs.aws.amazon.com/athena/latest/ug/querying-hudi.html`).

Apart from querying the data in S3, you can also query the data in other data stores such as Amazon CloudWatch Logs, Amazon DynamoDB, Amazon DocumentDB, and Amazon RDS, and JDBC-compliant relational data sources, such as MySQL and PostgreSQL, under the Apache 2.0 license using Amazon Athena Federated Query feature (`https://docs.aws.amazon.com/athena/latest/ug/connect-to-a-data-source.html`). Prebuilt Athena data source connectors exist for these sources. You can also deploy your own connector to connect to a data source (`https://docs.aws.amazon.com/athena/latest/ug/connect-to-a-data-source-lambda.html`).

Amazon Athena is also used in combination with AWS Step Functions (`https://aws.amazon.com/step-functions/`) to create a data processing pipeline that is orchestrated in AWS Step Functions and processed using Amazon Athena. These data processing pipelines can use **User-Defined Functions (UDFs)** (`https://docs.aws.amazon.com/athena/latest/ug/querying-udf.html`) in Amazon Athena for reusable and standardized processing that has to be used multiple times within the same pipeline or across multiple pipelines. Additionally, the same **USING EXTERNAL FUNCTION** syntax that was used with UDFs can be used to run the ML inference using Amazon SageMaker (`https://aws.amazon.com/sagemaker/`). Now, let's look at some of the Amazon Athena features that can help us be more efficient in querying data.

You can create views in Athena to simplify the querying process for less SQL-savvy resources and to ensure consistent results for common queries.

Often, data exploration requires parsing nested structures and arrays. Amazon Athena supports both of these and can also parse a JSON object. This flexibility to parse complex structures helps Amazon Athena enable data exploration on less-than-perfect data. Amazon Athena also supports queries on **geospatial data**. The input data should be in **WKT** (**Well-known text**) format or JSON-encoded geospatial data format. Amazon Athena can also be configured to query AWS CloudTrail logs, Amazon CloudFront logs, Classic Load Balancer logs, Application Load Balancer logs, Amazon VPC flow logs, and Network Load Balancer logs.

Additionally, you can parameterize the queries that are used more often. This is done using the PREPARE and EXECUTE statements (`https://docs.aws.amazon.com/athena/latest/ug/querying-with-prepared-statements.html`). You also have the option to save the queries per workgroup.

Querying in Athena

In the previous section, we learned about the various features of Athena that can help to simplify data exploration in AWS. In this section, we will look at a simple example for querying the data.

Run the following query in your Athena console. You should be able to see the data inserted in the **01 - Seed data job for Data Analysis Chapter** job in the *Creating a dataset* section:

```
SELECT * FROM "AwsDataCatalog"." chapter-data-analysis-glue-
database"."employees" order by emp_no;
```

You will see the following output:

Figure 13.11 – Query output in the Athena console

So far, we have created sample data using Glue jobs, we have learned about the various features for data exploration in Athena, and we have also queried our sample data through Athena.

Next, we will look at another tool for exploring data in Amazon S3.

Amazon Redshift Spectrum

Redshift Spectrum is a feature within the Redshift toolset. It is a mechanism used to query S3 data by employing massive parallelism to query the data on a big data scale. The feature also enables Redshift to offload a part of the query compute such as aggregation and filtering to the spectrum layer. Just like Athena, Amazon Redshift Spectrum can query data from the AWS Glue Data Catalog or an external Hive metastore. So, tables created in the AWS Glue Data Catalog can be accessed within Redshift through Redshift Spectrum using an external schema. Later in this section, we will check a related example.

Users can also partition the data, and the intelligent spectrum layer can prune those partitions when users query for the specific data within the partitions. Because the data lives externally, the same data can be accessed in multiple Redshift clusters through the spectrum layer. Other big data technologies such as Hudi can be used to create a transactional data lake. Redshift supports **Copy-on-Write (CoW)** Hudi tables (https://hudi.apache.org/docs/concepts.html#copy-on-write-table). Check out the documentation (https://docs.aws.amazon.com/redshift/latest/dg/c-spectrum-external-tables.html#c-spectrum-column-mapping-hudi) for supported Hudi versions. We will discuss Hudi tables in more detail in the following sections. Updates to the CoW Hudi tables are available in Redshift through the Spectrum layer.

Additionally, you can query Delta Lake (`https://delta.io/`) tables through Redshift Spectrum. The data from Redshift Spectrum can be joined with the data maintained within Redshift. You can also use data handling options (`https://docs.aws.amazon.com/redshift/latest/dg/t_setting-data-handling-options.html`) to define Spectrum's behavior when it finds unexpected values in the columns of external tables. Spectrum supports the row and column level rules that have been set up for your data lake security for governed tables. Additionally, data in S3 accessed via Spectrum can be used to hydrate the materialized views in Redshift (`https://docs.aws.amazon.com/redshift/latest/dg/materialized-view-overview.html`).

One of the major improvements in Spectrum, which was introduced a few years ago, was the support for bloom filters. A **bloom filter** is a probabilistic, memory-efficient data structure that accelerates join queries. Redshift decides on its own whether to use the bloom filter for a query at runtime. Spectrum supports modern BI tools by enabling you to query for complex and nested data types (`https://docs.aws.amazon.com/redshift/latest/dg/tutorial-query-nested-data.html`) such as structs, arrays, or maps in S3 data.

Now that we understand Redshift Spectrum, let's create an external schema in Redshift to query the table that we created using the **01 - Seed data job for Data Analysis Chapter** job in the *Creating a dataset* section.

The CloudFormation template shipped with this chapter creates a role called `HandsonSeriesWithAWSGlueRSRole` and a Redshift cluster to enable us to use Amazon Redshift Spectrum to query the data from S3. Please navigate to the IAM console (`https://console.aws.amazon.com/iamv2/home#/roles/details/HandsonSeriesWithAWSGlueRSRole`) and check out the definition of this role to ensure that it is compliant with your organization. This role will be used by Amazon Redshift to access the AWS Glue Data Catalog:

1. Go to the Redshift SQL workbench console (`https://console.aws.amazon.com/sqlworkbench/home?#/client`).

2. Click on the Redshift cluster created by the CloudFormation template. You can get this from the `RedshiftClusterId` key of the **Outputs** tab of the CloudFormation stack:

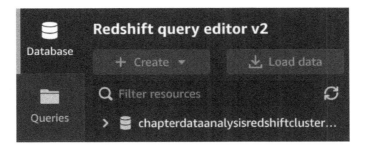

Figure 13.12 – The Redshift cluster in Redshift query editor v2

3. Select the **Database user name and password** option and enter the username and password entered during the creation of the CloudFormation stack. You can keep the default value of dev for the **Database** field. Click on the **Create connection** button:

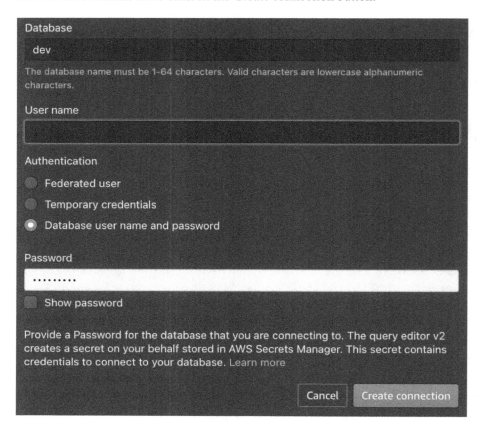

Figure 13.13 – The Database username and password options

4. Make sure that the dev database has been selected at the top:

Figure 13.14 – Selecting the dev database

5. Enter the following command, and click on the **Run** button:

    ```
    create external schema chapter_data_analysis_schema
    from data catalog database 'chapter-data-analysis-glue-
    database' region '<region>'  iam_role 'arn:aws:iam::<aws_
    account_id>:role/HandsonSeriesWithAWSGlueRSRole';
    ```

    Replace region and aws_account_id in the preceding command.

    Here, database is the AWS Glue Data Catalog database. This database was created through the CloudFormation stack.

6. Now, expand the dev database. You should notice the chapter_data_analysis_schema schema underneath it. Now you should be able to see the employees table created in the *Creating a dataset* section:

Figure 13.15 – Expanding the dev database option

7. Run the following SELECT query to see the data loaded into the Glue Data Catalog table:

    ```
    SELECT * FROM "dev"."chapter_data_analysis_
    schema"."employees" order by emp_no;
    ```

The output is as follows:

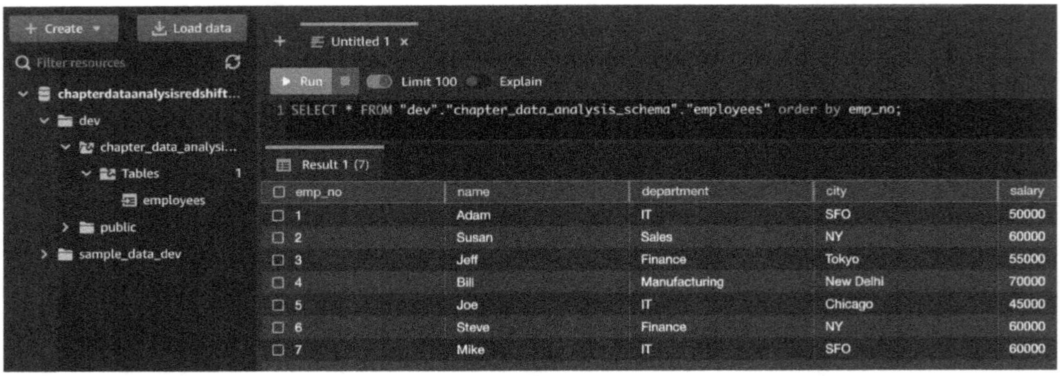

Figure 13.16 – Data in the Glue Data Catalog table

Alright, so we saw how the data written in S3 can be accessed by both Redshift and Athena for analysis. But what if the data had to be updated? One mechanism is to overwrite, that is, truncate and then load the table. In some cases, this approach can be expensive. We can probably come up with a more cost-optimized approach where we, first, partition the table and then only overwrite a partition. However, this approach comes with its own drawbacks.

For this approach to work, the newer updates will have to be limited to only a few of the partitions because if the newer updates are across partitions, then all of the partitions will have to be overwritten. As you might have noticed, creating a logic to upsert data in a data lake can become quite complex very quickly. An alternative is to use open source solutions such as Hudi and Delta Lake to make the data lake more transactional. Solutions such as Hudi bring additional benefits, such as the ability to create **Merge on Read** (**MoR**) or **CoW** (tables along with the ability to only query the incremental data and time travel.

In order to simplify the process of using these open source technologies, the AWS Glue team came up with AWS Glue custom connectors (https://aws.amazon.com/about-aws/whats-new/2020/12/aws-glue-launches-aws-glue-custom-connectors/).

In this chapter, we will use quite a few Marketplace Glue connectors. Previously, you created Apache Hudi, Delta Lake, and OpenSearch connections in the *Creating Marketplace connections* section. Now we will use Apache Hudi and Delta Lake connections for upserting data in the S3 data lake.

Creating and updating Hudi tables using Glue

Apache Hudi is an open source data management tool that was initially developed by Uber. Its superpower is enabling incremental data processing in a data lake. The Apache Hudi format is supported by a wide range of tools on AWS such as AWS Glue, Amazon Redshift, Amazon Athena, and Amazon EMR.

The CloudFormation template, for this chapter, creates two Hudi batch jobs. They are `02 - Hudi Init load for Data Analysis Chapter` and `03 - Hudi Incremental load for Data Analysis Chapter`. Both of these jobs use the Hudi connection created in the *Creating the Marketplace connections* section. Additionally, these jobs accept the target bucket as an input parameter. This input parameter is prepopulated by the CloudFormation template. Navigate to the job details page of the `02 - Hudi Init load for Data Analysis Chapter` job (`https://console.aws.amazon.com/gluestudio/home?#/editor/job/02%20-%20Hudi%20Init%20load%20for%20Data%20Analysis%20Chapter/details`) to check out the configurations for the job.

Now we will execute the Glue Hudi jobs to create Hudi tables in the Glue Data Catalog:

1. Navigate to the AWS Glue Studio console (`https://console.aws.amazon.com/gluestudio/home?#/jobs`), check the checkbox next to **02 - Hudi Init load for Data Analysis Chapter**, and click on the **Run Job** button.

2. Now you can go to the AWS Glue Studio monitoring page (`https://console.aws.amazon.com/gluestudio/home?#/monitoring`) and check the status of the job. You might see a lag of a few seconds for the execution to show up on the monitoring page:

Figure 13.17 – Viewing the job status

3. After the job finishes, this job will create a Hudi table, and you will be able to query it in Athena using the following query:

```
SELECT emp_no, name, department, city, salary FROM
"AwsDataCatalog"."chapter-data-analysis-glue-
database"."employees_cow" order by emp_no;
```

The results are as follows:

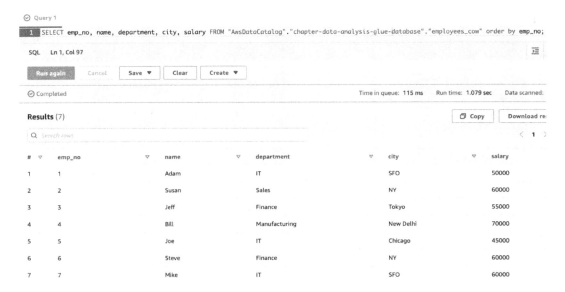

Figure 13.18 – The query results for the Hudi table

4. Now, let's say that **Jeff** got a raise along with a transfer to **Cincinnati**. Additionally, let's say that Jeff's new salary is 75,000. Run the **03 - Hudi Incremental load for Data Analysis Chapter** job just as you ran the previous one. This job will help to update the information in the `employees_cow` table. Note that the value of `salary=75000` and `city=Cincinnati` for `emp_no=3` is hardcoded in this job.

5. Go to the go the AWS Glue Studio monitoring page (`https://console.aws.amazon.com/gluestudio/home?#/monitoring`) and check the status of the job. You might see a lag of a few seconds for the execution to show up on the monitoring page:

Job runs (4) Info

Job name	Type	Start time	End time	Run status	Run time	Capacity	Worker type	DPU hours
03 - Hudi Incremental load for Data Analysis Chapter	Glue ETL	06/12/2022 18:25:27	-	Running	-	2	Standard	
02 - Hudi Init load for Data Analysis Chapter	Glue ETL	06/12/2022 18:16:04	06/12/2022 18:18:16	Succeeded	2 minutes	2	Standard	0.06
01 - Seed data job for Data Analysis Chapter	Glue ETL	06/12/2022 17:13:34	06/12/2022 17:14:40	Succeeded	1 minute	2	Standard	0.03

Figure 13.19 – Monitoring the status of the 03 - Hudi Incremental load for Data Analysis Chapter job

6. After the successful completion of the job, run the query on the `employees_cow` table in Amazon Athena again. You will notice that the record has been updated:

    ```
    SELECT emp_no, name, department, city, salary FROM
    "AwsDataCatalog"."chapter-data-analysis-glue-
    database"."employees_cow" order by emp_no;
    ```

The results are as follows:

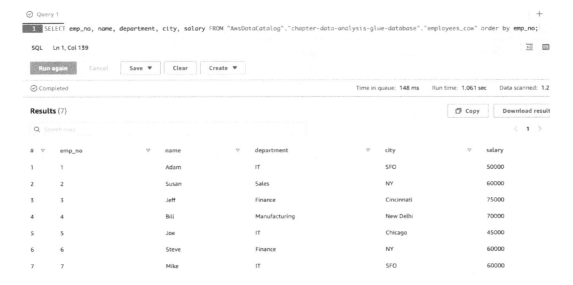

Figure 13.20 – The updated table

We just saw the use of Apache Hudi for upserting the data in a lake and querying the upserted data in Athena. Now we will try to upsert the data using the Delta Lake connection created in the *Creating Marketplace connections* section.

Creating and updating Delta Lake tables using Glue

Delta Lake is also an open source framework that was initially developed by Databricks. Similar to Hudi, Delta Lake is also supported by Spark, Presto, and Hive among many others.

We will now execute the **04 - DeltaLake Init load for Data Analysis Chapter** job to create a Delta Lake table. The **04 - DeltaLake Init load for Data Analysis Chapter** job was created by the CloudFormation template executed earlier:

1. Run the Glue job: **04 - DeltaLake Init load for Data Analysis Chapter**. Notice in the job script that we are using Spark SQL to create a table definition in the Glue Catalog for the Delta Table. Here is the Spark SQL statement from the code of the **04 - DeltaLake Init load for Data Analysis Chapter** job:

```
spark.sql("CREATE TABLE `chapter-data-analysis-
glue-database`.employees_deltalake (emp_no int,
name string, department string, city string,
salary int) ROW FORMAT SERDE 'org.apache.hadoop.
hive.ql.io.parquet.serde.ParquetHiveSerDe' STORED
AS INPUTFORMAT 'org.apache.hadoop.hive.ql.io.
SymlinkTextInputFormat' OUTPUTFORMAT 'org.apache.hadoop.
hive.ql.io.HiveIgnoreKeyTextOutputFormat' LOCATION
'"+tableLocation+"_symlink_format_manifest/'")
```

Also, notice that we have put /tmp/delta-core_2.12-1.0.0.jar in the **Python lib path** argument. This can be seen in the following screenshot:

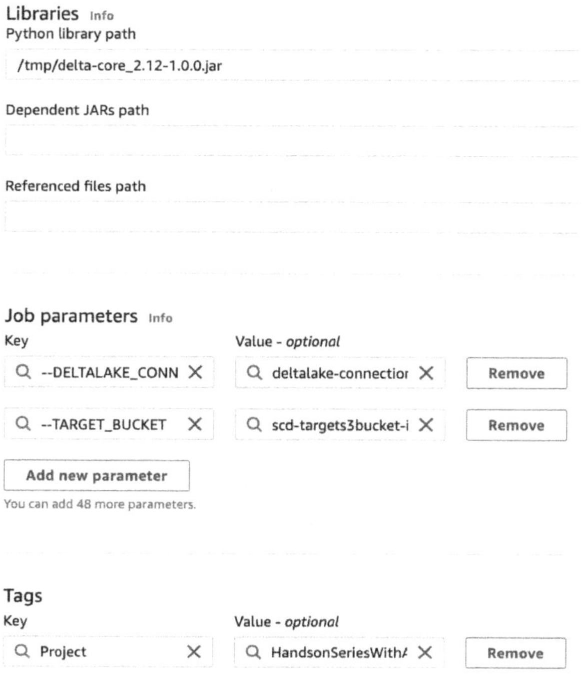

Figure 13.21 – Running the 04 - DeltaLake Init load for Data Analysis Chapter job

Additionally, we generate `symlink_format_manifest` from within the Glue job. This helps us to read the table from Athena or Presto.

2. Go to the AWS Glue Studio monitoring page (`https://console.aws.amazon.com/gluestudio/home?#/monitoring`) and check the status of the job. Once the job is complete, go to Athena, and execute the following statement:

```
SELECT * FROM "AwsDataCatalog"."chapter-data-analysis-
glue-database"."employees_deltalake" order by emp_no;
```

You will notice that the data has been inserted into the Glue Catalog table and can be queried through Athena, as shown in the following screenshot:

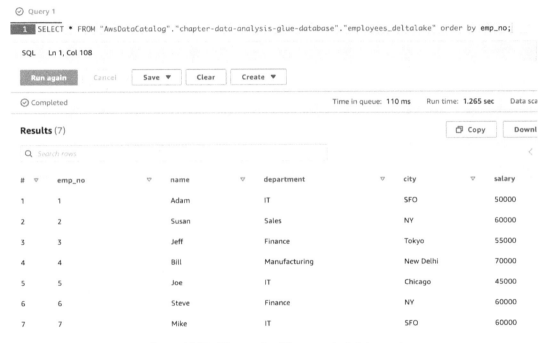

Figure 13.22 – The result of the executed statement

3. Now, let's say that we want to update `city` to `Cincinnati` and `salary` to `70000` for `emp_no = 3`. Run the **05 - DeltaLake Incremental load for Data Analysis Chapter** job and let it finish. The value of `salary=75000` and `city=Cincinnati` for `emp_no=3` is hardcoded into this job.

4. Run the following query in Athena and notice that the data for `emp_no = 3` has changed:

```
SELECT * FROM "AwsDataCatalog"."chapter-data-analysis-
glue-database"."employees_deltalake" order by emp_no;
```

The results are as follows:

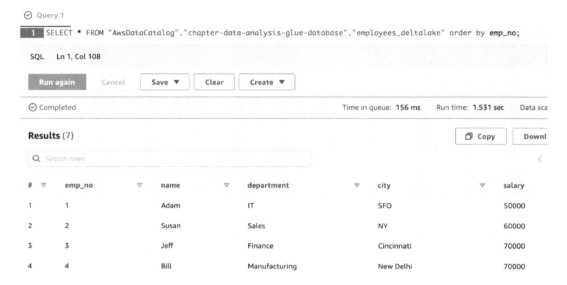

Figure 13.23 – The updated data for emp_no = 3

In this section, we saw how we can create and update tables and data in the Glue Data Catalog using Delta Lake. Now we will look at how we can insert data into governed tables.

Inserting data into Lake Formation governed tables

Governed tables are packed with a lot of features such as ACID transactions, automatic data compaction for faster query response times, and time travel queries. Now we will go through the process of creating Lake Formation governed tables using Glue jobs:

1. Go to the **Outputs** tab of the CloudFormation stack and grab the S3 path for the `LakeFormationLocationForRegistry` key.

2. Go to AWS Lake Formation (https://console.aws.amazon.com/lakeformation/home) and register the S3 location, from *step 1*, with Lake Formation, as shown in the following screenshot:

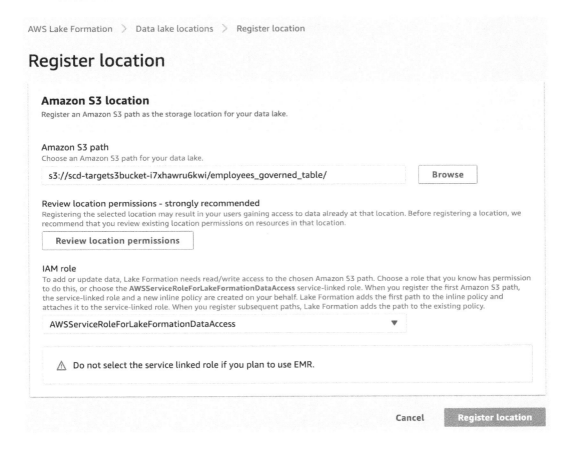

AWS Lake Formation > Data lake locations > Register location

Register location

Amazon S3 location
Register an Amazon S3 path as the storage location for your data lake.

Amazon S3 path
Choose an Amazon S3 path for your data lake.

| s3://scd-targets3bucket-i7xhawru6kwi/employees_governed_table/ | Browse |

Review location permissions - strongly recommended
Registering the selected location may result in your users gaining access to data already at that location. Before registering a location, we recommend that you review existing location permissions on resources in that location.

Review location permissions

IAM role
To add or update data, Lake Formation needs read/write access to the chosen Amazon S3 path. Choose a role that you know has permission to do this, or choose the **AWSServiceRoleForLakeFormationDataAccess** service-linked role. When you register the first Amazon S3 path, the service-linked role and a new inline policy are created on your behalf. Lake Formation adds the first path to the inline policy and attaches it to the service-linked role. When you register subsequent paths, Lake Formation adds the path to the existing policy.

AWSServiceRoleForLakeFormationDataAccess ▼

⚠ Do not select the service linked role if you plan to use EMR.

Cancel   Register location

Figure 13.24 – Registering the location

The format of this path is `s3://<target_s3_bucket>/employees_governed_table/`. Make sure that you register it in the same region where you created the Cloud Formation stack.

Note that you should use the `AWSServiceRoleForLakeFormationDataAccess` role. This role has been granted access to the KMS key so that we can query the governed table successfully.

3. Go to **Data locations** tab in Lake Formation and grant privileges from `s3://<target_s3_bucket>/employees_governed_table/` to HandsonSeriesWithAWSGlueJobRole. You will have to paste the `s3://<target_s3_bucket>/employees_governed_table/` path inside the **Storage locations** textbox and select **HandsonSeriesWithAWSGlueJobRole** from the **IAM users and roles** drop-down list:

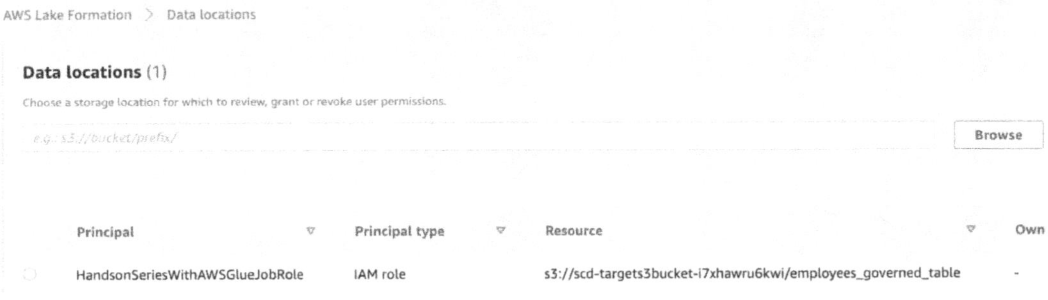

AWS Lake Formation > Data locations

Data locations (1)

Choose a storage location for which to review, grant or revoke user permissions.

e.g. s3://bucket/prefix/			Browse

Principal	▽	Principal type	▽	Resource	▽	Own
HandsonSeriesWithAWSGlueJobRole		IAM role		s3://scd-targets3bucket-i7xhawru6kwi/employees_governed_table		-

Figure 13.25 – The Data locations tab

4. Run the **06 - Governed Table Create Table for Data Analysis Chapter** job from Glue Studio, just as you ran the previous jobs. This job will create `employees_governed_table` in `chapter-data-analysis-glue-database`. After the job has been successfully completed, you should be able to see the table in Athena.

5. Now we will load this table with data. Execute the **07 - Governed Table Init Load for Data Analysis Chapter** job. This code starts a transaction, loads the data, and then commits it.

6. After the job finishes, you will now be able to query the data in Athena. Run the following command:

```
SELECT * FROM "AwsDataCatalog"."chapter-data-analysis-
glue-database"."employees_governed_table" order by emp_
no;
```

The following screenshot shows the data in the `employees_governed_table` table:

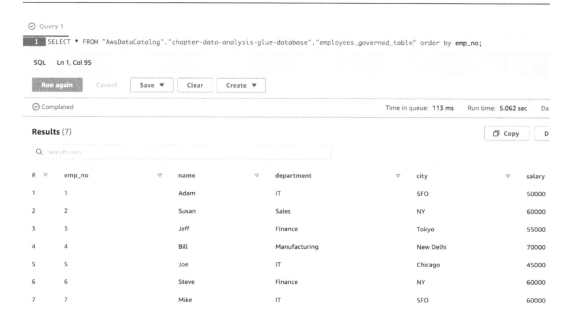

Figure 13.26 – Data in the employees_governed_table table

In this section, we saw how governed tables can be used to ingest data in a data lake. The Glue job used to ingest the data ran as a batch. In fact, in this chapter, all of the jobs that have been executed so far have been batch jobs. These jobs include the jobs related to both Hudi and Delta Lake. Next, we will look at how to stream ingestion jobs.

Consuming streaming data using Glue

Now that we understand how Glue works in batch mode, let's understand the process of updating the data coming through a stream.

The CloudFormation stack creates a **Managed Streaming for Apache Kafka (MSK)** cluster for this purpose. You will have to create a Glue connection for this MSK cluster. It is important that you name this connection as `chapter-data-analysis-msk-connection`. This connection is used in the jobs that follow. These jobs get the Kafka broker details from the connection.

Creating chapter-data-analysis-msk-connection

We will execute Glue jobs to load data into an MSK topic and also consume data from the topic. Both of these jobs require broker information and other details about the MSK cluster. Now we will create an MSK connection in Glue. Please ensure that you put the name of the connection as `chapter-data-analysis-msk-connection`. This is because the Glue jobs have been preconfigured to use this name as the connection name:

1. Navigate to the **Connections** page in the AWS Glue console (`https://console.aws.amazon.com/glue/home?#catalog:tab=connections`), and then go to the **Connections** section.

2. Click on the **Add connection** button.

3. Set **Connection type** as **Kafka** and put **Connection name** as `chapter-data-analysis-msk-connection`. Select the MSK cluster created using CloudFormation in the **Select MSK cluster** drop-down list and ensure that the **Require SSL connection** flag has been checked. Click on **Next**:

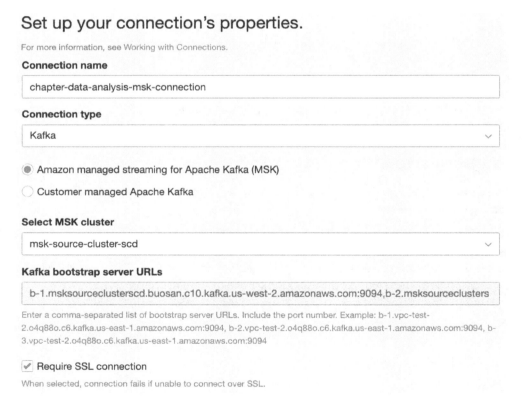

Figure 13.27 – Setting up the properties

4. Select the VPC ID, one of the private subnet IDs, and a security group, and click on **Next**. You should be able to get all of these values from the **Outputs** tab of the CloudFormation stack. Click on **Finish**:

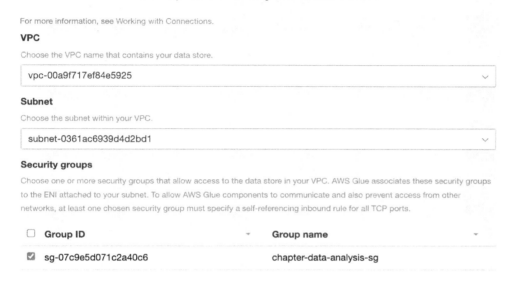

Figure 13.28 – Setting up access

Now that we have created an MSK connection in Glue, we will load data into a topic in the MSK cluster. Later, we will consume data from the topic through Glue streaming jobs.

Loading and consuming data from MSK using Glue

Run the Python shell's **08 - Kafka Producer for Data Analysis Chapter** job. This job will use `chapter-data-analysis-msk-connection`, as created in the preceding section, and load data into the MSK cluster.

This job uses the AWS wrangler `whl` file and the `kafka-python` `whl` file to read the data from the S3 path and load it into Kafka. Both of these `whl` files have been copied in the S3 bucket of your account through the CloudFormation template and have been configured in the Glue Python shell job. This job creates a `chapter-data-analysis` topic and then loads data into it.

After the job has successfully finished, you will have the data in the MSK cluster. Now we should execute the Glue streaming jobs to consume the data from the topic.

Glue streaming job as a consumer of a Kafka topic

First, we will check out the traditional micro-batch pattern that is commonly employed to consume streaming data using Glue.

Start the **09 - Kafka Consumer for Data Analysis Chapter** job. This is a Spark streaming job that consumes data from the `chapter-data-analysis` topic. It micro-batches the processing using the `forEachBatch` (https://docs.aws.amazon.com/glue/latest/dg/aws-glue-api-crawler-pyspark-extensions-glue-context.html#aws-glue-api-crawler-pyspark-extensions-glue-context-forEachBatch) method of `GlueContext`.

The `forEachBatch` method micro-batches the streaming dynamic frame. In the **09 - Kafka Consumer for Data Analysis Chapter** job, the micro-batch is 10 seconds. Each micro-batch is processed in the `processBatch` method. In the **09 - Kafka Consumer for Data Analysis Chapter** job, we write the micro-batch into a Hudi table just as we had written one in the batch operation.

Notice that the processing of these micro-batches was no different from the processing of the Hudi batch jobs shared earlier. Essentially, this means that the process can be applied to consume a stream in other formats such as Delta Lake using the batch code for the Delta Lake shared earlier.

After a couple of minutes of execution, you should see the `employees_cow_streaming` table under `chapter-data-analysis-glue-database`. You should be able to query it in Athena using the following query:

```
SELECT emp_no,name,department,city,salary FROM
"AwsDataCatalog"."chapter_data_analysis"."employees_cow_
streaming" order by emp_no;
```

The results are as follows:

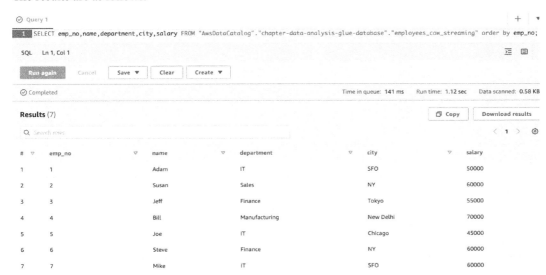

Figure 13.29 – The results of the query

In this section, we used a streaming Glue job to consume data from a Kafka topic. In the next section, we will use the Hudi DeltaStreamer (https://hudi.apache.org/docs/hoodie_deltastreamer/) utility to consume data from the same Kafka topic.

Hudi DeltaStreamer streaming job as a consumer of a Kafka topic

Now that we have seen the traditional micro-batch method used to consume streaming sources in Hudi tables using Glue, let's look at the mechanism of using Hudi DeltaStreamer to consume streaming data.

Run the **10 - DeltaStreamer Kafka Consumer for Data Analysis Chapter** job. Just as in the previous job, this also uses the Hudi connection. However, notice that the dependent JARs path has been set to /tmp/*. This is required to ensure that the right classes are available on the classpath. While some of the configurations are similar to the configurations of the Hudi jobs that we created up till now, the DeltaSteamer job requires the schema files for the source and target. Since our use case is about replicating data, our source and target schema has the same file. The structure of this avro schema file is as follows:

```
{
  "type":"record",
  "name":"employees",
  "fields":[{
```

```
        "name": "emp_no",
        "type": "int"
    }, {
        "name": "name",
        "type": "string"
    }, {
        "name": "department",
        "type": "string"
    },{
        "name": "city",
        "type": "string"
    },{
        "name": "salary",
        "type": "int"
    },{
        "name": "record_creation_time",
        "type": "float"
    }
  ] }
```

This file is written into your S3 bucket through CloudFormation, and the **10 - DeltaStreamer Kafka Consumer for Data Analysis Chapter** job is configured to use this avro schema file.

Once the job has been executing for 2–3 minutes, you should be able to see and query the employees_ deltastreamer table in chapter-data-analysis-glue-database, in Athena, using the following query:

```
SELECT emp_no,name,department,city,salary FROM
"AwsDataCatalog"."chapter-data-analysis-glue-
database"."employees_deltastreamer" order by emp_no;
```

The result is as follows:

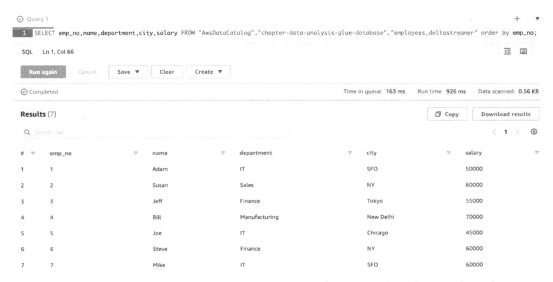

Figure 13.30 – The 10 - DeltaStreamer Kafka Consumer for Data Analysis Chapter job results

Now we have our traditional Glue streaming and DeltaStreamer jobs running. This means that if we add new data to our MSK topic, the data will be consumed by both of our jobs. Now we will load CDC data into our topic. Our streaming jobs should consume and process the data. Our query through Athena should be able to show the updated data in the processed tables since our streaming jobs are using Hudi.

Creating and consuming CDC data through streaming jobs on Glue

Now, we will load CDC data into the MSK topic.

Run the **11 - Incremental Data Kafka Producer for Data Analysis Chapter** job. This job adds the following CDC data to the `chapter-data-analysis` topic:

```
{"emp_no": 3,"name": "Jeff","department": "Finance","city":
"Cincinnati","salary": 70000,"record_creation_time":now}
```

This job uses the same `whl` files as the **08 - Kafka Producer for Data Analysis Chapter** job.

As soon as the job finishes, you should be able to see the update in both the `employees_deltastreamer` and `employees_cow_streaming` tables. The following screenshot shows the result in the `employees_deltastreamer` table:

Figure 13.31 – The results of the employees_deltastreamer table

The following screenshot shows the result in the `employees_cow_streaming` table:

Figure 13.32 – The results of the employees_cow_streaming table

Since our Glue streaming jobs are configured to consider emp_no as the record key, it will automatically update city and salary to the new values.

> **Note**
>
> Please shut down the Glue Streaming job so that you do not incur any additional charges.

Now we will discuss the process of loading the Amazon OpenSearch domain using Glue.

Glue's integration with OpenSearch

Now, let's focus on a search use case. Let's say that you were interested in searching through log data. Amazon OpenSearch could be your answer to that. Originally, it was forked from Elasticsearch and comes with a visualization technology called **OpenSearch Dashboards**. OpenSearch Dashboards has been forked from **Kibana**. OpenSearch can work on petabytes of unstructured and semi-structured data. Additionally, it can auto-tune itself and use ML to detect anomalies in real time. Auto-Tune analyzes cluster performance over time and suggests optimizations based on your workload.

For the purpose of this chapter, we will use our employee data as the source and show how we can load the data into OpenSearch. Then, we will visualize the data in OpenSearch Dashboards.

The CloudFormation template creates a secret that stores the OpenSearch domain's user ID and password. The Marketplace connection created by you using the OpenSearch connector should have this secret configured in it. This is because the Glue job will use this secret to authenticate against the OpenSearch domain. Now we will set the secret in the Glue OpenSearch connection:

1. Navigate to the **Connectors** tab of the AWS Glue Studio console (https://console.aws.amazon.com/gluestudio/home?#/connectors) and then go to the OpenSearch connection that you created earlier. This connection should be in the **Connections** section.

2. Click on the **Edit** button:

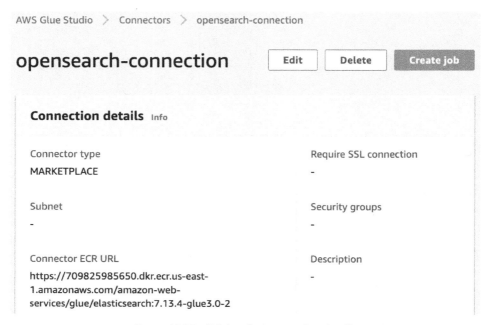

AWS Glue Studio > Connectors > opensearch-connection

opensearch-connection   [Edit]   [Delete]   [Create job]

Connection details Info

Connector type   Require SSL connection
MARKETPLACE   -

Subnet   Security groups
-                                       -

Connector ECR URL   Description
https://709825985650.dkr.ecr.us-east-   -
1.amazonaws.com/amazon-web-
services/glue/elasticsearch:7.13.4-glue3.0-2

Figure 13.33 – Editing the connection details

3. Go to the **Connection access** section and select **ChapterDataAnalysisOSSecret** from the drop-down list. Then, click on the **Save changes** button. **ChapterDataAnalysisOSSecret** is created by the CloudFormation template. The values of the OpenSearch master user and password supplied during the Cloud Formation stack have been stored in this secret:

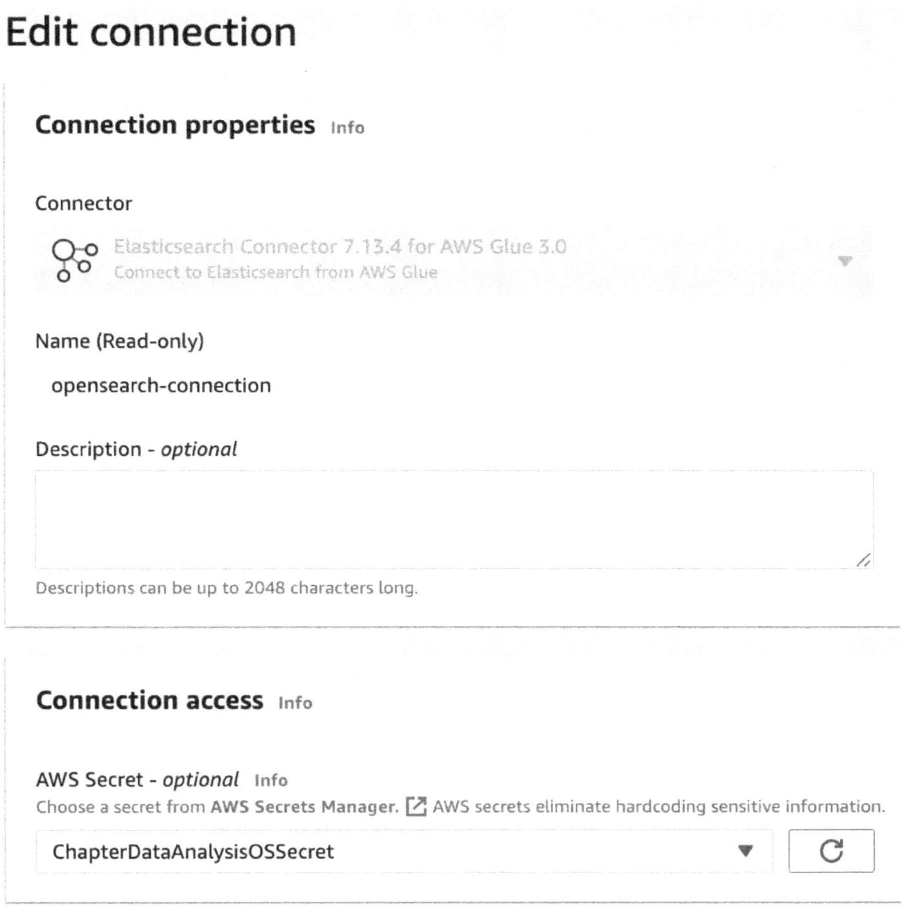

Edit connection

Connection properties Info

Connector

Elasticsearch Connector 7.13.4 for AWS Glue 3.0
Connect to Elasticsearch from AWS Glue

Name (Read-only)

opensearch-connection

Description - *optional*

Descriptions can be up to 2048 characters long.

Connection access Info

AWS Secret - *optional* Info
Choose a secret from **AWS Secrets Manager.** AWS secrets eliminate hardcoding sensitive information.

ChapterDataAnalysisOSSecret

Figure 13.34 – Filling in the connection properties

4. Run the **12 - OpenSearch Load for Data Analysis Chapter** job. On the successful completion of this job, the employee information will be available in the `employees` index of the OS domain.

5. Now that we have data in our OS domain, it's time to access that. The CloudFormation template has created a Windows EC2 instance for you to check the data. First, you will need the password to the EC2 instance. Run the following command to retrieve the password:

```
aws ec2 get-password-data --instance-id <instance_id_
of_windows_ec2_instance> --priv-launch-key <key_file_
selected_during_the_creation_of_the_cloudformation_stack>
--query PasswordData | tr -d '"'
```

You can get the instance ID from the `InstanceIDOfEC2InstanceForRDP` key in the **Outputs** tab of the CloudFormation stack.

6. Now, navigate to your remote desktop client and use the public IP address of the EC2 instance. Use the password from the preceding step and `Administrator` as the username to log in. You can get the public IP address of the EC2 instance from the `PublicIPOfEC2InstanceForRDP` key in the **Outputs** tab of the CloudFormation stack.

   If you had keyed in the correct IP address of your laptop in the `ClientIPCIDR` parameter of the CloudFormation stack, then a security group rule to allow a **Remote Desktop Protocol (RDP)** connection from your laptop on port `3389` should already be in place.

7. Install your favorite browser on the EC2 instance after logging in, and then open the OpenSearch Dashboards URL. You can get this URL from the `OpenSearchDashboardsURL` key in the **Outputs** tab of the CloudFormation stack.

8. Use the username and password entered for the OpenSearch domain during the creation of the CloudFormation stack. Additionally, you can also retrieve it from the `ChapterDataAnalysisOSSecret` secret in the AWS Secrets Manager (`https://console.aws.amazon.com/secretsmanager/home`).

9. Click on the **Explore on my own** link, select the **Private** radio button in the **Select your tenant** popup, and then click on the **Confirm** button:

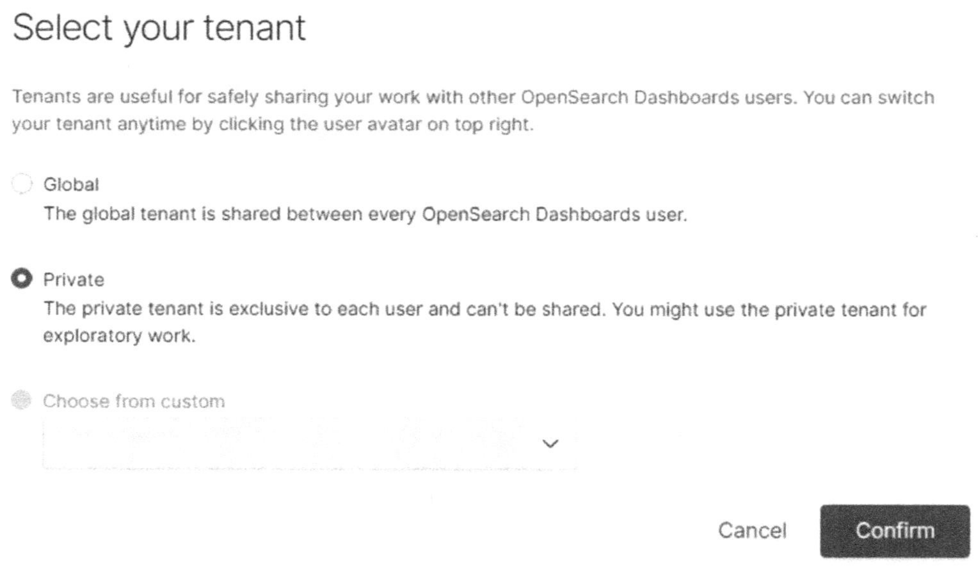

Select your tenant

Tenants are useful for safely sharing your work with other OpenSearch Dashboards users. You can switch your tenant anytime by clicking the user avatar on top right.

○ Global
The global tenant is shared between every OpenSearch Dashboards user.

● Private
The private tenant is exclusive to each user and can't be shared. You might use the private tenant for exploratory work.

○ Choose from custom

⌄

Cancel   **Confirm**

Figure 13.35 – Selecting the private tenant option

10. Click on **Query Workbench** from the left-hand pane:

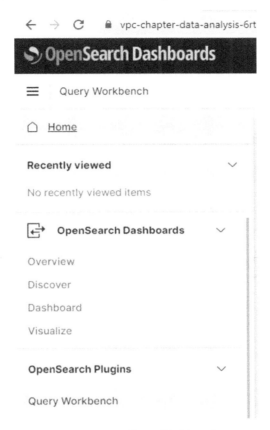

Figure 13.36 – Query Workbench

11. Run the following query in the **Query editor** window. You will notice that the data is available in OpenSearch to use:

```
select * from employees order by emp_no;
```

The following screenshot shows the data:

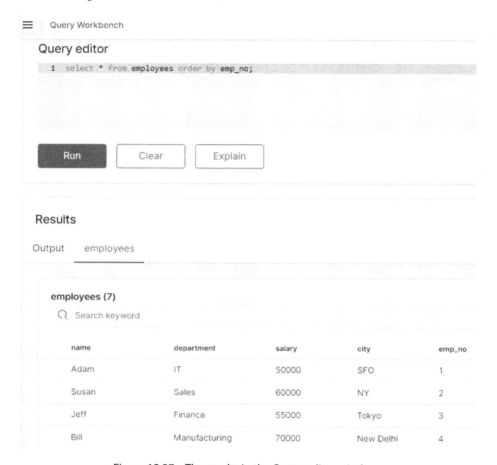

Figure 13.37 – The results in the Query editor window

In this section, we inserted data from Glue into OpenSearch and then queried it from OpenSearch Dashboards.

Cleaning up

Delete the CloudFormation stack and remove the registration of the S3 location in AWS Lake Formation along with the *Data locations* permissions that were granted manually for the governed tables part.

Summary

In this chapter, we learned how data in the data lake can be consumed through both Athena and Redshift. Then, we saw how we can create transactional lakes using technologies such as Hudi and Delta Lake. We then checked various mechanisms for consuming streaming sources in Glue using the `forEachBatch` method and Hudi DeltaStreamer. Finally, we checked how the ElasticSearch connector from the AWS Glue connector offerings can be used to push data into an OpenSearch domain and consumed through OpenSearch Dashboards. This chapter familiarized you with the most common patterns of data analysis and ETL using AWS Glue.

In the next chapter, we will learn about ML. We will find out more about the strengths and weaknesses of SparkML and SageMaker and when to use each of those tools.

14
Machine Learning Integration

Machine learning (**ML**) is one of the cornerstones of today's computing for any software-related company. ML models are capable of making predictions or deductions based on past experience, provided as training data. This enables a wide variety of applications with large benefits to any organization.

Because it relies on training data, ML is closely tied to data mining, data processing, and, in general, any kind of **extract, transform, load** (**ETL**) process. Training data must be properly cleaned, formatted, and classified before it can be fed to a model – a process that greatly affects the effectiveness of the model itself. Because of this, services such as AWS Glue offer ML-specific features and integrations, catered to making ML easier and more effective to use.

Training data preparation is not the only relationship ML has with ETL processes – it can also be used to enhance and provide new transformations within the processes themselves, enabling new capabilities that were not possible before. ML models can be used, for instance, to automatically detect duplicate data or to tag columns in datasets based on specific properties.

In this chapter, we will cover the following topics:

- Glue ML transformations
- SageMaker integration
- Developing ML pipelines with Glue

By the end of this chapter, you will understand how ML transformations work with Glue, how to combine Glue and SageMaker effectively to power all your ML needs, and how to deploy an ML pipeline in the AWS cloud using Glue.

Technical requirements

For this chapter, the only requirement is that you have access to this book's GitHub repository page (`https://github.com/PacktPublishing/Serverless-ETL-and-Analytics-with-AWS-Glue`).

Glue ML transformations

As mentioned previously, ML is not just an entity that reads the output data from ETL processes, but also one that powers its transformations. ML models enable a wide variety of operations that were not possible before due to computer intelligence limitations.

Because of this, Glue started to offer ML powered-operations with specific purposes under the **ML transforms** feature. As the name suggests, ML transforms are specific kinds of Glue transforms that are powered by ML models but must be trained and prepared before they can be used. Once they are ready, they can be called from your ETL job's code, just like other Glue transforms.

At the time of writing, Glue has only released one ML transform, FindMatches, which will automatically find duplicated records within a dataset. Even though this seems like a simple task (most ETL engines could provide this by simply comparing records and checking if they are equal, or if they share a primary key), ML allows for duplicate detection, even in scenarios where records don't have the same identifier or primary key, or when all the fields are not the same.

The FindMatches operation enables use cases that were not possible or considerably harder before, such as fraud detection (where a user may have created a duplicate account while trying to avoid a ban) or finding duplicates in a product catalog (where two entries may have different capitalization or spelling but refer to the same product).

As mentioned earlier, ML transforms must be trained, which means you will need sample training data, but the transformation must be fine-tuned to the specifics of your dataset and use case. ML transforms also abstract most of the logic that goes into training an ML model, enabling data engineers to take advantage of ML without necessarily being experts on it.

In this section, we'll go through the life cycle of an ML transform. We'll learn how to create one, train it, and use it in ETL jobs.

Creating an ML transform

Before the training stage, an ML transform must be created and configured according to the desired results. To create an ML transform, you must provide the following:

- **Job configuration parameters**: Just like with ETL jobs, the transform will need an IAM role, resource configuration, and security configurations.

- **Source dataset parameters**: The dataset to be read, plus the column within it to be used as a primary key.

- **ML tuning parameters**: As with other ML models, the transformation can be configured to favor precise or non-strict results, where the non-strict configuration would give more results but also a larger number of false positives. A Glue ML transform also allows the user to decide between spending extra resources to make the transform more accurate or saving costs by having fewer results at the cost of less accuracy.

Once these have been set, the transform is created and is set to the **Needs training** state. Transforms in this state cannot be used in ETL jobs, as they are required to go through the model training process first.

Training an ML transform

ML transforms follow a supervised learning mechanism called **labeling**. Within ML, data labeling is a mechanism by which a human actor provides context to a dataset so that a machine can learn and understand it. For instance, when creating an image object recognition ML model, a human could take a set of pictures and label them based on the object shown in them (for example, "car", "bicycle", or "orange"). This labeling can be as simple or as complex as required, and with it, the ML model can understand the context of what it is trying to recognize based on the labels.

The same mechanism applies here. When training an ML transform, Glue will take a sample of the records of the specified input dataset and provide it in a pre-formatted CSV file in an S3 path of the user's choice. The user can then download the file, inspect the records, and label them accordingly by filling out the **label** column in the CSV file. For instance, let's say we have the following sample records:

label	book_id	book_title	authors
	12367126	1-Safe Algorithms for Symmetric Site Configurations	John Hayes, Richard B. Bauchman
	08272651	1-Safe Algorithms for Symmetric Site Configurations	John Hayes, Richard B. Bauchman
	71616223	2003 SIGMOD Innovations Award Speech	Martha Smith
	12637181	2Q: A Low Overhead High-Performance Buffer Management	Elena Garcia
	72521341	2Q: A Low Overhead High-Performance Buffer Management	Elena Garcia

Figure 14.1 – Sample records with an empty label column

The labeling should look like this, considering we are trying to identify duplicated records:

label	book_id	book_title	authors
0	12367126	1-Safe Algorithms for Symmetric Site Configurations	John Hayes, Richard B. Bauchman
0	08272651	1-Safe Algorithms for Symmetric Site Configurations	John Hayes, Richard B. Bauchman
1	71616223	2003 SIGMOD Innovations Award Speech	Martha Smith
2	12637181	2Q: A Low Overhead High-Performance Buffer Management	Elena Garcia
2	72521341	2Q: A Low Overhead High-Performance Buffer Management	Elena Garcia

Figure 14.2 – Sample records with a filled label column

As you can see, the labeling process consists of setting the same label identifier for records that refer to the same book, even if the records are slightly different in terms of title or authors. This process is quite literally teaching the ML model how different records refer to the same entity.

There are several considerations regarding how this works:

- The value of the label column can be anything (a number, a letter, or a word), so long as it is consistent and the same for equal rows.

- The file will also contain a second column called labeling_set_id. This column identifies different groups of rows with their own, separate labeling. Label values can be repeated across different labeling sets without causing a match.

- The file you upload to S3 for Glue to take as labels must be a UTF-8-encoded CSV file, the columns of which must be the same as the source dataset's, plus the label and labeling_set_id columns.

Once the labeling file is ready, it can be uploaded to S3 and provided to Glue so that it can train the transform's model. Upon being uploaded, Glue can perform two calculations:

- **Transform quality estimation**: This is an estimation of how good the transform is at doing its job, as specified by several percentage values.

- **Column importance**: This calculation determines how relevant the columns in the dataset are to the success of the transform. Irrelevant columns can be omitted and the transform would still be able to find matches.

The labeling process is repeatable and can be done an unlimited number of times. If the results of the quality estimation process (or the results of your ETL jobs that rely on the ML transform) are not good enough, the labeling process can be repeated to improve the accuracy of the model through human curation.

Using an ML transform

Once the transform has been trained at least once, it will change status to **Ready for use**. A transformation in this state can be used within ETL jobs. The **FindMatches** ML transform can be used in two modes: regular and incremental. Let's start by looking at a regular invocation:

```
findmatches_result = FindMatches.apply(
frame = my_dynamic_frame,
transformId = "tfm-d03f274ad2f0136dacc5bcb54deced1eea54371a",
transformation_ctx = "findmatches")
```

As you can see, the transform simply needs the DynamicFrame to apply the transformation to, as well as the ID of the trained transformation. The result of this operation (findmatches_result) will be a DynamicFrame with the same schema as the input one, but with two added columns:

- match_id: If the ML model considers two rows are the same, they will have the same value for this column – for instance, two matching rows may have a match_id value of 2, whereas a different pair of matching rows may have a value of 3.

- match_confidence_score: This represents a number between 0 and 1 that estimates the quality of the decision made by the model.

Using these two columns, a pipeline could automatically cull duplicated records, provided that the confidence score is high enough, for example.

Using FindMatches in this way lets users detect duplicates in a dataset. However, it can cause challenges. If new records were to come in and had to be matched against the previous ones, they would have to be added to the already-existing table, and the transform would have to be executed against the entire dataset. This approach is doable but will increase the execution time and resource consumption as the dataset becomes larger. Because of this, Glue provides an incremental way of using a transform:

```
findincrementalmatches_result = FindIncrementalMatches.apply(
existingFrame = my_dynamic_frame,
incrementalFrame = my_incremental_data,
transformId = "tfm-d03f274ad2f0136dacc5bcb54deced1eea54371a",
computeMatchConfidenceScores = true,
transformation_ctx = "findincrementalmatches")
```

When using `FindIncrementalMatches`, several parameters must be provided:

- `existingFrame`: This represents the already existing and cleaned dataset.

- `incrementalFrame`: This represents the batch of new records that must be matched against the already-existing ones.

- `transformId`: This is the ID of the trained transformation.

- `computeMatchConfidenceScores`: A Boolean value that determines whether the `match_confidence_score` column should be generated or not.

Using `FindIncrementalMatches` allows for faster, easier, and less resource-intensive match detection and should always be used for incremental setups.

Running ML training tasks and ML ETL jobs

Training an ML model and using ML-based transformations is a resource-intensive task that often requires additional memory. Because of this, it is always recommended to use larger EC2 instance types, or in the case of Glue resources, worker types. When training an ML transformation or running an ML-based ETL job, we recommend always using the G.2X worker type unless you know the task is simple and small in advance.

SageMaker integration

Amazon SageMaker is AWS's primary service for ML development. It provides a set of tools and features that lets users handle all the stages of the ML development pipeline, from data collection and preparation to model deployment and hosting.

Just like any other ML tool, SageMaker relies on the concept of model training to get models up to the accuracy level expected from them. And as we mentioned previously, training ML models usually requires large amounts of data to be prepared and processed. Because of this, SageMaker offers native integration with Apache Spark (`https://docs.aws.amazon.com/sagemaker/latest/dg/apache-spark.html`), which provides model-training capabilities using an AWS-tailored version of Spark.

One of the most important features SageMaker offers is serverless notebooks (`https://docs.aws.amazon.com/sagemaker/latest/dg/nbi.html`). A notebook instance is a serverless EC2 instance that runs Jupyter (`https://jupyter.org`), a web-based code execution service that lets users run code and visualize results interactively through the concept of notebooks. Code running in notebooks can be written in a variety of languages and use as many external libraries and frameworks as necessary, including Apache Spark. That said, the code within the notebook is usually executed locally unless a framework provides the capabilities to do otherwise.

To execute SageMaker features using Apache Spark jobs in a proper cluster environment, SageMaker offers AWS Glue integration. This allows users to execute Spark code written in a SageMaker notebook in a Glue Development Endpoint rather than locally within the notebook instance – which is always recommended to take advantage of Spark's concurrent execution model.

Glue-integrated SageMaker notebooks have the following limitations and considerations:

- They can only be launched from the Glue web console.
- The Development Endpoint they attach to must be launched in a VPC.
- Just like with ETL jobs, the security group attached must contain a self-referencing inbound rule that allows all traffic. This ensures that communication between the notebook and the endpoint, as well as between all the nodes of the endpoint, is possible.

Once a notebook has been launched, the **Sparkmagic** kernel can be used to run code within the Development Endpoint. Even though this feature was originally designed to run the SageMaker Spark library, you can also use it to interactively run and debug your regular ETL job code in a notebook easily.

In the next section, we'll discuss ways to orchestrate the elements we discussed previously into a pipeline using Glue and SageMaker.

Developing ML pipelines with Glue

The combination of SageMaker's model-hosting features and libraries, plus Glue's data preparation and orchestration features, allow you to create complex and highly-configurable ML pipelines. In this architecture, each service is responsible for different roles:

- Glue handles data handling and orchestration. Data handling includes extraction, processing, preparation, and storage. Orchestration refers to the overall execution of the pipeline itself.
- SageMaker handles all ML-related tasks such as model creation, training, and hosting.

Several components are critical to this, as follows:

- **Glue workflows** are the main form of orchestration in Glue. Workflows allow users to define graph-based chains of crawlers, ETL jobs, and triggers, and to see their execution visually in the web console.
- **Python Shell jobs** are a sub-class of Glue ETL jobs that are designed to run plain Python scripts instead of PySpark ones. They are similar to AWS Lambda functions but come with fewer restrictions since they don't have a time or memory limit. Python Shell jobs are typically used to automate tasks in an ETL pipeline using the AWS SDK or to run any code that does not need the capabilities of Spark in a cheaper, faster-to-launch environment.

- **SageMaker Model hosting** allows users to create an ML model and host it in the AWS cloud. Users don't have to worry about managing hardware or infrastructure to hold the model, and SageMaker provides tools to train and access the model in different ways.

A Glue-based ML pipeline would consist of a workflow where the following steps take place:

1. **Data extraction**: A Spark-based Glue ETL job obtains data from a source and stores it in intermediate, temporary storage, such as S3.

2. **Data preparation**: A second Spark-based Glue ETL job takes the output of *Step 1* and prepares the dataset for ML usage using ETL transformations.

3. **Model creation and training**: Using the AWS SDK, a Glue Python Shell job creates an ML model hosted in SageMaker and starts a SageMaker training job using the dataset that was created in *Step 2*. Once the model has been trained, a SageMaker inference endpoint is created to let other applications use the model.

Interaction with the workflow (starting it and notifying its completion) can be handled with Amazon SQS queues and messaging (https://aws.amazon.com/sqs/).

Parts of the pipeline could be replaced by other services if Glue's capabilities are not enough, although orchestration would have to be handled with a different feature since Glue workflows only orchestrate Glue resources. The following are some examples:

- The data extraction phase could be handled by any other ETL service in AWS, such as Amazon EMR, AWS Batch, or AWS Data Exchange.

- The data preparation phase could potentially be handled better by AWS Glue DataBrew, a service specifically designed for visual data preparation. Alternatively, you could also use Amazon EMR or AWS Batch.

- Pipeline orchestration can be handled by AWS Step Functions, CloudWatch events, or even Lambda functions.

With this, we've discussed everything about ML data pipelines using Glue.

Summary

In this chapter, we discussed all aspects of ML within AWS Glue. We talked about Glue ML transforms, what they are, how they are trained, and how they can be used. We also discussed AWS SageMaker and how it can integrate with Glue resources to accelerate the execution of ML code in notebooks. Finally, we analyzed reference architectures and services for ML pipelines using AWS Glue and SageMaker.

These concepts should have given you a complete overview of how Glue can be used for ML purposes, and how Glue can fit into your ML architecture in the AWS cloud. In the next chapter, we will talk about the data lake architecture and designing use cases for real-world scenarios.

15

Architecting Data Lakes for Real-World Scenarios and Edge Cases

We are now well versed in the concept of a **data lake**, a *centralized repository* that allows you to store all your structured and unstructured data at any scale. Since a data lake primarily focuses on storage, it does not require as much processing power as other methods (such as the data warehouse), making it easier, faster, and more cost-effective to scale up as data volumes grow.

The data lake is not just a repository – it requires a well-designed data architecture, along with proper planning and management. As it is driven by a data-based design, it helps you rapidly ingest raw data before any business requirements come into the picture. There are a variety of tools you can use for ingesting raw data into a data lake, including ETL tools such as Ab Initio, Informatica, and DataStage.

This chapter mainly covers practical examples of real-world data problems that exhibit certain bottlenecks and how to overcome these. By the end of this chapter, you should be familiar with common data problems, such as various ETL optimization techniques you can apply with AWS Glue to handle large volumes of data, handling a large number of small files, common performance issues with join operations involving fact and dimension tables, and how you design a data layout for highly selective queries with AWS Glue.

In this chapter, we're going to cover the following main topics:

- Running a highly selective query on a big fact table using AWS Glue
- Dealing with Join performance issues with big fact and small dimension tables in ETL workloads
- Solving Join problems involving big fact and big dimension tables using AWS Glue

- Reducing time on read operations involving large-dimension tables using AWS Glue grouping

- Solving S3 eventual consistency problems and faster writes to Amazon S3 for large fact table datasets using AWS Glue

Technical requirements

To follow along with the examples in this chapter, you will need the following:

- Access to GitHub, S3, and the AWS console (specifically AWS Glue, AWS Lake Formation, and Amazon S3).

- A computer with Chrome, Safari, or Microsoft Edge installed and the **AWS command-line interface** (**AWS CLI**):

  - Regarding the AWS CLI, you can use not only the AWS CLI but also AWS CLI version 2. In this chapter, the AWS CLI (not version 2) will be used. You can set up the AWS CLI (and version 2) by going to `https://docs.aws.amazon.com/cli/latest/userguide/cli-chap-getting-started.html`.

- An AWS account and an accompanying IAM user (or IAM role) with sufficient privileges to complete this chapter's activities. We recommend using a minimally scoped IAM policy to avoid unnecessary usage and making operational mistakes. You can get the IAM policy for this chapter from this book's GitHub repository, which can be found at `https://github.com/PacktPublishing/Serverless-ETL-and-Analytics-with-AWS-Glue`. This IAM policy includes the following access:

  - Permissions to create a list of IAM roles and policies for creating a service role for an AWS Glue ETL job

  - Permissions to read, list, and write access to an Amazon S3 bucket

  - Permissions to read and write access to Glue Data Catalog databases, tables, and partitions

- An S3 bucket for reading and writing data by AWS Glue. If you haven't done so yet, you can create one via the AWS console (`https://s3.console.aws.amazon.com/s3/home`). You can also create a bucket by running the following AWS CLI command:

  ```
  aws s3api create-bucket --bucket <your_bucket_name>
  --region us-east-1
  ```

Running a highly selective query on a big fact table using AWS Glue

We will start with one of the common data processing use cases, where you would end up scanning a large volume of data but it returns a selected value as a result. For example, if you want to find out the city with the highest population within the US, it would end up scanning data for more than 19,000 cities and then returning only one city as a result. Working with a large volume of data comes with the challenges of high amounts of processing costs and spending a lot of time scaling them. You should know the right techniques for data filtering to avoid any kind of data processing bottlenecks.

In this section, you will learn how to handle highly selective queries with AWS Glue. Let's say that you have a use case to query a big fact table that consists of humongous clickstream data stored in Amazon S3 that contains billions of records. The clickstream data stores information that's been collected about a user while they browse through a website or use a web browser. You are looking to query the dataset to check how much time a specific customer had spent on a given website at a specific time or how many views were generated for a specific product for a given timeframe. These are considered highly selective queries. It can be an intense operation and creates a bottleneck when it comes to scanning billions of records and returning the data. Under the hood, the Apache Spark driver splits the overall query into tasks and sends these tasks to executor processes on different nodes of the cluster. To improve query performance, one strategy is to reduce the data that is read by the Spark executors. One way to prevent loading data that is not needed is to use Glue partition indexes, which reduce the data movement and query processing time. This becomes even more important if the executors are not on the same physical machine as the data.

In the next section, you will learn how to use AWS Glue to implement a solution to be able to run highly selective queries efficiently. We will demonstrate how to use the Glue partition indexing technique. The solution provided uses an AWS Glue crawler to crawl an S3 table and then conduct an analysis using Spark SQL queries with Glue interactive sessions, as shown in the following diagram:

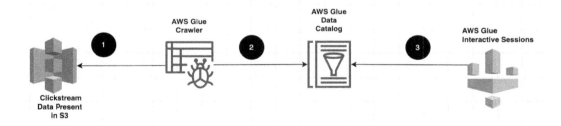

Figure 15.1 – Handling a highly selective query with AWS Glue

As illustrated in the preceding diagram, the solution works as follows:

1. An AWS Glue crawler parses the schema from AWS S3 and registers the table in AWS Glue Data Catalog with the metadata.

2. The Glue interactive sessions use Spark SQL and AWS Glue Data Catalog as their external schema stores for the newly written data in Amazon S3. You can perform highly selective queries on the data by using interactive sessions to query Amazon S3 directly using SQL.

So far, we have understood how you can handle running highly selective queries with AWS Glue with Glue's partition index feature. Now, let's get hands-on by preparing the test data and running some of the sample highly selective queries.

Hands-on tutorial

In this tutorial, we will use the AWS CLI to prepare the test data, create a Glue database and Glue crawlers, and define partition indexing on tables to see how it works and helps with highly selective queries. Follow these steps:

1. **Prepare the test data**: We will use a partitioned dataset from a sample clickstream data source to work with partition indexing with Glue Data Catalog. You can execute the following commands using the AWS CLI. It will take about 6 to 8 minutes to copy the data. Notice that we loaded data for customers 1, 2, and 3. Execute the following commands after replacing `${YOUR_BUCKET_NAME}` with your respective AWS S3 bucket:

   ```
   aws s3 sync s3://packt-serverless-etl-glue/chapter-15/
   uservisits-src-data/customer=1/ s3://${YOUR_BUCKET_NAME}/
   input/clkstreamdata/customer=1/ --exclude "*" --include
   "visitYearMonth=1998*"

   aws s3 sync s3://packt-serverless-etl-glue/chapter-15/
   uservisits-src-data/customer=2/ s3://${YOUR_BUCKET_NAME
   }/input/clkstreamdata/customer=2/ --exclude "*" --include
   "visitYearMonth=1998*"

   aws s3 sync s3://packt-serverless-etl-glue/chapter-15/
   uservisits-src-data/customer=3/ s3://${BUCKET_NAME}/
   input/clkstreamdata/customer=3/ --exclude "*" --include
   "visitYearMonth=1998*"
   ```

2. **Create the database**: The following command will create a Glue database called `serverless_glue`. This database is created using the AWS CLI for the partition index:

   ```
   aws glue create-database \
       --database-input "{\"Name\":\"serverless_glue \"}"
   ```

> **Note**
>
> You can also use the Glue console to create partition indexed tables. For more information, go to `https://docs.aws.amazon.com/glue/latest/dg/partition-indexes.html#partition-index-creating-table`.

Go to the AWS Glue console and click **Databases** on the left. You will see a database called `serverless_glue`.

3. **Create the crawler**: The following command will create a Glue crawler called `crawl-table-without-partition-index`, a database called `serverless_glue`, and an S3 path called `s3://${BUCKET_NAME}/input/clkstreamdata/` Execute below command after replacing the `${YOUR_BUCKET_NAME}` with your respective AWS s3 bucket and `${YOUR_GLUE_SERVICE_ROLE}` with the IAM role you would like to attach to Glue Crawler:

```
aws glue create-crawler \
--name crawl-table-without-partition-index \
--role ${YOUR_GLUE_SERVICE_ROLE} \
--database-name serverless_glue \
--table-prefix tbl_without_index_ \
--targets "{\"S3Targets\": [{\"Path\": \"s3://${YOUR_
BUCKET_NAME}/input/clkstreamdata /\"} ]}"
```

The following command will create another Glue crawler called `crawl-table-with-partition-index`, a database called `serverless_glue`, and an S3 path called `s3://${BUCKET_NAME}/input/clkstreamdata/`:

```
aws glue create-crawler \
--name crawl-table-with-partition-index \
--role ${YOUR_GLUE_SERVICE_ROLE} \
--database-name serverless_glue \
--table-prefix tbl_with_index_ \
--targets "{\"S3Targets\": [{\"Path\": \"s3://${YOUR_
BUCKET_NAME}/input/clkstreamdata /\"} ]}"
```

Verify that the crawlers have been created successfully in the Glue console.

> **Note:**
>
> While creating a crawler via CLI you need to ensure the IAM role that is being used, has the required permission to perform CreateCrawler operation and it should be able to assume the provided role in the CLI command.

4. **Start the crawlers**: Now that we have created and verified both crawlers, we will run the crawlers using the AWS CLI. It will take a minute or two for each crawler to run. You can also start them in the Glue console, as you did earlier. You can run them one by one or in parallel:

```
aws glue start-crawler --name crawl-table-without-
partition-index
aws glue start-crawler --name crawl-table-with-partition-
index
```

Once the crawlers have finished running and are in a ready state, we can view the results by clicking **Tables** on the left of the page. We should see two new tables that were created by the crawlers: `tbl_with_index_clkstreamdata` and `tbl_without_index_clkstreamdata`.

Click on the `tbl_with_index_clkstreamdata` and `tbl_without_index_clkstreamdata` tables – you will see the table schema that was automatically generated by the crawler based on the Parquet files. Notice the partition columns that were identified by both crawlers in both table definitions. In this case it will have two partitions, which are the `customer` and `visityearmonth`.

5. **Add a partition index to a table**: You can define partition indexes for a given table in AWS Glue Data Catalog at any point in time. You can use the `CreateTable` API with a required list of `PartitionIndex` objects to create a brand-new table in AWS Glue Data Catalog. For an existing table in AWS Glue Data Catalog, you can use the `CreatePartitionIndex` API to add partition indexes. In total, you can have three partition indexes on a table. For the `tbl_with_index_clkstreamdata` table, the possible index that was identified by the crawler is (`customer`, `visityearmonth`).

Run the following command in the CLI to add a partition to the `tbl_with_index_clkstreamdata` table. We will add a partition index using both partition columns that were identified by the crawler:

```
aws glue create-partition-index \
--database-name serverless_glue \
--table-name tbl_with_index_clkstreamdata \
--partition-index
Keys=customer,visityearmonth,IndexName=idxbycustvym
```

Click `tbl_with_index_clkstreamdata` to review the table schema. On the schema page, click the **Partitions and indices** button at the top right to validate the partition index that was created in the previous step.

To ensure that the partition index has been created, you can check the status column. First, it will show its status as **Creating**. This process can take some time, depending on how many partitions are present for the given table. Once the status is **Active**, you can test the partition index using a Glue interactive session (a Spark notebook).

Testing the partition index via a Spark notebook

Now that our partition index has been created, we will query the dataset in AWS Glue Data Catalog using a Glue interactive session. We have learned about Glue Interactive sessions in *Chapter 2, Introduction to important AWS Glue Features*. Please refer to the *Glue Interactive session* section in Chapter 2 if you run into any issues with the setup Follow these steps to run `spark-sql` queries against the tables that were created in the previous section:

1. Open an interactive session notebook from the AWS Glue console and initiate a `SparkSession`:

   ```
   Spark
   ```

2. Run `select` against the `serverless_glue.tbl_without_index_clkstreamdata` table. This table has no partition index on it. Capture the time to run using the `%%time` Spark magic command:

   ```
   %%time
   %%sql
   select count(*)
   from serverless_glue.tbl_without_index_clkstreamdata
   where customer = 2 and visityearmonth = 199812
   ```

```
%%time
%%sql
select count(*) from serverless_glue.tbl_without_index_clkstreamdata where customer = 2 and visityearmonth = 199812

CPU times: user 52.6 ms, sys: 19.1 ms, total: 71.7 ms
Wall time: 29.7 s

Type:    Table        Pie


count(1)

7342094
```

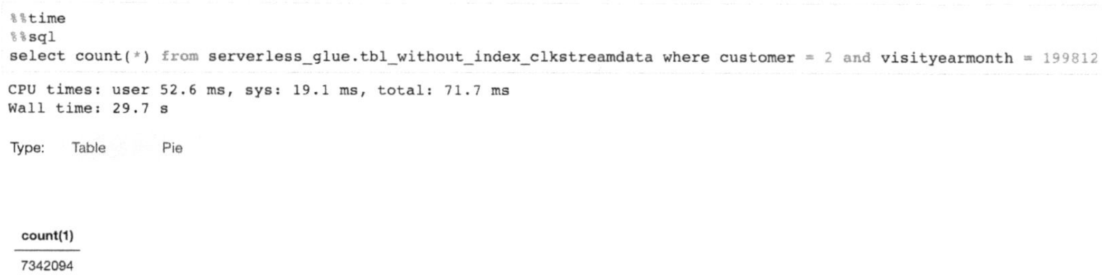

Figure 15.2 – Highly Selective Query on table without partition index

3. Run `select` against the `serverless_glue.tbl_with_index_clkstreamdata` table. This table has a partition index on it. Capture the time to run using the `%%time` Spark magic command:

   ```
   %%time
   %%sql
   select count(*)
   from serverless_glue.tbl_with_index_clkstreamdata
   where customer = 2 and visityearmonth = 199210
   ```

```
%%time
%%sql
select count(*) from serverless_glue.tbl_with_index_clkstreamdata| where customer = 2 and visityearmonth = 199812

CPU times: user 42.2 ms, sys: 2.21 ms, total: 44.4 ms
Wall time: 9.43 s

Type:    Table      Pie


count(1)

7342094
```

Figure 15.3 – Highly Selective Query on table with partition index

4. Notice the difference between the wall time for the same query targeting two tables.

5. The results of the query are the same. The queries have the same filter applied to the same dataset on S3 but the execution times it took to run the queries are different. The query against the serverless_glue.tbl_with_index_clkstreamdata table was completed much faster than the query against the serverless_glue.tbl_without_index_clkstreamdata table.

6. In a scenario where no partition index is present on the serverless_glue.tbl_without_index_clkstreamdata table, AWS Glue will make a GetPartitions API call with all the partitions present in the table and then filter the partitions that were used in the query expression. These highly selective queries without partition indexes can result in higher I/O because the number of partitions typically increases over time, whereas using partition indexes makes a great performance optimization technique. In our test case, the serverless_glue.tbl_without_index_clkstreamdata table was loaded with three times more partitions than when querying serverless_glue.tbl_with_index_clkstreamdata. This becomes even more evident when you have more partitions than what was provided by our sample dataset.

In this section, you learned how to perform highly selective queries on a big fact table using Glue's partition index feature. In the next section, we will cover another real-world use case scenario that deals with performance issues when it comes to performing Join operations between a big fact and a small dimension table in ETL pipelines.

Dealing with Join performance issues with big fact and small dimension tables in ETL workloads

In a scenario where you are joining a big fact table with a small dimension table, Spark can apply the join operation using two different join techniques – it can use a Sort Merge/Shuffle Hash join if both tables are bigger or a Broadcast join if one of the datasets for the underlying table is small enough to be stored in the Spark memory of all executors.

A broadcast join can significantly increase performance and helps with optimizing join operations. A join operation can result in a large data shuffle across the network between the different executors running on multiple workers. This leads to **out-of-memory (OOM)** errors or data spilling to physical disks on the respective workers. While using a broadcast join, you must ensure the smaller table is broadcasted to the executors running on the worker nodes. By doing so, each of the executors running on the workers will be capable enough to handle these join operations between the big fact table and the small dimension table. A broadcast join will be automatically applied if the dimension table is smaller than 10 MB. You can still enforce a broadcast join and let Spark know which table it needs to be applied to.

The following code shows how to use a join operation between a big fact table and a small dimension table and ensure the broadcast join is used:

```
val ClickstreamFactDF = ClickstreamFactRDD.toDF
val SessionDimensionDF = SessionDimensionRDD.toDF

// Applying Broadcast
val tmpSessionDimension = broadcast(SessionDimensionDF.
as("SessionDimension"))

val joinedDF = ClickstreamFactDF.
join(broadcast(tmpSessionDimension),
   $"Session_key" === $"S_key",  // join by ClickstreamFact.
depID == SessionDimension.id
   "inner")

// Show the explain plan and confirm the table is marked for
broadcast
joinedDF.explain()

== Physical Plan ==
*BroadcastHashJoin [Session_key#14L], [S_key#18L], Inner,
BuildRight
:- *Range (0, 100, step=1, splits=8)
+- BroadcastExchange HashedRelationBroadcastMode(List(input[0,
bigint, false]))
   +- *Range (0, 100, step=1, splits=8)
```

Now, let's look at an example to understand whether you should use broadcasting or not.

In this exercise, you will be joining two tables: `clickstream_fact_table` and `session_dimension_table`. First, let's see how big they are:

```
clickstream_fact_table.count // #rows 1,201,233,333
Session_dimension_table.count // #rows 2,922,556
```

Now, we can try to perform a join operation without broadcasting to see how long it takes:

```
val t0 = System.nanoTime()

// Create the Execution Plan
clickstream_fact_table = clickstream_fact_table.join(session_dimension_table,
             clickstream_fact_table.col("session_key")  ===
session_dimension_table.col("s_key"))

// Perform an action to run the execution
Clickstream_fact_table.count

val t1 = System.nanoTime()
println("Elapsed time: " + (t1 - t0)/10e8 + "s")
Output: Elapsed time: 215.115751969s
```

Now, what happens if we broadcast the dimension table? By making a simple addition to the join operation – that is, replacing the `dimension_table` variable with the broadcast (`dimension_table`), we can force Spark to handle our tables using a broadcast:

```
val t0 = System.nanoTime()

// Create the Execution Plan
Clickstream_fact_table = clickstream_fact_table.
join(broadcast(session_dimension_table),clickstream_fact_table.
col("session_key") === session_dimension_table.col("s_key"))

// Perform an action to run the execution
Clickstream_fact_table.count

val t1 = System.nanoTime()
println("Elapsed time: " + (t1 - t0)/10e8 + "s")
Output: Elapsed time: 61.1358s
```

Using the broadcast join between a big fact table and a small dimension table resulted in *70%* faster execution. When you use a broadcast join with a small-sized table, you need to ensure it will remain small to medium in size in the future so that you don't run into OOM exceptions or make your application code problematic.

In this section, you learned how to deal with Join performance issues when it comes to performing join operations between a large fact table dataset and a small dimension table dataset. We explained the concept of a broadcast join and how useful it can be in these scenarios since it saves a lot of execution time and cost. In the next chapter, we will learn how to solve an edge-case problem involving using a join operation between a large fact table and a large-dimension table.

Solving Join problems involving big fact and big dimension tables using AWS Glue

Whether you are a data engineer, big data architect, or business analyst, one thing you need to do is scale your data processing and ETL batch workloads. In this section, we are going to talk about one of Glue's Spark runtime optimization features: workload partitioning with bounded execution. This can help you handle join operations between a large fact table and a dimension table. We will also provide a hands-on tutorial to demonstrate the difference this feature can make concerning performance. This feature works in conjunction with AWS Glue bookmarks, which we discussed in *Chapter 2, Introduction to Important AWS Glue Features*. It can help you break down your complex and humongous workloads by bounding the execution of the respective Spark applications. In layman's terms, you can partition your ETL workloads by putting a restriction in place for each of these independent workloads to process a certain number of files sequentially or in parallel. The following diagram depicts an ETL architecture for this use case:

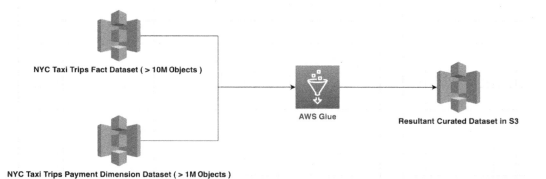

Figure 15.4 – Handling join operations between a large fact table and dimension tables

Keeping this architecture in mind, let's talk about a scenario where you want to process and perform join operations between a big fact table (NYC taxi trips fact dataset) and a dimension table (payments dimension dataset) and write the resultant data to the target – Amazon S3. In our experience, some of the common errors you can primarily run into while executing this use case are OOM issues with a Spark driver as a result of listing billions of files in Amazon S3 for the fact table, or OOM issues with a Spark executor as a result of data skew in the fact table. We will show you how to handle these edge case scenarios using workload partitioning. This can help you avoid these problems by setting up the bounded execution for Spark applications.

In the next section, we will demonstrate how workload partitioning feature performs when it comes to perform join operations between large fact and a large dimension table and how it can be optimized. You can follow through the series of steps provided from beginning below to learn about the problem and how to deal with it using Glue workload partitioning technique.

Hands-on tutorial

In the following section, we will use AWS CLI to create a Glue database, Glue crawlers and define workload partitioning while reading data using Glue Dynamic frames to see how it works and helps with join operations between a large fact and a dimension table.. In this demonstration we are going to use public dataset, that is, NYC TLC data available here `https://registry.opendata.aws/nyc-tlc-trip-records-pds/`.We did some pre-processing on the original dataset to create our test datasets for fact and dimension table. Our setup included two large datasets . One for NYC Taxi Trips Fact Dataset containing 1.3 million objects totaling 42 GB and other for NYC Taxi Trips Payment dataset containing 1.3 million objects with total size - 17 GB.

Now, let's dive into some hands-on and follow the step by step instructions provided below and learn about Glue workload partitioning implementation:

1. **Prepare the test data**: We will use a partitioned dataset from a sample NYC taxi data source to work with workload partitioning. You can execute the following commands using the AWS CLI. It will take about 6 to 8 minutes to copy the data. Notice that we loaded data for a couple of years from these sample datasets. Execute the following commands after replacing `${YOUR_BUCKET_NAME}` with your respective AWS S3 bucket:

   A. **Copy the fact table data**: To create a sample dataset, you may run the below command three times to copy the data for multiple years ranging from `year=2011` to `year=2013`. Doing so the total number of objects that will be copied are 423,255 and total size will be approximately 13.7 GB:

   ```
   aws s3 cp --recursive s3://packt-serverless-etl-
   glue/chapter-15/trips_fact_data/ s3://${YOUR_BUCKET_
   NAME}/input/trips_fact_data/ --exclude "*" --include
   "year=2011/*"
   aws s3 cp --recursive s3://packt-serverless-etl-
   ```

```
glue/chapter-15/trips_fact_data/ s3://${YOUR_BUCKET_
NAME}/input/trips_fact_data/ --exclude "*" --include
"year=2012/*"

aws s3 cp --recursive s3://packt-serverless-etl-
glue/chapter-15/trips_fact_data/ s3://${YOUR_BUCKET_
NAME}/input/trips_fact_data/ --exclude "*" --include
"year=2013/*"
```

B. **Copy the dimension table data:** To create a sample dataset, you may run the below command to copy the data for `year=2012`. We are copying only single partition data for hands-on exercise. Doing so the total number of objects that will be copied are 142,965 and total size will be approx. 4.7 GB:

```
aws s3 cp --recursive s3://packt-serverless-etl-glue/
chapter-15/payments_dim_data/ s3://${YOUR_BUCKET_NAME}/
input/payments_dim_data/ --exclude "*" --include
"year=2012/*"
```

2. **Create the database:** The following command will create a Glue Database `workload_partitioning`. This database is created using AWS CLI for demonstrating workload partitioning. Run the following command in AWS CL.

Go to the AWS Glue console, click **Databases** on the left. You will see a database with name `workload_partitioning`. Replace profile and endpoint per the region where you are running this command:

```
aws glue create-database \
--database-input "{\"Name\":\"workload_partitioning\"}" \
```

3. **Create the crawler:** The following commands will create Glue Crawler with Name: `crawl-nyc-trips-taxi-fact-table`, Database: `workload_partitioning` and S3 path: `s3://${BUCKET_NAME}/input/trips_fact_data /`. Execute below command after replacing the `${YOUR_BUCKET_NAME}` with your respective AWS s3 bucket and `${YOUR_GLUE_SERVICE_ROLE}` with the IAM role you would like to attach to Glue Crawler:

```
aws glue create-crawler \
--name crawl-nyc-trips-taxi-fact-table \
--role ${YOUR_GLUE_SERVICE_ROLE} \
--database-name workload_partitioning \
--targets "{\"S3Targets\": [{\"Path\": \"s3://${YOUR_
BUCKET_NAME}/input/trips_fact_data/\", ,\"SampleSize\":
1}]}"
```

The following command will create Glue Crawler with `Name` : `crawl-nyc-trips-payments-dim-table`, `Database`: `workload_partitioning` and `S3 path`: `s3://${BUCKET_NAME}/input/payments_dim_data/`. Execute below command after replacing the `${YOUR_BUCKET_NAME}` with your respective AWS s3 bucket and `${YOUR_GLUE_SERVICE_ROLE}` with the IAM role you would like to attach to Glue Crawler:

```
aws glue create-crawler \
--name crawl-nyc-trips-payments-dim-table \
--role ${YOUR_GLUE_SERVICE_ROLE} \
--database-name workload_partitioning \
--targets "{\"S3Targets\": [{\"Path\": \"s3://${YOUR_
BUCKET_NAME}/input/payments_dim_data/\", ,\"SampleSize\":
1} ]}"
```

Verify that the crawlers were created successfully in the Glue console.

4. **Start the crawlers**: Once we have created both crawlers, we will run the crawlers using the CLI. It will take a minute or two for each crawler to run. You can also start them in the Glue console, as you did earlier. You can run them one by one or in parallel:

```
aws glue start-crawler --name crawl-nyc-trips-taxi-fact-
table
aws glue start-crawler --name crawl-nyc-trips-payments-
dim-table
```

Once the crawlers have finished running, you can view the results by clicking **Tables** on the left of the page. You should see that two new tables were created by the crawlers: `payments_dim_data` and `trips_fact_data`.

Click on the `trips_fact_data` and `payments_dim_data` tables to see the table schema that was automatically generated by the crawler based on the Parquet files. Notice that the partition columns that were identified by both crawlers are in both table definitions.

5. **Run a Glue job**: By now you already know how to create a Glue job. Once the job is created, we can enable AWS Glue Job bookmarks to use with AWS Glue DynamicFrame to take advantage of incremental processing. Here is our sample spark application code - `https://github.com/PacktPublishing/Serverless-ETL-and-Analytics-with-AWS-Glue/tree/main/Chapter15` in the GitHub repository we used for our setup. During our benchmarking for demonstration purpose, we ran total 3 iterations. Starting with the code which does not use bounded execution and perform join operations between a large fact and a dimension table. We will apply the bounded execution after the first iteration, hence executing it using workload partitioning to conclude the demo.

> **Note:**
>
> As a part of hands-on exercise for creating and running the glue job you don't have to run all the iterations , you can directly run iteration 3. For Iteration 3 the PySpark code is available in the GitHub repository: `https://github.com/PacktPublishing/Serverless-ETL-and-Analytics-with-AWS-Glue/blob/main/Chapter15/Workload_Partitioning_Itr_3.pyx`

During our benchmarking, when we executed the first iteration of this code as is, without bounded execution, the spark driver struggled with the memory due to a large number of objects in both fact and dimension tables, that is, 1.6 million objects in `trips_fact_data` where as `payments_dim_data` had approximately 1 million objects. In this scenario, spark driver has to keep a track of large number of objects in its memory and at the same time it need to keep a track of number of spark tasks. It eventually failed with a Spark driver OOM. To conclude you can check the spark driver and executor memory profile using AWS Glue job metrics. The Job metric graph will looks as follows:

Figure 15.5 – Spark driver memory peaked above 50% and led to OOM

To conclude, you can check the Spark driver and executor memory profile using AWS Glue job metrics. The job metric graph will look like what's shown in the preceding screenshot.

Solution

To overcome this Spark driver OOM error, we modified the previously written code so that it uses workload partitioning and includes the `boundedFiles` parameter as `additional_options`. We will only process 95,000 files from the `trips_fact_data` and `payments_dim_data` data sources. As the Spark application will only process 95,000 files, hence it will put less pressure on the Spark driver. Bounded execution keeps track of the files and partitions with the specified bound concerning the number of files. One thing to note here is bounded execution works well along with job bookmarks and we need to ensure it's enabled beforehand. As we already know, job bookmarks keep track of already processed files and partitions from the source data based on the timestamp and path.

The following code snippet shows the changes that can be made in the code to implement workload partitioning with bounded execution:

```
tripsfactDyf = glueContext.create_dynamic_frame.from_
catalog(database = "trips_fact_data", table_name = "trips_fact_
data", transformation_ctx = "datasource0", additional options =
{"boundedFiles" : "95000"} )

paymentDimDyf = glueContext.create_dynamic_frame.from_
catalog(database = "workload_partitioning", table_name =
"payments_dim_data", transformation_ctx = "datasource0"",
additional options = {"boundedFiles" : "95000"})
```

Once the above changes have been made, we can rerun our spark application using Iteration-2. We will find that the spark driver memory consistently stayed below 50% with a peak of 25% but that the spark executors struggled with heavy memory usage. This caused the job to eventually fail with an executor OOM:

Memory Profile: Driver and Executors

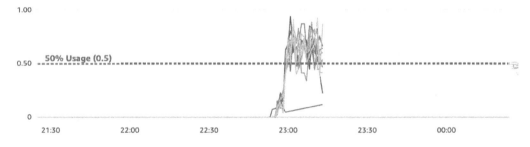

Figure 15.6 – Spark driver memory peaked below 50% but executors are stuggling with memory

So finally we decided to run iteration-3 and use more conservative bound for the number of files to be processed in a given iteration. We changed the boundedFiles value to 45,000 files:

```
tripsfactDyf = glueContext.create_dynamic_frame.from_
catalog(database = "trips_fact_data", table_name = "trips_fact_
data", transformation_ctx = "datasource0", additional_options =
{"boundedFiles" : "45000"} )

paymentDimDyf = glueContext.create_dynamic_frame.from_
catalog(database = "workload_partitioning", table_name =
"payments_dim_data", transformation_ctx = "datasource0"",
additional_options = {"boundedFiles" : "45000"})
```

This time, the Spark application ran without any driver or executor memory problems. Based on your use case, you can perform benchmarking to find the optimal value for `boundedFiles` that would work for your workloads. To conclude with an optimal value, you may require multiple iterations of the Spark application to be executed. One of the other great advantages of using workload partitioning is that it allows you to execute multiple Spark applications in parallel for the same dimension and fact datasets. Let's assume that, in your production environment, you have a strict SLA to meet for data processing. You can optimize this problem by creating more than one copy of the Glue job. Then, to process a subset of the data from the input data sources, you can take advantage of Glue's push-down predicate with bounded execution.

The following are two pieces of code from two different Glue jobs that are processing data from the same `trips_fact_data` and `payments_dim_data` tables. However, each of these are from different input partitions – that is, 2020 and 2021, respectively:

- The following code is for Glue job 1:

```
tripsfactDyf = glueContext.create_dynamic_frame.
from_catalog(database = "trips_fact_data", table_
name = "trips_fact_data", transformation_ctx =
"datasource0",push_down_predicate=("year=2020"),
additional_options = {"boundedFiles" : "45000"} )

paymentDimDyf = glueContext.create_dynamic_frame.from_
catalog(database = "workload_partitioning", table_name =
"payments_dim_data", transformation_ctx = "datasource0"",
,push_down_predicate=("year=2020"), additional_options =
{"boundedFiles" : "45000"})
```

- The following code is for Glue job 2:

```
tripsfactDyf = glueContext.create_dynamic_frame.
from_catalog(database = "trips_fact_data", table_
name = "trips_fact_data", transformation_ctx =
"datasource0",push_down_predicate=("year=2021"),
additional options = {{"boundedFiles" : "45000"} )

paymentDimDyf = glueContext.create_dynamic_frame.from_
catalog(database = "workload_partitioning", table_name =
"payments_dim_data", transformation_ctx = "datasource0"",
,push_down_predicate=("year=2021"), additional options =
{{"boundedFiles" : "45000"})
```

Once these jobs have been created, you can use AWS Glue workflows to execute them in parallel.

In this section, you learned how to perform a join operation between a large fact table and a large-dimension table using Glue's workload partitioning feature. You also learned how to divide large workloads into partitioned workloads so that you can read from a single data source in parallel, thus reducing the overall time for such workloads. In the next section, you will learn how to process the data in a large-dimension table, which can contain millions of small files, in Amazon S3.

Efficient spark reads with large dimensions using AWS Glue Grouping

Let's assume you have an edge use case where you have over 1 billion rows in one of your dimension table data sources available in Amazon S3 and that you have written some ETL code in a Glue job. This code reads millions of small files with billions of rows with a standard Glue worker, does some file conversion, and writes the files back to S3. In this section, you will learn how to deal with expensive Spark read operations, especially while reading the data from large-dimension tables with AWS Glue.

As we know, Glue manages provisions and manages the resources that are required to perform ETL for you. That being said, when you encounter OOM exceptions thrown by the Spark driver, we need to understand how Spark works to resolve them. Once the Glue job is executed, the Glue console provides you with the ETL metrics and memory profiles for each job run you execute, which helps you identify job abnormalities and performance issues, similar to the one shown here:

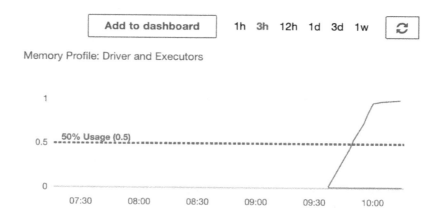

Figure 15.7 – ETL metrics showing performance issues while reading millions of small files

As you can see, the memory of the Spark driver (the blue line) exceeds the threshold of 50%, and once it reaches 100, the job fails with an OOM exception and is killed. The executors (the green line) haven't even started to use any memory yet. In this instance, the transformation that was performed in ETL isn't the problem here. There is no data movement and the Spark tasks haven't been distributed to the executors yet.

The problem is that Spark tries to make large recursive calls to the S3 list method. The S3 list method becomes too expensive in this scenario since there are too many small files in the S3 dataset – in this case, your large-dimension table. As we know, the Spark driver's job is to record the file metadata it reads and store it in the driver's memory. This leads to OOM errors with the Spark driver.

The best practice for solving this problematic use case while dealing with large-dimension tables is to use the grouping feature from Glue. When enabled and used with a Glue DynamicFrame, it allows Spark to form a group for multiple small files and assign this group of files to a single Spark task rather than individual files. Using this feature, you can significantly reduce the memory pressure on the Spark driver as it stores significantly less information in memory about fewer tasks. This reduces the probability of OOM exceptions while reading from these large-dimension tables. The downside of not using this feature is that Spark would process individual files using a single Spark task. Eventually, the Spark driver would get the status of each of these Spark tasks individually, which would overwhelm the Spark driver and lead to an OOM error.

Now, let's learn how to configure this feature within a Glue ETL job and take advantage of it. The following boilerplate code examples for Scala and PySpark use the AWS Glue DynamicFrame API in an ETL script with the configuration that is required to enable the AWS Glue grouping feature:

```
#Scala Example when you are reading directly from Amazon S3
glueContext.getSourceWithFormat(
    connectionType = "s3",
    options = JsonOptions(Map("paths" -> s3Paths,
    "groupFiles" -> "inPartition",
    "useS3ListImplementation" -> true)),
    format = "xml",
    formatOptions = JsonOptions(Map("rowTag" -> "our-row-tag"))
  ).getDynamicFrame()
#PySpark Example when you are reading directly from Amazon S3
 df = glueContext.create_dynamic_frame.from_options(
   "s3", {'paths': ["s3://s3path/"],
   'recurse':True,
   'groupFiles': 'inPartition',
   'groupSize': '1048576'},
    format="json")
```

If you want to read the data directly from AWS Glue Catalog, you can set grouping configuration parameters in the following two ways:

- Edit the table definition and provide these parameters as key values:

Figure 15.8 – Editing table details in Glue Data Catalog

- Provide these parameters while creating a DynamicFrame within the ETL script:

```
#PySpark Example
datasource = Gluecontext.create_dynamic_frame_from_
catalog(
        database= "many_files_dataset",
```

```
                    table_name ="Json_2k_million",
                    additional_options= {"groupsize"
    :1024*1024*1024},"groupFiles": "acrossPartition")
```

In the preceding boilerplate scripts, there are some important things to note:

- The configurations you can tweak per your use case are groupFiles, groupSize, and recurse. AWS Glue enables grouping when you have more than 50,000 input files in the Amazon S3 data source by default.

- You can still set groupFiles to inPartition if you want to group a large number of small files in Amazon S3 data sources and perform benchmarking.

- groupSize is purely an optional configuration that allows Spark tasks to process a certain amount of data while reading and then process it as a single AWS Glue DynamicFrame partition. Be careful when using a considerably small or large groupSize values because it can result in significant task parallelism or underutilization of the resources in the Glue environment.

- You can use the recurse config with grouping, which allows you to recursively read all the files in the subdirectories for the Amazon S3 path provided.

- You can use useS3ListImplementation along with grouping to help resolve OOM exceptions.

In this section, you learned how to solve a problematic use case – that is, optimizing read operations for the large-dimension table. You experimented with using the AWS Glue grouping technique, which helps solve small file problems that can run into OOM issues. You learned how to use this configuration while reading the data from Amazon S3 and Glue Data Catalog. Using grouping, you can ensure the ETL pipelines do not run into these corner cases and build a scalable ETL pipeline.

Solve S3 eventual consistency and faster spark writes to S3 for large fact tables

Let's assume you have a use case where you are dealing with writing huge data into Amazon S3 – that is, you have a clickstream fact table dataset in Parquet format but the Spark application fails with an exception *File not found* error. When running Spark jobs on Amazon S3, Spark writes the output to a _TEMPORARY prefix in S3, then moves the data from _TEMPORARY to its final destination. In S3, a move is a rename operation. If the move happens immediately after the write operation, there is a chance of eventual consistency, which causes this move operation to fail. You will see that it failed due to a *Rename failed* or *File not found* error message. In this section, you will learn how to handle these problematic scenarios and fix them in the long term. The following diagram shows the S3 eventual consistency model:

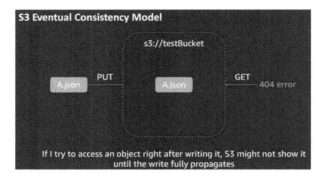

Figure 15.9 – S3 eventual consistency model

First, let's understand how the S3 eventual consistency model works. The important concept to understand here is that the file rename process in a POSIX-based filesystem is a metadata-only operation. Only the pointer changes and the file remain as is on disk. For example, I have a file called abc.txt and I want to rename it xyz.txt. This is an instantaneous and atomic process. The xyz.txt file's last modified timestamp remains the same as the abc.txt file's last modified timestamp. On the other hand, in AWS S3 (the object store), the file that was renamed under the hood is a copy followed by a delete operation. The source file is copied to the destination and then the source file is deleted. So, aws s3 mv changes the last-modified timestamp of the destination file, unlike what happens in the POSIX filesystem. The metadata here is a key-value store where the key is the file path and the value is the content of the file. There is no such process as changing the key and getting this done immediately. The renaming process depends on the size of the file. If there is a directory rename (there is nothing called *directory* in S3 so for simplicity, we can assume a recursive set of files is a directory), then it depends on the number of files inside the directory, along with the size of each file. So, in a nutshell, renaming is a very expensive operation in S3 compared to this being done in a normal filesystem. S3 comes with two kinds of consistency: read after write and eventual consistency. In some cases, it results in a *File not found* error, files being added and not listed, or files being deleted or not removed from the list.

To deal with this problem, you have two options:

- Use glueparquet.
- Use an S3-optimized output committer.

Let's look at each of these options in detail.

Using glueparquet

Using `glueparquet` as the output format will internally change the output committer to `DirectOutputCommiter`, which does not do renames. To understand this concept better, let's see what the PARQUET file looks like under the hood:

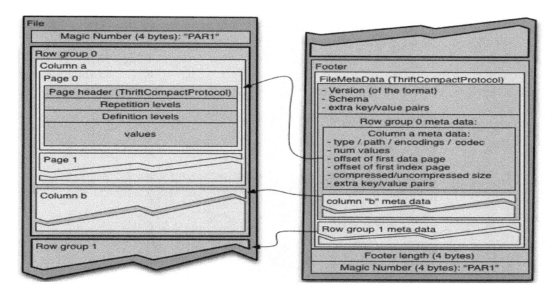

Figure 15.10 – Parquet file format internals

Let's talk about the preceding figure. The following are the generic properties for a Parquet file format that you should be familiar with. They help you choose between the different file formats that are available in the big data ecosystem:

- Parquet file uses a magic number (4 bytes) that acts as a separator and helps identify the beginning and end of the file.

- Following the first magic number, there are several row groups and then a footer.

- `FileMetaData` is placed in the footer because the Parquet file metadata is written once the actual data is written. The row groups contain the data.

- The Parquet file contains three types of metadata: **file metadata**, **column** (chunk) **metadata**, and **page header metadata**.

Apache Parquet format is typically faster for reads compared to writes because it has the columnar storage layout and also offers a precomputed schema that is written along with the data. Now, let's understand why a regular Parquet file format can cause issues while writing large fact table datasets in Amazon S3. The problem is that, when using the Parquet file format with a Glue DynamicFrame, Spark does not know the complete schema. Hence, the number of buffers is unknown. This leads to an additional pass over the dataset, which is an expensive operation and leads to more time in the overall execution. This may lead to failures during the write operation. By introducing glueparquet when it's time to write each executor, you can compute the schema of the data it has in memory. Doing a pass over the data in memory is way faster than in disk, so there are no performance issues. The executor creates one buffer for each column group. As rows are being written, if a row with a new column comes in, new buffers are created. When the buffers are full, data is written to disk:

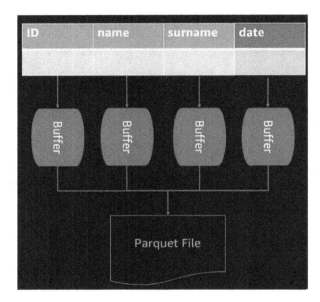

Figure 15.11 – Illustration of a glueparquet write

It uses a different committer called DirectOuputCommitter that also does not do any rename operations, hence saving a significant amount of I/O processing while performing write operations in Spark. When the data is available to write, the writer computes and merges the schema dynamically at runtime, leading to faster job runtimes. In comparison to a regular Apache Parquet writer, it does not perform an extra scan over the input dataset to infer the schema. It enables schema evolution by allowing you to add and delete new columns. The following is a boilerplate PySpark script example that demonstrates how to use the glueparquet format option:

```
glueContext.write_dynamic_frame.from_options(
frame = dyFrame,
```

```
    connection_type = "s3",
    connection_options = {"path": output_dir},
    format = "glueparquet",
    format_options = {"compression": "snappy",
    blockSize = block_size,
    pageSize = page_size})
```

The following default values are set for the configurations:

- `compression` is `"snappy"`
- `blockSize` is 128 MB
- `pageSize` is 1 MB

`blockSize` specifies the size of a row group that is buffered in memory, while `pageSize` specifies the size for compression that will be used. When the read operation is executed, each page can be decompressed independently. It is the smallest unit in a Parquet file that must be read fully to access a single record.

S3-optimized output committer

Now, let's talk about the last option that we have to deal with this problem. You can use the EMRFS S3-optimized committer, which is an alternative `OutputCommitter` implementation that is optimized for writing Parquet files to Amazon S3 when using EMRFS. Glue can use EMR's S3-optimized output committer in Spark applications that use Spark SQL, DataFrames, or datasets to write Parquet files. In terms of performance, it is considered better than `DirectOutputCommiter`, which we discussed in the previous section.

Now, let's discuss the problems that the S3-optimized output committer can address:

- List and rename operations are considered expensive in Spark applications while doing write operations. This committer helps improve a Spark application's performance by avoiding both of these operations during the job and task commit phases.

- Glue ETL jobs run on the Apache Spark framework, which writes all the output to a temporary directory in S3 by default. When all the executors have finished writing, the files are moved from this temporary directory to your selected destination path.

- S3 does not use the concept of directories (everything is a named prefix), so this move operation is just a rename to change the file's prefix. Sometimes, the jobs fail while writing the data to S3.

The best way to address this is to enable the EMRFS S3-optimized committer, which is available in Glue. It removes such errors by using optimized S3 write logic. Whenever you or one of your applications writes a file to S3, there's a very small time window where the file needs to be propagated throughout S3's backend system. If you try to access that file within that window of time (as in, immediately after writing it), there's a chance the file has not finished propagating and S3 will return an error.

To enable this feature with AWS Glue, you can supply the necessary key pair value via the AWS Glue console when creating or updating an AWS Glue job. Setting the value to `true` enables the committer. By default, the flag is turned off:

```
Key       :    --enable-s3-parquet-optimized-committer.
Value : true
```

This feature is available with Glue 2.0 onward and is used by default in Glue 3.0. There are certain scenarios where this committer will not be used, even if you have enabled it. Please check `https://docs.aws.amazon.com/emr/latest/ReleaseGuide/emr-spark-committer-reqs.html` for more details.

In this section, you learned how to solve Amazon S3's eventual consistency problem and enable faster write operations, which involves writing data to Amazon S3 for large fact table datasets. You learned about what the S3 eventual consistency model is all about and how it works. Then, we explained the solutions to this problem by using either the `glueparquet` file format or an S3-optimized committer with AWS Glue. For both these options, you understood the different file committers that are involved to help you choose one over the other in real-world data processing problems.

Summary

In this chapter, you learned about some of the best practices for dealing with real-world problems and how to run highly selective queries on big fact tables. After that, you learned how to run highly selective queries by experimenting with the Glue partition indexing technique, which allows you to query humongous fact tables and make data retrieval smooth. Next, you learned how to deal with join performance issues between a large fact table and a small dimension table. Here, you learned how to use the broadcast mechanism to optimize the join operation.

After that, you learned how to deal with dimension tables when something goes wrong and you don't have a way to partition the workloads into smaller workloads. Here, you applied a Glue bounded execution with Glue bookmarks to restrict the number of files that can be processed with incremental workloads. For the edge case scenario, where you read a large-dimension table, you learned how to configure Glue jobs to use the grouping technique, which put less pressure on the Spark driver, thus avoiding OOM issues. Lastly, you learned how to deal with S3 eventual consistency and the best practice to handle faster writes to Amazon S3 for large fact tables.

Index

Packt.com

Subscribe to our online digital library for full access to over 7,000 books and videos, as well as industry leading tools to help you plan your personal development and advance your career. For more information, please visit our website.

Why subscribe?

- Spend less time learning and more time coding with practical eBooks and Videos from over 4,000 industry professionals

- Improve your learning with Skill Plans built especially for you

- Get a free eBook or video every month

- Fully searchable for easy access to vital information

- Copy and paste, print, and bookmark content

Did you know that Packt offers eBook versions of every book published, with PDF and ePub files available? You can upgrade to the eBook version at packt.com and as a print book customer, you are entitled to a discount on the eBook copy. Get in touch with us at customercare@packtpub.com for more details.

At www.packt.com, you can also read a collection of free technical articles, sign up for a range of free newsletters, and receive exclusive discounts and offers on Packt books and eBooks.

Other Books You May Enjoy

If you enjoyed this book, you may be interested in these other books by Packt:

Simplify Big Data Analytics with Amazon EMR

Sakti Mishra

ISBN: 9781801071079

- Explore Amazon EMR features, architecture, Hadoop interfaces, and EMR Studio
- Configure, deploy, and orchestrate Hadoop or Spark jobs in production
- Implement the security, data governance, and monitoring capabilities of EMR
- Build applications for batch and real-time streaming data analytics solutions
- Perform interactive development with a persistent EMR cluster and Notebook
- Orchestrate an EMR Spark job using AWS Step Functions and Apache Airflow

Time Series Analysis on AWS

Michaël Hoarau

ISBN: 9781801816847

- Understand how time series data differs from other types of data
- Explore the key challenges that can be solved using time series data
- Forecast future values of business metrics using Amazon Forecast
- Detect anomalies and deliver forewarnings using Lookout for Equipment
- Detect anomalies in business metrics using Amazon Lookout for Metrics
- Visualize your predictions to reduce the time to extract insights

Packt is searching for authors like you

If you're interested in becoming an author for Packt, please visit authors.packtpub.com and apply today. We have worked with thousands of developers and tech professionals, just like you, to help them share their insight with the global tech community. You can make a general application, apply for a specific hot topic that we are recruiting an author for, or submit your own idea.

Share Your Thoughts

Now you've finished *Serverless ETL and Analytics with AWS Glue*, we'd love to hear your thoughts! Scan the QR code below to go straight to the Amazon review page for this book and share your feedback or leave a review on the site that you purchased it from.

https://packt.link/r/1-800-56498-8

Your review is important to us and the tech community and will help us make sure we're delivering excellent quality content.